M000222170

Radio Tracking and Animal Populations

Radio Tracking and Animal Populations

Edited by

Joshua J. Millspaugh

The School of Natural Resources
Department of Fisheries and Wildlife Sciences
University of Missouri
Columbia, Missouri

John M. Marzluff

College of Forest Resources
Wildlife Science Group
University of Washington
Seattle, Washington

ACADEMIC PRESS

A Harcourt Science and Technology Company

San Diego   San Francisco   New York   Boston   London   Sydney   Tokyo

Cover photo credit: © Milo Burcham

This book is printed on acid-free paper.

Copyright © 2001 by ACADEMIC PRESS

All Rights Reserved.
No part of this publication may be reproduced or transmitted in any form or by any
means, electronic or mechanical, including photocopy, recording, or any information
storage and retrieval system, without permission in writing from the publisher.

Requests for permission to make copies of any part of the work should be mailed to:
Permissions Department, Harcourt Inc., 6277 Sea Harbor Drive,
Orlando, Florida 32887-6777

Academic Press
A Harcourt Science and Technology Company
525 B Street, Suite 1900, San Diego, California 92101-4495, USA
http://www.academicpress.com

Academic Press
Harcourt Place, 32 Jamestown Road, London NW1 7BY, UK
http://www.academicpress.com

Library of Congress Catalog Card Number: 00-110674

International Standard Book Number: 0-12-497781-2

Transferred to Digital Printing in 2009

CONTENTS

3 Effects of Tagging and Location Error in Wildlife Radiotelemetry Studies

John C. Withey, Thomas D. Bloxton, and John M. Marzluff

PART **III**

Equipment and Technology

4 Recent Telemetry Technology

Arthur R. Rodgers

PART **IV**

Animal Movements

5 Analysis of Animal Space Use and Movements

Brian J. Kernohan, Robert A. Gitzen, and Joshua J. Millspaugh

PART **V**

Resource Selection

12 High-Tech Behavioral Ecology: Modeling the Distribution of Animal Activities to Better Understand Wildlife Space Use and Resource Selection

John M. Marzluff, Steven T. Knick, and Joshua J. Millspaugh

PART **VI**

Population Demographics

13 Population Estimation with Radio-Marked Animals

Gary C. White and Tanya M. Shenk

14 Analysis of Survival Data from Radiotelemetry Studies

Scott R. Winterstein, Kenneth H. Pollock, and Christine M. Bunck

PART **VII**
Concluding Remarks

CONTRIBUTORS

Number(s) in parentheses indicates page on which the contribution begins.

RICHARD K. BAYDACK (167), Natural Resources Institute, University of Manitoba, Winnipeg, Manitoba R3T 2N2, Canada

THOMAS D. BLOXTON (43), Division of Ecosystem Science, College of Forest Resources, University of Washington, Seattle, Washington 98195

CLAIT E. BRAUN[1] (167), Colorado Division of Wildlife, Fort Collins, Colorado 80526

CHRISTINE M. BUNCK (351), U.S. Geological Survey, Biological Resources Division, Madison, Wisconsin 52711

L. MIKE CONNER (275), Joseph W. Jones Ecological Research Center, Newton, Georgia 31770

ANDREW B. COOPER[2] (243), Quantitative Ecology and Resource Management, University of Washington, Seattle, Washington 98195

WALLACE P. ERICKSON (209), Western EcoSystems Technology, Inc., Cheyenne, Wyoming 82001

EDWARD O. GARTON (15, 291), Department of Fish and Wildlife Resources, University of Idaho, Moscow, Idaho 83844

KENNETH G. GEROW (209), Statistics Department, University of Wyoming, Laramie, Wyoming 82071

[1]Present address: Grouse Inc., Tucson, Arizona 85750
[2]Present address: Living Oceans Program, National Audubon Society, Islip, New York 11751

xi

ROBERT A. GITZEN (125), Wildlife Science Group, University of Washington, Seattle, Washington 98195

CHRISTIAN A. HAGEN[3] (167), Natural Resources Institute, University of Manitoba, Winnipeg, Manitoba R3T 2N2, Canada

SHAY HOWLIN (209), Western EcoSystems Technology, Inc., Cheyenne, Wyoming 82001

BRUCE K. JOHNSON (15, 291), Oregon Department of Fish and Wildlife, La Grande, Oregon 97850

NORM C. KENKEL (167), Department of Botany, University of Manitoba, Winnipeg, Manitoba R3T 2N2, Canada

ROBERT E. KENWARD (3), Centre for Ecology and Hydrology, Winfrith Technology Center Dorchester, Dorset DT2 8ZD, England

JOHN W. KERN (209), Spectrum Consulting Services, Inc., Pullman, Washington 99163

BRIAN J. KERNOHAN (125), Boise Cascade Corporation, Boise, Idaho 83728

JOHN G. KIE (291), Pacific Northwest Research Station, La Grande, Oregon 97850

STEVEN T. KNICK (309), U.S. Geological Survey, Boise, Idaho 83706

MICHAEL A. LARSON (397), Department of Fisheries and Wildlife Sciences, University of Missouri, Columbia, Missouri 65211

FREDERICK A. LEBAN[4] (15, 291), Department of Fish and Wildlife Resources, University of Idaho, Moscow, Idaho 83844

JOHN M. MARZLUFF (43, 309, 383), Wildlife Science Group, University of Washington, Seattle, Washington, 98195

TRENT L. MCDONALD (209), Western EcoSystems Technology, Inc., Cheyenne, Wyoming 82001

JOSHUA J. MILLSPAUGH (125, 243, 309, 383), Department of Fisheries and Wildlife Sciences, University of Missouri, Columbia, Missouri 65211

---

[3]Present address: Division of Biology, Kansas State University, Manhattan, Kansas 66506
[4]Present address: Youth With A Mission, Highfield Oval, Harpenden, Hertfordshire AL5 4BX, United Kingdom

RICHARD M. PACE III[5] (189), U.S. Geological Survey, Biological Resource Division, Louisiana Cooperative Fish and Wildlife Research Unit, Louisiana State University, Baton Rouge, Louisiana 70803

BRUCE W. PLOWMAN[6] (275), Joseph W. Jones Ecological Research Center, Newton, Georgia 31770

KENNETH H. POLLOCK (351), Department of Statistics, North Carolina State University, Raleigh, North Carolina 27695

ARTHUR R. RODGERS (79), Ontario Ministry of Natural Resources, Centre for Northern Forest Ecosystem Research, Thunder Bay, Ontario P7B 5E1, Canada

TANYA M. SHENK (329), Research Section, Colorado Division of Wildlife, Fort Collins, Colorado 80526

DAVID J. WALKER (167), Department of Botany, University of Manitoba, Winnipeg, Manitoba R3T 2N2, Canada

GARY C. WHITE (329), Department of Fishery and Wildlife Biology, Colorado State University, Fort Collins, Colorado 80523

SCOTT R. WINTERSTEIN (351), Department of Fisheries and Wildlife, Michigan State University, East Lansing, Michigan 48824

MICHAEL J. WISDOM (15, 291), Pacific Northwest Research Station, La Grande, Oregon 97850

JOHN C. WITHEY (43), Division of Ecosystem Science, College of Forest Resources, University of Washington, Seattle, Washington 98195

[5]Present address: Northeast Fisheries Science Center, Protected Species Branch, Woods Hole, Massachusetts 02543
[6]Present address: USDA, APHIS, Wildlife Services, Montpelier, Vermont 05602

PREFACE

Radiotransmitters provide convenient and cost-effective means of remotely monitoring the physiology, movements, resource selection, and demographics of wild animals. During the past 10 years, efficient global positioning system collars, smaller, longer lasting, and more reliable radiotransmitters, and tools such as geographic information systems have become widely available. These developments have changed the way radio-tracking data are collected and analyzed and have produced a stimulating environment for users of radio-telemetry. As we grappled with the latest analytical techniques and techno-logical advancements in our own studies, we sensed the need for a consolidated and updated synthesis of design and analytical techniques for wildlife radio-tracking studies.

The use of radiotelemetry in wildlife studies has been the focus of several books. However, the design and analysis of radio-tracking studies are gener-ally of secondary consideration to technological advancements. A notable exception is *Analysis of Wildlife Radio-Tracking Data* by Gary C. White and Robert A. Garrott published in 1990. This important book was the first attempt to consolidate the analytical techniques and computer programs available for wildlife radio-tracking studies into one reference. The purpose of this book is similar: to provide an up-to-date assessment of design and analytical techniques for wildlife radio-tracking studies. In doing so, we hope to update and refocus the concepts presented in White and Garrott (1990).

This book is organized into seven sections designed to encompass research design and aspects of animal ecology that are most commonly evaluated using radio-tracking data. These sections include experimental design, equipment and technology, animal movements, resource selection, and population demographics. Each section begins with an introductory chapter that reviews the topic and summarizes current and future directions in that subject area. Subsequent chapters within each section highlight recent analytical approaches. As a result, the book highlights historical and contemporary advancements in the design and analysis of radio-tracking studies, provides practical advice for the field biologist conducting telemetry studies, and supplies directions and ideas for future research. This book is intended for conservation and wildlife biologists, managers, and students interested in the design and analysis of wildlife radio-tracking studies. It could thus be used as a primary text for advanced undergraduate and graduate students studying the analysis of radio-tracking data and would also be appropriate as a supplemental text in courses dealing with quantitative assessment of wildlife populations. Because of the quantitative focus of the book, readers should have basic knowledge of introductory statistics, with some chapters requiring more advanced mathematical training.

Clearly, peer review is critical to an edited volume such as this one. We are grateful to the following for their timely and helpful reviews: Gary Brundige, Mike Conner, Andy Cooper, Wally Erickson, Mark Fuller, Oz Garton, Bob Gitzen, Christian Hagen, Mike Hubbard, Robert Kenward, John Kern, Brian Kernohan, John Kie, Steve Knick, Mike Larson, Fred Leban, Trent McDonald, Richard Pace, Art Rodgers, Gary Roloff, Steve Sheriff, Brian Washburn, Gary White, and Scott Winterstein. Several people graciously reviewed two manuscripts. We thank students in QSCI 477 and 555 at the University of Washington for stimulating our thinking about the design and analysis of radio-tracking studies. Gary White provided constructive and valuable comments on the production of this book and we appreciate his efforts. Bob Garrott offered useful suggestions in the early stages of book development. We thank Darren Divine, Mark Fuller, and the GIS/RS/Telemetry working group of The Wildlife Society for their support, assistance, and many contributions. The School of Natural Resources, University of Missouri, and the College of Forest Resources, University of Washington, supported production of this book. Lorna Gilliland at the University of Missouri is gratefully acknowledged for her help in formatting the chapters and day-to-day management of this effort in the final few weeks. Our thanks are extended to Carol Washburn for her constructive and meticulous editorial reviews of the entire book. We thank the National Science Foundation, Custer State Park, South Dakota Department of Game, Fish, and Parks, University of Washington, University of Missouri, Boise Cascade Corporation, Washington Department of Natural Resources, Rayonier, U.S. Fish and Wildlife Service, U.S. Forest Service, the Idaho Army National Guard, USGS Biological Resources Division, W. S. Seegar, and the Rocky Mountain Elk Foundation for support of research that has been the impetus for our interest in the use of radiotelemetry in wildlife studies. We are sincerely grateful to all of these people and organizations.

Joshua J. Millspaugh
John M. Marzluff

PART **I**

Introduction

CHAPTER 1

Historical and Practical Perspectives

ROBERT E. KENWARD

The First 20 Years
The Third Decade
The 1990s
The Future

The turn of the millennium is a good time to take stock of developments in the radio-tagging of wildlife. This book comes 40 years after publication of the first papers on the use of radio-tags in wildlife (Le Munyan et al. 1959, Eliassen 1960). This book, conceived and edited by Josh Millspaugh and John Marzluff, follows by one decade a first thorough review in the *Analysis of Wildlife Radio-Tracking Data*, by Gary White and Robert Garrott (1990). It also coincides with a second review of the whole field, including the technology and field techniques in *A Manual for Wildlife Radio Tagging* (Kenward 2001), which in turn follows *Wildlife Radio Tagging: Equipment, Field Techniques and Data Analysis* (Kenward 1987). All of these books come from Academic Press, thanks to the wisdom of editor Andy Richford.

Andy Richford's doctorate was in ornithology at Oxford University, where some of the earliest European radio-tagging was started, producing papers in *A Handbook on Biotelemetry and Radio Tracking* (Amlaner and Macdonald 1980). Other useful reviews over the years have come in a concise early book by L. David Mech (1983), as chapters in manuals from The Wildlife Society (Brander and Cochran 1971, Cochran 1980, Samuel and Fuller 1994), by R. Harris et al. (1990), and in many conference proceedings.

Radio Tracking and Animal Populations
Copyright © 2001 by Academic Press. All rights of reproduction in any form reserved.

THE FIRST 20 YEARS

Radio-tags small enough for wildlife depended on the development of the transistor, which reduced the power and hence the battery mass needed for radiotransmitters. The first wildlife tags followed the radiotelemetry of physiological data from human pilots in the early days of aerospace activity. Continuous signals from early tags revealed the heart rate of chipmunks (*Tamias striatus*) in the United States (Le Munyan et al. 1959) and heart rate with wing beat from mallards (*Anas platyrynchos*) in Norway (Eliassen 1960). Despite the early European success, radio-tracking of wildlife was developed primarily in North America, where it was quickly realized that frequent transmission of short signal pulses was more efficient than continuous transmission. This permitted early tags to transmit for many days from small game species (Cochran and Lord 1963, Mech et al. 1965) as well as from relatively large mammals (Marshall et al. 1962, Craighead and Craighead 1970) and eagles (Southern 1964).

Radio-tags are used mainly to study physiology, behavior, and demography. Studies of physiology generally require tags to be implanted, and often anchored in place, which requires veterinary supervision and usually reduces detection distances. Although implants have been used productively to study temperature regulation of ectotherms (e.g., Swingland and Frazier 1980), and in a few cases to measure heart rate or other parameters (e.g., Ball and Amlaner 1980, Butler 1980, Gessamen 1980), most physiological telemetry has been confined to cages or other enclosures.

The value of radio-tags in demographic studies, especially for recording mortality and recovering fresh carcasses for *post mortem* examination, was recognized at an early stage (Cook et al. 1967, Mech 1967). Tags were also used to help record productivity, by locating mammal dens (Craighead and Craighead 1970) or finding bird nests as well as recording mortality of the wearers (Dumke and Pils 1973). This approach was extended to register survival rates (Trent and Rongstad 1974). However, the poor reliability of early radio-tags, due to the construction techniques and components (especially batteries), was liable to bias survival rate estimates. If signals vanished, had the tag failed on a surviving animal? Or had death and tag failure occurred together, through the impact of a predator's bite, a vehicle, or even lead shot? In the latter cases, an assumption that tag failure was independent of death would overestimate survival. Yet if tags were unreliable, an assumption that animals with failed tags had died would greatly underestimate survival.

A particular strength of radio-tags has always been to permit "sampling on demand" and therefore potentially without bias. Even if sampling of survival might be biased by death-associated failures, unbiased data on behavior and movements could be gathered before tags failed. Animals could be located and observed when required, as opposed to when vigorous movement or presence in open habitats made them especially conspicuous. The ability to locate and observe elusive species produced new ideas about social behavior of elusive predators (Macdonald 1979). Such "radio-surveillance" also gave detailed data on predatory behavior and kill rates (Mech and Korb 1977, Fritts and Mech 1981). Predation rates of individual goshawks (*Accipiter gentilis*) were combined with mark-resighting estimates of density, again based on radio-tags, to estimate total predation impacts (Kenward 1977, Kenward et al. 1981a). Long, reliable tag life was not necessary for studies of movements and behavior, and locations could even be recorded automatically (Cochran et al. 1965, Deat et al. 1980). For wildlife biologists, movements and behavior therefore became an emphasis of early radio-tagging.

However, the estimation of movements with radio-tags can be problematic too. As an extreme example, if a signal was lost, did this indicate a failed tag or long-distance dispersal? Even the methods used to estimate home ranges were found to be bias-prone. If home ranges were built up from grid cells of a size defined by the tracking accuracy (Heezen and Tester 1967), the area plotted was strongly dependent on the number of locations until there were very large numbers, unless rules were used to include extra cells (Voight and Tinline 1980). On the other hand, if outlines were plotted as developed for small numbers of trap records, the resulting convex polygons (Mohr 1947) or ellipses (Jennrich and Turner 1969) were greatly influenced by excursive movements (Dunn and Gipson 1977, Macdonald and Amlaner 1980). Therefore, estimated areas would be biased if excursions were inadequately recorded. Moreover, ellipses and polygons extended over areas unused by animals. Polygon edges could be restricted in length to avoid such areas (Harvey and Barbour 1965), but this involved an arbitrary decision about the edge length permitted in these concave polygons.

The first two decades also saw the introduction of equipment that would excel in later decades, including the use of satellites to track UHF (ultra high frequency) tags on wildlife (Buechner et al. 1971, Kolz et al. 1980) and circuits to give more stable performance from VHF (very high frequency) transmitters (Taylor and Lloyd 1978, Anderka 1980, Lotimer 1980). There was realization that systematic radio-tracking could be used to estimate relative use of different habitats (Kohn and Mooty 1971, Nicholls and Warner 1972, Marquiss and Newton 1982) provided that suitable maps were available. Johnson (1980) noted that use of resources could be examined at different levels, from global distribution (level 1) through placement of ranges in landscapes (level 2) and of radio locations within ranges (level 3) to, for example, height within a tree (level 4), and also introduced the concept of assessing habitat preference by ranking.

PLATE 1.1 A radio-tagged goshawk (*Accipiter gentilis*), released to reestablish the species in southern Britain, eats a rabbit (*Oryctolagus cunniculus*), its first kill.

Following the excitement of any new development comes anti-climax, and this began for radio-tracking at the end of its first two decades. Perhaps because of an early emphasis on recording behavior, early studies were deemed too qualitative for some observers (Lance and Watson 1980), except when radio-tags were used for the "hard-science," such as comparing experimental animals and controls in the field (Ims 1988, Kenward et al. 1993). For conservation biologists, radio-tags started to become useful for monitoring survival of animals in reintroduction programs (Hessler et al. 1970, Kenward et al. 1981b). Tags were mounted so that they would be shed from legs or tail feathers after monitoring successful dispersal or the making of a few kills that indicated successful transition to the wild (Plate 1.1). There was growing realization that tags themselves might affect survival if attached as harnesses to birds for longer studies.

THE THIRD DECADE

The first two decades can perhaps be seen as a "honeymoon" period for wildlife radio-tagging. In the following decade, radio-tags were used increasingly in large numbers. Much effort went to address analysis issues that had started to emerge in the early years, and analysis techniques proliferated.

Early estimates of survival rate had been based on assumptions that rates were constant through time. New methods could be applied (Trent and Rongstad 1974, Heisey and Fuller 1985) to estimate loss rates for bird clutches

(Mayfield 1961, 1975). This approach could be modified to include survival rates that were gradually changing at a constant rate (Lee 1980, Cox and Oakes 1984) because mortality was typically greatest in young animals. A mean rate could then be estimated across a number of time periods and parametric tests used to compare it with other estimates. The assumption that rates were constant or changing at a constant rate could be avoided with the Kaplan-Meier approach (Kaplan and Meier 1958): a survival rate was estimated for each consecutive time period and then compared between samples across each period in turn with nonparametric tests (Pollock et al. 1989a). The estimation of rates in separate intervals accommodated the addition of tagged animals during the course of a study and permitted animals with lost signals to be censored at the time of loss, with their survival included in each period before loss. Retrap-rates of animals with and without tags could be used to test for adverse effects on survival (Kenward 1982) in readily trapped species, such as squirrels (Plate 1.2). A comparison of re-recording rates after normal tag failure or unexplained loss could also be used to correct survival rates for the "loss = death?" ambiguity (Kenward 1993).

In home range analyses, problems with outliers could be avoided by returning to an earlier concept of home range as an area used by animals in the course of their "normal" activities (Burt 1943). Methods were developed for defining range "cores." A single polygon could be plotted around the locations closest to a center or where density was greatest (Hartigan 1987, Kenward 1987), but this did not suit multinuclear or curved patterns of range use. Multinuclear alternatives were developed to outline contours

PLATE 1.2   A gray squirrel (*Sciurus carolinensis*) being radio-tracked in Britain to study the stripping of tree bark by this introduced North American species.

across a matrix of density functions (Dixon and Chapman 1980, Worton 1989), to plot polygons around clusters of locations with minimal nearest-neighbor distances (Kenward 1987), or to create tiles based on neighbor distances and use the smaller of these tesselations as a range core (Wray et al. 1992).

The new analysis techniques required extensive computation, but software was becoming available for biologists to use on personal computers (Lewis and Haithcoat 1986, Stüwe and Blohowiak 1985, Ackerman et al. 1990, Kenward 1990). In a stock–taking of the many techniques that had appeared for analyzing demography and behavior of radio-tagged animals, White and Garrott (1990) also provided programs for survival analysis.

THE 1990s

Following a decade of analysis innovations, the 1990s saw substantial improvements in many types of technology for radio-tagging. Especially notable was the introduction for civilian use of the military global positioning system (GPS). A GPS receiver estimates its location from the time delay between signals reaching it from a network of 24 satellites. GPS was valuable for biologists to fix their own position when radio-tracking in poorly mapped areas, and receivers could also be used in tags to store animal locations for subsequent tag recovery or retransmission (see Chapter 4). The early GPS tags were relatively heavy (Rempel et al. 1995, Moen et al. 1996), but efforts to develop watch-sized units for humans reduced the size of potential animal tags to about 30 g (Hünerbein et al. 1997). With correction of the errors that have been introduced to stop the system guiding non-U.S. missiles, tracking resolution can be better than 100 m (Carrel et al. 1997).

This accuracy was an order of magnitude better than the tracking resolution being obtained for tags that transmitted to satellites in the Argos location system. The Argos system estimates a vector from a passing satellite to a platform transmitter terminal (PTT) by the Doppler shift of the PTT's carefully controlled frequency. If two or more vectors are estimated during a satellite pass, a tag's location can be estimated on either side of the satellite track, with the correct side determined by previous locations or natural history (e.g., whales remain in the sea). Whereas a GPS tag can estimate its altitude if signals are received from at least four satellites, accuracy of a PTT location is reduced if tag altitude is unknown (Keating 1995). Argos tracking is ideal for studying long-distance movements in remote areas, such as the foraging of albatrosses or whales (Jouventin and Weimerskirch 1990, Mate et al. 1997). With tags of 20 g now available, much is being discovered about migration routes and wintering areas of rare birds (e.g., Fuller et al. 1995, Higuchi et al. 1996,

Meyburg and Meyburg 1998), in one case revealing danger from pesticide use in wintering grounds (Woodbridge et al. 1995).

Another technology that developed rapidly in the 1990s is the storage of data in tags, for recovery with the tag or for retransmission. With appropriate data compression techniques, even video material can be transmitted to the Argos system (Seegar et al. 1996). Tags that have recorded movements of large pelagic animals can be released to transmit data from the ocean surface. If high-accuracy tracking is required in remote areas, GPS data can now be transmitted to the Argos system from tags as small as 70 g (Howey 2000).

For managing and conserving animals, knowledge of habitat use is becoming more and more important. Habitat studies typically require accurate tracking, and benefit from new mapping technologies. Geographic information systems (GIS) make it practical to create complex maps, combining data on topography, soils, vegetation, human activities, and climate. In small areas, contamination of animals with pesticides can be related to their use of fields with mapped pesticide treatments (Fry et al. 1999). Satellite images can be used to relate movements of radio-tagged animals to habitat maps over larger areas. Maps of landcover can now be prepared for whole countries (Fuller et al. 1994), so that animals can indicate which habitats they favor despite moving over wide areas (Marzluff et al. 1997a, Walls et al. 1999). New methods for assessing resource selection (Aebischer et al. 1993) and importance of habitats are based on variation in selection patterns (Mysterud and Ims 1998) or in areas used (Kenward et al. in review).

Despite advances in the use of satellites for radio-tracking, PTTs and GPS tags remain relatively expensive in comparison with the VHF tags used in most projects. VHF tags have benefited greatly in reliability from the introduction of new circuits, surface-mount construction techniques, and batteries designed to military specifications. High reliability and long tag lives mean that animal deaths greatly outnumber tag failures. The minimum survival rates, based on an assumption that lost animals are dead, are then reasonably close to the maximum survival rates that use only the recorded deaths (Fig. 1.1). Minimum rates provide conservative estimates for modeling impacts of human actions on rare species. Moreover, low loss rates enable logistic regression to be used for robust investigation of factors that may affect survival between seasons.

New techniques for careful fitting of harnesses have decreased impacts on foraging performance (Vekasy et al. 1996, Marzluff et al. 1997b) and survival of raptors (Kenward et al. 2000, in press). Raptor survival rates from banding data are especially vulnerable to age-specific finding bias, and for several species the juvenile survival estimates from radio-tagging are substantially higher than from banding (Kenward et al. 1999, 2000,

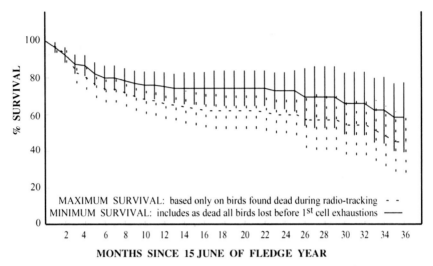

FIGURE 1.1 Survival estimates, with 95% Cox-Oakes confidence intervals, for radio-tagged common buzzards (*Buteo buteo*) in southern England. Maximum survival rates (solid lines) are based on recorded deaths alone; minimum survival rates (dotted lines) include birds whose signals were lost inexplicably. Divergence between estimates is minimized by the use of reliable tags and tracking techniques.

Nygård et al. 2000). Such data are leading to reassessments of raptor populations (Hunt 1998) at a time when radio-tagging of prey has helped to record unexpectedly high predation impacts on some game species (Thirgood and Redpath 1997).

THE FUTURE

Prediction is always a risky business. However, it may be reasonable to extrapolate a little from the way in which radio-tagging has developed to date. Improvements in automated data collection, including GPS tags and other new radio location systems (Burchardt 1989, Spencer and Savaglio 1995), will surely increase the volume of data available from individual animals. Microprocessors will be used increasingly in tags, both to enhance transmission life and to enable output from more sophisticated sensors. This will increase the quality of data from each animal. There is now general acceptance of the view that individual locations lack statistical independence (Kenward 1992, Aebischer et al. 1993, Otis and White 1999; see also Chapter 5), which focuses attention on collecting data from large numbers of individuals. Studies will therefore collect more data from more animals.

The need to process large volumes of data will increase the importance of software for processing and analysis. Without such software, projects can drown in data. However, sufficient software is already available to motivate comparative reviews of utility for analyzing home ranges (Larkin and Halkin 1994, Lawson and Rodgers 1997). Software is increasingly being used for analysis of large datasets to investigate when best to use different outline methods (Robertson et al. 1998, Kenward et al. in press b). The fundamental difference between all of these methods is in the degree of smoothing, along an axis from location-density methods with one center (ellipses), through density contouring, to methods based on linkages between pairs of locations that create abrupt polygon edges, including multinuclear outlines from cluster analysis of nearest-neighbor distances. Whereas ellipses give stable size estimates from few locations (see Chapter 2), and contours may give the most significant results when examining effects of range size in homogeneous habitats or for wide-ranging species in fine-grained habitats, cluster polygons often give strong results when habitats are coarse grained and for separating territories. Automated processing could save much time if consensus can be reached on *a priori* choice of estimator for different analyses. Such a choice will probably depend on the biological questions, numbers of locations, and habitat structure. (See Chapters 2, 5, 8, and 11).

Software is important for much more than analyzing home range. It is important for analyzing survival, estimating densities in ways that may involve radio-tagging (White and Garrott 1990), determining when animals disperse or stop dispersing, and generating indices of interaction with resources and other individuals (Kenward 2000). The latter interactions may, but need not, be based on range outlines. Movements of some species make home ranges hard to define (Gautestad and Mysterud 1995), and at other times it may be inconvenient to collect many locations. With tags that transmit reliably for several years, survival and productivity can be monitored by recording locations three to five times a year along animal life paths. Nests and other focal points estimated from locations between breeding seasons can then be used to obtain strong relationships between the performance of individuals (survival, persistence, breeding) and habitats in expanding circular buffers around the focal points (Kenward et al. in press c).

The linking of data from radio-tagged animals with sophisticated GIS data on their environments will enable the construction of increasingly complex population models. Models are needed in ecology to predict the responses of populations and eventually whole systems to changing landscapes and climates. Models that associate presence with habitat have already been built by radio-tracking (Lamberson et al. 1994), but more complex models will be based on simulating mechanisms of foraging, dispersal, and breeding of individuals. Such models are already being built with detailed data recorded visually from flocking birds (Goss-Custard 1996, Sutherland 1996, Pettifor et al. in press). Radio-tags could well be the key to obtaining appropriate data from the majority of animals, which are dispersed and elusive.

PART II

Experimental Design

CHAPTER 2

Experimental Design for Radiotelemetry Studies

EDWARD O. GARTON, MICHAEL J. WISDOM, FREDERICK A. LEBAN, AND BRUCE K. JOHNSON

A substantial literature is available to assist investigators in planning projects that use radiotelemetry. Amlaner and Macdonald (1980) edited one of the first collections of papers, published as *A Handbook on Biotelemetry and Radio Tracking*. Kenward (1987) offered excellent suggestions on study design, equipment, field methods, and analysis techniques for wildlife telemetry studies.

Radio Tracking and Animal Populations
Copyright © 2001 by Academic Press. All rights of reproduction in any form reserved.

White and Garrott (1990) provided detailed suggestions for designing radio-telemetry systems and studies, estimating home range, analyzing habitat use, estimating survival rates, and using radiotelemetry for population estimation. Priede and Swift (1992) edited a more recent collection of papers on technical aspects of radiotelemetry and satellite tracking, with extensive coverage of individual groups of animals from fish and crustacea to mammals and birds. Samuel and Fuller (1994) provided some of the most helpful suggestions for designing wildlife studies that use radiotelemetry; their suggestions included detailed but succinct sections on study design, telemetry equipment, field procedures, and analysis.

CRITICAL QUESTIONS FOR EXPERIMENTAL DESIGN

Designing an observational or manipulative study utilizing radiotelemetry requires that questions be answered regarding every aspect of the study, ranging from goals and objectives to selection of sample sizes and methods of analysis. We discuss some of the most important questions in the following sections. Answering these questions is impossible without careful thought, evaluation of previous studies, and analysis of data from a pilot survey or data gathered by earlier researchers.

WHAT IS THE PURPOSE OF THE STUDY?

Is the purpose of the work to study animal movement, migration and dispersal, space use, home range size, habitat use, or resource selection? Is the goal to estimate population size, density, survival rate, or fecundity? Answering these questions has strong bearing on the experimental design of radiotelemetry studies. For example, estimating simple movements to determine general migration patterns may require few locations of low accuracy during the migration period. By contrast, estimating fine-scale patterns of resource selection may require many locations of high accuracy taken across all times of day and seasons of interest. Consequently, the purpose of the radiotelemetry study should be outlined in a detailed proposal beforehand, with explicit identification as to which of the following objectives the study is designed to meet.

Movement, Migration, and Dispersal

Radiotelemetry is an essential tool in modern studies of movement, migration, and dispersal of most vertebrates. Use of radiotelemetry has dramatically

increased the amount and detail of information available for estimating movements of larger animals and provides extremely valuable additions to information from studies that use tagging, banding (ringing), and other forms of
marking for smaller animals. In addition, radiotelemetry studies are an important complement to recent approaches that use genetic markers.

Use of radiotelemetry can be especially helpful in evaluating movement
theory and models. For example, the simplest model of dispersal assumes that
animals move away from their point of birth through a homogeneous environment according to a simple random walk, analogous to Brownian motion
in physics (Skellam 1951). These simple assumptions lead to a normal distribution of individuals after a set length of time (Ricklefs and Miller 2000),
which is described by the population variance (σ^2). Unfortunately, the sampling distribution of variances from the well-known normal distribution is not
symmetrical and does not follow a normal or t distribution, but it can be
described using a chi-squared distribution (Zar 1984:115):

$$\chi^2 = \frac{vs^2}{\sigma^2} \tag{2.1}$$

Thus, the chi-squared distribution (χ^2) can be used to define a confidence
interval for dispersal distance where there is a $1 - \alpha$ chance of including the
population variance of dispersal distance (σ^2):

$$\sqrt{\frac{vs^2}{\chi^2_{(\alpha/2,\,v)}}} \leq \sigma \leq \sqrt{\frac{vs^2}{\chi^2_{(1-\alpha/2,\,v)}}} \tag{2.2}$$

where $v = n - 1$ degrees of freedom and $s^2 =$ sampling variance of dispersal
distances.

Other models of dispersal have been proposed and evaluated with radiotelemetry observations for a diversity of vertebrates (Murray 1967, Waser
1985, 1987). However, such models or analyses will produce biased estimates
if the observed dispersal distances are truncated through failure to detect long-
distance movements. Porter and Dooley (1993:2436) found that 13 of 15 field
studies they reviewed had used methods that sampled unevenly over distance.
Using simulation methods, they concluded that the effects of uneven sampling
on simple mathematical models of animal movement are profound. They
promulgated corrections to the field data to mitigate for distance-weighted
sampling and showed that "in only one of five studies do simple mathematical
models adequately fit the observed distribution of movements." Telemetry can
help to decrease the bias against detection of long-distance dispersal events by
increasing the detection distance by one or more orders of magnitude if
researchers are cognizant of this potential bias and adjust their searching
methods appropriately. Even when these adjustments are made it will

be virtually impossible to eliminate all bias against very long dispersal distances, so that analysis methods that correct for the bias must still be applied.

Space Use and Home Range Size

Movements of individual animals are typically summarized over time to describe the animal's pattern of space use or home range (see Chapter 5). Home ranges have been measured in a myriad of ways: bounded areas or polygons (Mohr and Stumpf 1966), probability areas such as ellipses (Jennrich and Turner 1969), and nonparametric probability contours (Worton 1989) are typical metrics. Space use also has been described in terms of grid densities (Siniff and Tester 1965) or nonparametric density surfaces (Dixon and Chapman 1980, Anderson 1982, Worton 1989).

Two sampling components have strong bearing on measurements of space use and home range size: number of radio-marked animals and number of locations collected per animal. We address the issue of number of animals later under the questions of sample unit and sampling intensity. The classic approach to estimating the required sample size within animals is to (1) assume a simple parametric model for the underlying distribution of the characteristic of interest, (2) obtain estimates for the variance of this distribution from the literature or a pilot survey, (3) choose a desired precision for the estimate, (4) and apply simple parametric estimators for the required sample size. For example, Dunn and Brisbin (1982) suggested the following formula to estimate the sample size (n) required to obtain a bound (B expressed as a proportion of the estimate, e.g., 0.1 for a 10% bound) on the 95% home range size estimated with the bivariate normal estimator:

$$n = \frac{z^2_{1-\alpha/2}}{B^2} + 2 \qquad (2.3)$$

where n = sample size, Z = tabled normal distribution value, α = confidence level, and B = bound expressed as a proportion of the estimate. Calculating the sample sizes required to obtain estimates of typically desired precision (Table 2.1) leads to the conclusion that much larger sample sizes are required to estimate home range size than are typically obtained in such studies.

We should not be surprised by the large sample size required because estimating home range size is analogous to estimating the variance of a distribution. Variance estimation is notoriously difficult for even simple distributions, such as the normal distribution, in comparison with estimating measures of central tendency, such as the mean or median. Applying home range estimators such as the kernel for more complex distributions will place

TABLE 2.1   Sample Sizes Required to Obtain Estimates of Home Range Size Using the Bivariate Normal Home Range Estimator

Confidence level (%)	Bound (%)	Sample size required
95	10	386
90	10	271
95	20	98
95	30	45

Data from Dunn and Brisbin (1982).

even larger demands on sample size within animals (see section on sampling intensity).

However, estimators such as the kernel have the great advantage of being able to describe the internal anatomy of the home range by identifying areas of particular significance, such as core areas (Samuel et al. 1985; see also Chapter 5). With the use of automated tracking systems, future researchers will have the capability to use more sophisticated home range estimators, which describe the details of the internal anatomy of home ranges, by estimating the density of large samples of radiotelemetry locations and assessing the precision of the estimators using the mean integrated squared error (Silverman 1986).

Resource Use and Selection

Habitat and other environmental characteristics (resources) associated with radiotelemetry locations are commonly used to characterize resource use and selection (Samuel and Fuller 1994; see also Chapters 8 and 12). A common approach is to describe habitats or resources using discrete categories of vegetation cover types (e.g., meadows, open forest, closed forest), vegetation habitat types [e.g., grand fir (Abies grandis); Daubenmire and Daubenmire 1968], topographic features (e.g., valley bottom, midslope, ridge top), and successional stage (e.g., clear-cut or burn, grass-shrub, tall shrub, pole-sized timber, mature timber, old growth forest; Irwin and Peek 1983). Most commonly, the distribution of the types used by radio-marked animals is summarized in univariate form as percent use and compared to the percentage of available cover types within the region, study area, or home range to identify resource selection (Alldredge and Ratti 1986, Johnson 1980). Such data are clearly multivariate in nature and should be evaluated more appropriately using log-linear models (Heisey 1985), selectivity indices (Manly 1974, Manly et al. 1993, Arthur et al. 1996), logistic regression (Manly et al. 1993), or multiple regression (see Chapter 12).

Alternatively, these same resource characteristics can be described as continuous variables at the point location (e.g., percent canopy cover, percent slope, elevation, distance to water, and distance to roads; Rowland et al. 1998, 2000), as percent cover of categorical cover types within a probability distribution of error that is centered on the point location (Findholt et al. in press), or as percent cover within the home range (Samuel and Fuller 1994). A variety of multivariate statistical approaches can be applied to these habitat characteristics when measured as continuous variables; approaches include principal components analysis (Rotenberry and Wiens 1980), discriminant analysis (Dueser and Shugart 1979), Mahalanobis distance (Clark et al. 1993), factor analysis (Capen 1981), logistic regression (Manly et al. 1993), polytomous logistic regression (North and Reynolds 1996), discrete choice modeling (Cooper and Millspaugh 1999, Chapter 9), and other distance measures (Petraitis 1979, 1981).

Population Abundance and Density

Monitoring animals with radiotelemetry can dramatically improve the estimates for closed-population mark-recapture experiments by ensuring that the assumption of a known, marked population is correct (Otis et al. 1978; see also Chapter 13). If the assumption of a closed population cannot be met, then methods for open populations, such as the Jolly-Seber (Seber 1982) must be used. Unfortunately, these estimates are typically much less precise (White and Garrott 1990) because of the need to estimate additional parameters related to survival. In general it is not possible to derive analytical equations for estimating sample size for complex mark-recapture studies. Instead, the standard approach is to perform Monte Carlo simulations, using the estimation programs, in advance of fieldwork and again after initial field data have been gathered. Individual capture and survival histories are chosen randomly from appropriate distributions (e.g., binomial, Poisson) and passed to the analysis software repeatedly. In many cases such a process can be automated with some simple programming, but in other cases this can be a challenging undertaking.

Another group of powerful approaches for estimating population abundance that benefits from use of radiotelemetry includes complete counts, sample counts on plots, and distance estimation. Aerial surveys of large mammals attempt a complete count of all members of a population or count animals on a sample of the area for extrapolation to the whole population, following procedures of probability sampling (Cochran 1977). Such procedures generally underestimate population abundance (Caughley 1977) because of visibility bias (Caughley 1974). Radio-marked animals can be used to develop models of visibility bias (Samuel et al. 1987; see also Chapter

13) that can be used to remove this bias (Steinhorst and Samuel 1989) and obtain unbiased estimates of population abundance and composition (Samuel et al. 1992). Line transect (Burnham et al. 1980) and variable circular plot (Roeder et al. 1987) methods of distance estimation (Buckland et al. 1993) require key assumptions that all animals directly on the line or plot center are seen and do not move away from the line or plot center before detection. These assumptions can be verified using radio-marked animals.

Survival

Monitoring animals with radiotelemetry can dramatically improve the precision and quality of estimates of survival rates compared to the classic approaches based on band recovery (Brownie et al. 1985; see also Chapter 14). One of the most powerful and flexible approaches, based on numerical optimization, has been developed extensively by White (1983). His computer program SURVIV (see Appendix A) can be used to construct likelihood functions for a variety of survival models to derive maximum likelihood estimates of the parameters as well as to construct likelihood ratio tests between models. SURVIV includes powerful procedures for designing survival studies that allow the user to perform Monte Carlo simulations of proposed experiments and evaluate sample size requirements (White 1983). Other approaches to survival estimation that are particularly applicable to radio-tagged animals include Heisey and Fuller's (1985) extension of the Mayfield (1961) method, Kaplan-Meier product limit estimators (Kaplan and Meier 1958), and Cox's proportional hazards approach (Cox and Oakes 1984).

Fecundity

Surprisingly few guidelines exist for estimating the reproductive rate for animal populations despite its great importance in determining individual fitness and population performance. Klett et al. (1986) made extensive suggestions for measuring nest success of ducks from nest visits and emphasized that timing of nest searches dramatically affects the estimates. Johnson (1994) also identified common biases associated with estimates of nest success; these biases were due to nest searches that occurred after nest abandonment or failure, resulting in overestimation of nest success. Radiotelemetry can be used to identify and correct these biases. However, radio-tags themselves can negatively affect nest success (Rotella et al. 1993), and this potential bias must be addressed (see Chapter 3). Perhaps the Heisenberg Uncertainty Principle applies to reproductive success as well as to electrons or smaller particles.

Is the Study Observational or Manipulative?

The kind and strength of conclusions that can be drawn from radiotele-metry studies are highly dependent on whether the experimental design is observational or manipulative. Manipulative studies measure responses to designed changes on experimental units through treatments, controls, and replicates that are structured as formal hypotheses to test cause-effect relations. By contrast, observational studies simply monitor the conditions of interest and document patterns without establishing cause-effect relations.

Observational studies can be descriptive or correlative (White and Garrott 1990). Descriptive studies use radio-tracking to observe natural processes of animal movement and behavior without formalizing the study design in terms of hypothesis testing. Correlative studies improve on descriptive studies in that hypotheses are formulated and tested, but no manipulations are con-ducted, which does not allow testing of cause-effect relations.

Unfortunately, most radiotelemetry studies have been observational in design. This lack of manipulative design severely limits the potential to make strong inferences about cause-effect relations, even though many obser-vational studies typically make such inferences despite the unreliability of such conclusions (see section on inference). Consequently, whether the study will be observational or manipulative is an important decision to make in the early phase of study design before fieldwork begins. This decision will ulti-mately define the resources needed to carry out the study, and the strength and kind of inferences that can be made from results.

Is the Study Designed to Test Hypotheses or Build Models?

One of the most important decisions is whether the radiotelemetry study will be used to test hypotheses or build models. This decision is not trivial, and should be made during the design phase of the project, well before fieldwork begins. Studies designed to evaluate biological relations postulated by a researcher are tested through the formal process of statistical hypothesis testing (Zar 1974). This process involves the *a priori* setting of null and alternative hypotheses, and the decision to reject or fail to reject the null hypothesis is based on statistical rules (Ratti and Garton 1994). In the simplest sense, hypothesis testing seeks confirmation of specific ideas through formal rules.

By contrast, model building is not a confirmatory evaluation but instead involves the analysis of data in a manner that provides better understanding of potential relations between predictor and response variables. For radio-telemetry studies, response variables are typically expressed in terms of animal response, such as resource selection, home range use, or survival. Predictor variables are typically the environmental conditions that presumably influence the response variables.

Ideally, the model-building exercise seeks to identify the most parsimonious model among a variety of plausible models that range from the simplest to the most complex. The most parsimonious model, which is defined as the model with "enough parameters to avoid bias, but not too many that precision is lost" (Burnham and Anderson 1992:16), can be measured by statistical indices, such as Akaike's Information Criterion (AIC) or Mallow's C_p (Burnham and Anderson 1992, 1998). The model with the smallest AIC is considered to be the most parsimonious model among a variety of plausible models tested.

Because most studies are not manipulative in design, such model-building approaches are more appropriate than hypothesis testing. Moreover, model-building approaches, because of their flexible, exploratory nature, may lead to greater insights about plausible relations of the radio-marked animals with their environment, which then can be validated through additional research (Burnham and Anderson 1998).

WHAT IS THE POPULATION TO WHICH INFERENCES WILL BE MADE?

To what population will the estimates apply? This question must be answered initially to develop a valid experimental design for any radiotelemetry study. If our goal is to make statements about all elk (*Cervus elaphus*) in North America, then our sample must be drawn from all of the elk in North America. If our sample consists of elk captured and radio-marked in the Clearwater Drainage of northern Idaho, then we can only make inferences to elk in that region.

Importantly, inferences made from radio-marked animals are valid only when animals are selected in a manner that yields unbiased samples, such as through simple random or stratified random sampling (Scheaffer et al. 1986); this issue is discussed later. Moreover, data from animals located during daytime only, or during specific seasons, cannot be used to make inferences about patterns of animal use during other periods of day or other seasons (Beyer and Haufler 1994). Similarly, estimates generated from data collected at one geographic scale, such as a watershed, may not provide reliable inference to a distinctly different geographic scale, such as an ecoregion.

Specification of the incorrect inference space, or lack of explicit identification of appropriate inference space and associated limitations, is a common problem among most radiotelemetry studies. For example, of 31 radiotelemetry studies that were published in *The Journal of Wildlife Management* during 1991–1996, only 8 (26%) described the population to which inferences were intended. Of the 23 studies that did not clearly identify inference space, all 23 made generalized inferences to geographic areas and populations beyond the boundaries of the study areas, and 12 (52%) made inferences to all populations over the entire range of the species that were studied.

WHAT IS THE SAMPLE OR EXPERIMENTAL UNIT?

Traditionally, each individual location was used as an independent observation from the population of interest (Neu et al. 1974), but more modern approaches correctly identify the individual animal as the experimental or sample unit (Johnson 1980, Aebischer et al. 1993) and individual locations as subsamples. Thus, the sample size becomes the number of animals from the population that are radio-marked (Otis and White 1999). For social species every effort should be made to radio-mark animals that do not associate within the same social group as only animals from different social groups can be treated as independent sample units (Millspaugh et al. 1998a).

Independence of sample units is motivated by the need to minimize pseudoreplication. Pseudoreplication refers to the dependence or correlation of sample units, and the artificially high estimates of precision that are generated from such a treatment of samples. Pseudoreplication is especially problematic when conducting statistical hypothesis testing (Hurlbert 1984). The lack of sampling independence increases the likelihood of falsely concluding that statistical differences exist among samples (type I errors). Concerns about pseudoreplication can be problematic for radiotelemetry studies because animal locations as well as animals are spatially and temporally correlated, and thus not independent (Millspaugh et al. 1998a).

Pseudoreplication can be minimized by using animals as the sample unit (Pendleton et al. 1998, Otis and White 1999) and by ensuring that radio-marked animals are randomly selected from the population. Importantly, pseudoreplication is not a problem for collection of subsamples, such as the use of radiotelemetry locations as subsamples to estimate each animal's movement or space use. A more important issue is whether the sample and subsamples are unbiased estimators of the response variables of interest, regardless of the correlation associated with these estimators. Assessing potential biases and minimizing them is challenging because it requires as much insight into the species biology as statistical or technological sophistication. Classic calculations of variance and standard errors only measure the precision of the estimators rather than addressing bias. Correct specification of the sample or experimental unit helps to provide reliable measures of precision for hypothesis testing or model building. Investigators must continually question whether the data gathered will provide unbiased estimates of the population characteristics that they wish to test or model.

WILL SAMPLING BE STRATIFIED, SYSTEMATIC, CLUSTERED, OR RANDOM?

Sample Design within Animals

The traditional design for obtaining observations within animals might be described as something between serendipity and systematic sampling. New technology for automated radiotelemetry allows intensive monitoring of animals, which allows us to draw truly systematic samples within animals. However, any system could draw approximately systematic samples by time, stratified samples by time of day, or samples clustered in time.

Aerial sampling with radiotelemetry has the great advantage that animals that could move substantial distances within short periods of time can still be located; that is, aerial surveys can cover large geographic areas in a short period. However, disadvantages of aerial sampling include the limited precision of location coordinates (large-error polygons based on speed of flight of receiving station), the inherent dangers of night-time sampling, and the limited detail of visually determined habitat data. By contrast, ground-based sampling has the advantage that detailed behavioral and habitat data can be collected simultaneously, but its disadvantages are substantial: bias against detection of long-distance movements, potential disturbance of animals, danger and difficulty of collecting night-time locations, and errors in estimating animal locations (size-of-error polygons and problems of signal bounce in rugged terrain).

Swihart and Slade (1985a) demonstrated that most methods of analyzing animal movements assume that an animal's position at time $t + 1$ is independent of its position at time t, and they provided a statistical test of this assumption. In a more recent simulation study, Swihart and Slade (1997) showed that exclusive use of independent observations is unnecessary but that it decreases the inherent negative bias in kernel and polygon estimators of home range size.

Estimating the serial correlation between successive observations of space and habitat use is important for designing a sampling interval that maximizes information content of each observation while ensuring adequate sample size to estimate use of rare habitat types. In that regard, systematic sampling (Scheaffer et al. 1986) can be designed to minimize serial correlation between successive locations by lengthening the time interval between locations. However, long time intervals between locations reduces the overall number of locations that can be gathered for a given radio-marked animal and thus may preclude collection of data about fine-scale movements and resource use in space and time.

By contrast, clustered sampling (Scheaffer et al. 1986) will result in high dependence of locations within each cluster but low dependence between clusters. Clustered sampling, therefore, can reveal fine-scale movements and resource use through collection of some of the data at narrow time intervals. However, collection of such fine-scale data may be at the expense of obtaining more data of relatively high independence if the available technologies or resources limit the overall number of locations.

Random sampling, while theoretically sound, often is inefficient and logistically infeasible without the use of automated tracking systems. Efficiency can be improved with the use of stratified random-sampling schemes (Scheaffer et al. 1986). In particular, stratified random sampling or stratified clustered sampling may be of high utility with the use of automated tracking systems: many locations per animal can be collected, allowing time-based strata that achieve relative independence of locations across strata, but assuring collection of finer scale data about animal movements and use within each stratum.

Sample Design across Animals

An equally important consideration is how sampling will be conducted across the radio-marked animals. This topic has not been evaluated in detail. However, the same concepts and relations logically apply to sampling across animals as those discussed above for sampling within animals.

In particular, two related issues appear to be important in designing the method for sampling across radio-marked animals. The first is that the radio-marked animal is the appropriate sample unit (Otis and White 1999), and sample units should be relatively independent to avoid pseudoreplication (Millspaugh et al. 1998a). Systematic or time-based stratified sampling can be used effectively to address this issue, as discussed above for sampling within animals. Second, an inherent trade-off exists between sampling schemes that emphasize collection of data on more radio-marked animals, with fewer locations collected per animal, vs. collection of more locations per animal but monitoring fewer radio-marked animals. We address this issue in the next section.

WHAT SAMPLING INTENSITY IS NEEDED?

How many radio-marked animals should be monitored, and how many locations per animal are needed to accurately estimate the true parameters of animal movement, resource use, home range use, or survival? This question is fundamental to appropriate design of radiotelemetry studies but often is ignored.

The classic approach to determining sample size requirements for experiments involves obtaining preliminary estimates of means and variances, or proportions, for the characteristics of interest, specifying the magnitude of the differences that are biologically meaningful and desirable to detect, choosing a desired power for the test, and calculating required sample sizes (Zar 1984). In a comparable way, survey design requires obtaining the same estimates, specifying the size of the desired precision for the estimates, choosing power, and calculating required sample sizes (Scheaffer et al. 1986).

This classic approach is complicated by the fact that many of the characteristics of interest, such as movement patterns, space use, and habitat selection, differ by time of day and season of year. Consequently, more sophisticated designs are often required that incorporate stratification in sample designs or blocking in experimental designs (Ratti and Garton 1994). These sophisticated designs require preliminary estimates of means and variances (or proportions) within strata (or blocks); these estimates will be available only from preliminary data obtained during a pilot survey or during initial stages of sampling. Such a situation may require an adaptive approach to the design in which the initial design based on guesses or pilot surveys will need modification after the first full field season (or set of experiments).

We evaluated the effects of sample size on the error in estimating home range size and resource selection for a large sample of adult female elk monitored intensively at the Starkey Experimental Forest and Range (Starkey), northeast Oregon (Rowland et al. 1997, 1998). Specifically, we estimated size of home ranges for 42 radio-marked elk from Starkey, each of which had been intensively monitored by collecting a large subsample of 300 or more locations per animal during spring–summer 1994. These 42 elk composed approximately 15% of the adult female elk population at Starkey and, combined with the intensive sampling of each radio-marked animal, lend themselves to evaluations of errors in parameter estimates under different levels of sampling intensity. Our evaluation was motivated by a problem common to all radiotelemetry studies. That is, given limited resources, a radiotelemetry study must allocate effort among two types of choices: increasing the number of animals marked or increasing the number of locations obtained per animal.

Using these elk data from Starkey under a simulation method, Leban et al. (Chapter 11) and Leban (1999) suggested that a minimum of 20 radio-marked animals, with at least 50 locations per animal, were needed to obtain accurate estimates of resource selection of elk for a given season. Leban's conclusions were based on simulations that assessed effects of varying the number of radio-marked animals and the number of locations per animal on accuracy of estimated resource selection under four methods of analysis comparing used and available habitats: χ^2 goodness-of-fit test (Neu et al. 1974), Friedman (1937) test of differences in ranks, Johnson's (1980) comparison of average difference in ranks, and compositional analysis (Aebischer et al. 1993). Alldredge and Ratti (1986, 1992) used similar methods and came to similar conclusions regarding the sampling intensity needed to accurately estimate resource selection. Overall Leban et al. (Chapter 11) recommended use of compositional analysis because of its correct treatment of animals as the sample/experimental unit and its modest sample size requirements.

In a similar manner, we used resampling simulations to test how well the size of home ranges could be estimated under varying sampling intensities, using the same data from Starkey and the same methods described and used by Leban et al. (Chapter 11) and Leban (1999). Results from these simulations suggested that up to 200–250 locations per radio-marked animal, combined with radio-tracking of at least 20 animals, may be needed for a given season to estimate size of home range to within ±10% of the true size, based on the adaptive kernel estimator (Fig. 2.1).

By contrast, simulation results using the bivariate normal estimator suggested that estimation to within ±10% of the true size of the home range for the population could be achieved with as few as 10 locations per animal, combined with radio-tracking of at least 30 animals for a given season. Alternatively, our results suggested that estimation to within ±10% of the true size of the home range could be achieved with radio-tracking of 20 animals with 20 locations per animal when using the bivariate estimator (Fig. 2.1).

How could the sample size requirements for the adaptive kernel and bivariate normal estimators differ so markedly when applied to the same population of elk? The kernel estimator is a nonparametric estimator that approximates a bivariate normal estimator with small sample sizes but also approximates more complicated multimodal distributions at larger sample sizes because it makes relatively few assumptions about the shape of the distribution except that it is continuous (see Chapter 5). At large sample sizes it provides a very detailed description of the pattern of use and distribution of observations in space. The bivariate normal estimator is a parametric estimator that assumes an underlying bivariate normal distribution of observations of use. Such an estimator provides an excellent *generalized* measure of space use because it is unbiased with respect to sample size; however, it does not provide a good description of actual areas used at any location unless the

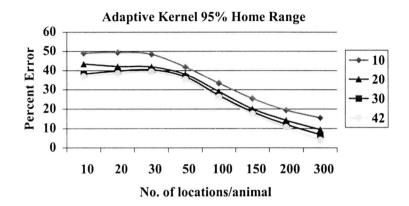

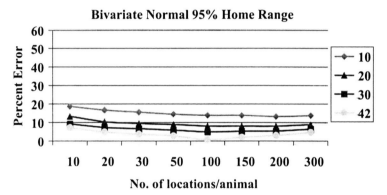

FIGURE 2.1   Mean error of home range size in relation to the true size, using the adaptive kernel and bivariate normal estimators, under varying number of resampled locations and animals, using data from 42 radio-marked elk whose movements were monitored in spring-summer, 1994, at the Starkey Experimental Forest and Range, northeast Oregon, USA. Resampling procedures follow methods described by Leban et al. (Chapter 11) and Leban (1999). Number of radio-marked animals was varied from 10, 20, 30, and 42 in the simulations (shown as diamonds, triangles, squares, and circles, respectively) and number of locations per radio-marked animal was varied from 10 to 300, as shown on the x axis.

underlying pattern of use is described by a bivariate normal distribution. As a parametric estimator the bivariate normal is highly efficient and would be expected to provide good estimates at small sample sizes in comparison with a nonparametric estimator, such as the kernel.

However, our results for the bivariate normal estimator differ substantially from the much higher sample sizes calculated from the theoretical equation of

Dunn and Brisbin (1982) when using the bivariate normal estimator to achieve relatively high accuracy of home range size estimates for individual animals (Table 2.1). Our results also differ from the lower sample size recommendations of Seaman et al. (1999) using kernel estimators; they concluded that bias in individual home range estimates reached an asymptote in many cases at 50 observations, leading them to recommend at least 50 locations per animal to obtain accurate home range estimates for individual animals. In contrast to their general recommendation, Seaman et al. (1999) showed that bias actually was quite large for complex distributions and declined substantially as sample sizes surpassed 50. Neither the theoretical approach of Dunn and Brisbin (1982) nor the simulation approach of Seaman et al. (1999) evaluated the needs for sample size from the population perspective, nor did they evaluate more powerful sampling approaches than simple random sampling. Both of these earlier groups of authors evaluated sample size requirements for estimating home range size for individual animals rather than for populations as we did. Seaman et al. (1999) did not evaluate trade-offs between number of animals sampled vs. number of locations collected per animal, as done here. Moreover, Seaman et al. (1999) did not use empirical data for their simulations but generated independent random observations from underlying unimodal or multimodal bivariate normal distributions. Consequently, parameters associated with such simulations are likely to be different from the parameters inherent to our use of empirical data for simulation. These differences in methods of simulation between our work and that of Seaman et al. (1999) may account for the difference in sample sizes calculated for accurate estimation of home range.

Powell et al. (2000b) also used simulations to estimate the sampling intensity required to obtain accurate estimates of survival and movement of radio-marked animals. Powell et al. (2000b) concluded that at least 25 radio-marked animals per season and geographic stratum were needed to estimate survival, movement, and capture rates accurately with the use of radiotelemetry.

The composite results of the above studies suggest that at least 20–25 radio-marked animals are needed to obtain accurate estimates of resource selection and survival, with at least 50 locations per animal per season of interest needed to estimate resource selection accurately. By contrast, accurate estimation of home range size may require 50–200 or more locations per animal per season of interest, combined with monitoring of 20 or more animals, depending on the type of home range estimator and the manner in which the requirements are calculated (Fig. 2.1, Table 2.1; Seaman et al. 1999).

Unfortunately, almost no radiotelemetry studies published to date have achieved a minimum sampling intensity of both 20 radio-marked animals and 50 locations per animal that may be required for accurate estimation of resource selection. Similarly, home range studies typically contain far fewer locations than 200 per radio-marked animal. These potentially high sample sizes needed to obtain accurate parameter estimates for radiotelemetry studies warrant further research to better understand the relations among number of radio-marked animals, number of locations collected per animal, and the accuracy of the associated response variables.

Interestingly, automated tracking systems based on GPS often monitor few animals intensively rather than monitoring a higher number of animals less intensively. The tendency to monitor few animals is due to the high cost of GPS collars vs. the relatively low cost of collecting a high number of locations per animal fitted with such a collar. One exception was the LORAN-C automated system described by Rowland et al. (1998), which monitored up to 118 animals per month and collected an average of more than 235 locations per animal per month, 1992–1996 (Johnson et al. 2000, Rowland et al. 2000).

Importantly, the tendency of studies that use GPS tracking technology to monitor few animals intensively would make such studies useful for estimating home ranges of the specific radio-marked animals, but probably insufficient to estimate resource selection, survival, or home range use for a larger population to which inferences are intended. Referring to this problem in relation to GPS technology, Otis and White (1999:1042) stated that "collection of a large number of locations from a few individuals does not imply that the entire population is behaving the same as those sampled, and inferences to the population are made from a small sample of individuals." Accordingly, we suggest that use of automated tracking systems will improve our estimates of resource selection, survival, and home range use only if a minimum of 20–25 radio-marked animals are monitored.

ARE RADIO-MARKED ANIMALS AN UNBIASED SAMPLE?

One of the most overlooked factors in the design of radiotelemetry studies is the question of whether the radio-marked animals are an unbiased sample of the population to which inferences will be made. Two major sources of sampling bias can exist: (1) effects of nonrandom capture location and (2) marking-induced or relocation-induced changes in animal behavior and survival.

Nonrandom capture location can bias the estimation of animal distribution, resource selection, and a myriad of other activity patterns. For example, traps used to capture and radio-mark animals often are placed along roads or other access points. Radio-marking animals captured along roads may oversample the portion of the population that inhabits areas near roads and undersample the portion that occurs far from roads. If the study is designed to assess resource selection in relation to roads or human activities, this bias could pose a serious flaw in study design. Moreover, if other resource variables have substantially different values near roads than far from roads, estimates of resource use and selection may be biased.

Effect of marking-induced or relocation-induced changes on animal behavior and resultant estimates of movement, resource use, home range use, and survival also is a potential flaw in study design (see Chapter 3). White and Garrott (1990) described the quandary of attempting to determine whether radio-tagged animals were behaving in a manner similar to nontagged animals. Independent sources of data are needed to assess behavior of radio-marked vs. non-radio-marked animals, and the assessment often is confounded by lack of power. Moreover, the methods used to assess behavior and survival of non-radio-marked animals may also be biased or function as a suitable alternative to radiotelemetry methods. Also, the effects of radio-marking on behavior and survival vary strongly by species and the methods of radio-marking (e.g., Paquette et al. 1997, Hubbard et al. 1998, Rotella et al. 1993, Schulz et al. 1998; see also Chapter 3). Finally, changes in behavior and survival of radio-marked animals can occur as a result of the relocation methods. For example, attempts to find and relocate avian nests can result in nest abandonment or increase the vulnerability of such nests to predation (MacInness and Misra 1972, Picozzi 1975). Consequently, researchers should conduct a thorough literature review of potential and documented effects of radio-marking and relocation on behavior and survival for the species and taxon being studied before the study is initiated, and adjust study design accordingly. (see Chapter 3).

ARE LOCATION ERRORS ADDRESSED?

Treating radiotelemetry locations as "error-free" point estimates can result in falsely concluding that differences exist in animal movements, resource selection, home range use, and survival (type I errors). Moreover, inaccurate locations reduce the power to detect real differences in animal movements, resource selection, home range, and survival, causing high rates of type II errors (i.e., falsely concluding that no differences exist among samples). These problems are further confounded by inaccurate estimation of resource variables that are associated with the animal locations. To avoid high rates of type I and type II errors, users of radiotelemetry must identify and account for the accuracy of the animal locations as well as the estimates of the resource variables that underlie the locations. Factors affecting telemetry error and techniques to estimate it are discussed in Chapters 3 and 4 of this book. These reviews conclude that detailed study of error is a critical part of any telemetry study using remote estimation of animal location and must be a substantial element of the experimental design. Consult these reviews in addition to the brief comments below.

Estimating Location Error

White and Garrott (1990) described methods to identify and account for location error that exists for a variety of radiotelemetry systems. Location error can vary widely, depending on methods of relocation. For example, direct observation of radio-marked animals is as accurate as the researcher's capability to map the animal's location, and error can be as small as 1 m when hand-held GPS units are used to map locations. By contrast, triangulation error can be as large as 500 to 1000 m (White and Garrott 1990).

New automated systems also can have relatively large location errors, depending on the technology and the environment in which the system operates. Findholt et al. (1996) described a LORAN-C automated system having a mean location error of 53 m (SE = 5.9). Rempel et al. (1995) documented median location errors of 51–74 m with a GPS tracking technology. Britten et al. (1999) reported errors as large as 3500 m with the use of small satellite transmitters.

Estimating location error typically involves the placement of radiotransmitters in known locations, and estimating each transmitter's location with the methods used for the study (White and Garrott 1990; see also Chapter 3). To estimate location error accurately, radiotransmitters must be placed in a variety of environments that span the conditions of the study area. Often, location error will vary spatially with differences in topography, slope, canopy closure, and other environmental factors. For example, steep canyons or dense vegetation may cause radiotelemetry signals to be reflected (bounce) or diffracted (spread), and substantially increase location error (Samuel and Fuller 1994). These problems illustrate the need to estimate location error for the unique conditions of the area in which the radiotelemetry study is conducted, and with the specific methods to be used for sampling the radio-marked animals.

Accounting for Location Error

Once location errors have been estimated and mapped (Fig. 2.2), a detailed analysis of their implications on the study's goals and inferences should be conducted. Findholt et al. (in press) and Samuel and Kenow (1992) mapped the spatially explicit probability distributions of location error in relation to resource values of interest (Fig. 2.3). Results were expressed in terms of the frequency at which estimated resource use was misclassified based on location error.

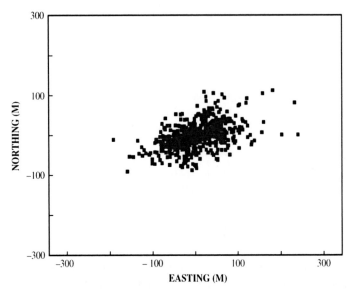

FIGURE 2.2 Error distributions of bias-corrected radiotelemetry locations from 39 collars for an automated telemetry system at Starkey Experimental Forest and Range, northeast Oregon (Findholt et al. 1996, in press). Location error was determined by comparing locations determined from the automated telemetry system to differentially corrected GPS locations of each collar.

Because most resource variables are spatially correlated, Findholt et al. (in press) found that point estimates of radiotelemetry locations, which always were nested within the highest probability values of the true location, typically provided accurate estimates of resource use (Fig. 2.3). Moreover, accuracy of resource use did not improve by estimating resource use under all or part of the probability distribution of location error.

Samuel and Kenow (1992) suggested that random points be assigned within the location error distribution to estimate the most likely resource values associated with each animal location. Similarly, the actual probability distribution of location error (or 1 minus the probability of error) can be mapped and placed in a grid of landscape pixels in relation to the habitat types of interest, allowing the researcher to calculate habitat use associated with each animal location as a weighted area estimate, with the probabilities used as weights (Findholt et al. in press; Fig. 2.3). When location error also varies by time of day or season, these adjustments to estimate resource use or movement must be made for each time period in which animals are sampled.

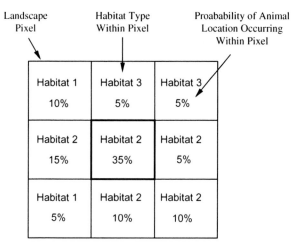

FIGURE 2.3 Conceptual relation between habitat types occurring within mapped landscape pixels and the probability of a radiotelemetry location occurring within the pixels, based on calculation of the probabilities from maps of location error, such as those generated by Samuel and Kenow (1992) and Findholt et al. (1996, in press). Notably, the central pixel contains the point estimate of the radiotelemetry location and logically has the highest probability of the true location. Moreover, because habitat types are spatially correlated, habitat types surrounding the point location (central pixel) are most often the same habitat type (Habitat 2 in this case). Consequently, the sum of probabilities across habitat types reveals that Habitat 2 has the highest probability of containing the location (75%), followed by Habitat 1 (15%) and Habitat 3 (10%). Importantly, the conclusion reached about which habitat type is associated with the highest probability of the true animal location (Habitat 2) is the same regardless of whether the conclusion is made based on the point estimate (central pixel) or the probability distribution encompassing all pixels. This would likely be the typical case for most radiotelemetry studies of habitat selection.

Accounting for Errors in Estimating Resource Values

By convention, estimates of resource values, such as maps of habitat types, typically are assumed by wildlife researchers to be measured without error. This assumption is not defensible even though it is a critical assumption of almost all analysis methods. For example, habitat maps can have classification errors of 50% or higher, and such errors may not be normally distributed (Goodchild and Gopal 1992). Failure to account for such mapping errors can cause high rates of type I or type II errors in estimating resource use and selection, depending on the type of errors and their distributions.

To account for such mapping errors, users of radiotelemetry should map, classify, and validate the accuracy of estimated resource values (e.g., Rowland et al. 1998). By convention, classification accuracy of 70% or higher often is considered acceptable for mapping habitats and other resources at landscape scales. However, a variety of methods are available for conducting such error analysis, and little agreement exists about the most appropriate methods (Goodchild and Gopal 1992). Complicating the process is the accumulation of mapping errors when a large number of environmental layers are mapped and overlaid (Newcomer and Szajgin 1984). Notably, the true implications of resource mapping errors must be analyzed by mapping such errors in relation to the animal location errors, and generating a joint probability distribution of the combined effects of both sources of error on the probability of committing type I and type II errors. This type of approach has received little attention in the design of radiotelemetry studies but would be a logical extension of the methods used by Samuel and Kenow (1992) and Findholt et al. (in press) to assess location errors in relation to the underlying distribution of resource values.

ARE BIASES IN OBSERVATION RATE IDENTIFIED AND CORRECTED?

Radiotelemetry studies have typically focused on location error as the main source of bias. However, additional spatial errors are often associated with the probability of successfully obtaining a location when a location is attempted, which has been defined as observation rate (Rempel et al. 1995, Johnson et al. 1998). Observation rate can vary spatially leading to biased estimates of home range, movements, and resource selection for either manual or automated tracking systems. Observation rate has been studied most extensively for automated tracking systems.

Observation rates of automated tracking systems can range from 0% to 100%, depending on the unique environmental characteristics of the study area, the temporal sampling scheme used, and performance of the technologies. Rempel et al. (1995) documented observation rates of 10–92% for GPS collars, with rates significantly lower in forested stands of dense canopy or high basal area. Moen et al. (1996) documented a similar bias in observation rate of GPS collars placed in stands of dense canopy. Rumble and Lindzey (1997) also found that observation rate of GPS collars was progressively lower with increasing density of trees, with rates typically less than 10% in the most dense stands, and often more than 50% in stands of lowest density. Johnson et al. (1998) documented observation rates of a LORAN-C automated system, with observation rate biased downward with increasing canopy closure, increasing slope, and concave topography. Johnson et al. (1998) found that the spatial bias in observation rate in relation to these environmental conditions could be predicted reliably with the use of kriging models and spatial statistics (Fig. 2.4).

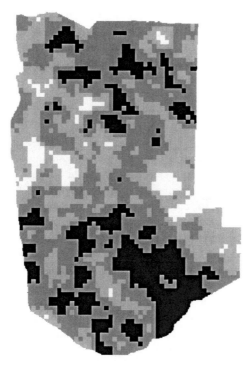

FIGURE 2.4   Predicted observation rate of an automated radiotelemetry system from 1995, Starkey Experimental Forest and Range, northeast Oregon (from Johnson et al. 1998). Observation rate was predicted from the spatial variance of the exponential isotropic semivariogram of observed observation rates of elk, mule deer, and cattle associated with a representative sample of 180 × 180 m pixels in the study area. The four gray-scale colors, from light to dark, correspond to observation rates of >70%, 60–70%, 50–60%, and <50%.

Identifying and Correcting Biased Observation Rates

Documenting and correcting the spatially explicit bias in observation rates involves the following steps, as described by Johnson et al. (1998). First, radiotransmitters are placed throughout the study area in a random or stratified random manner, such that locations of transmitters span the spatial and environmental conditions that are representative of the study area. Second, locations are attempted, using the same sampling scheme that is planned for locating radio-marked animals; that is, locations are attempted with the same type of sampling approach (e.g., cluster, systematic, or random sampling) and frequency and timing of sampling (e.g., one location attempt per hour per 24 hours per season) as that planned for sampling radio-marked animals in the study area. Third, observation rates are mapped and evaluated with the use of statistical models, such as regression or kriging approaches (Rumble and Lindzey 1997, Johnson et al. 1998). Fourth, performance of models to predict bias in observation rates (e.g., Fig. 2.4) are tested and validated in a spatially explicit manner. And fifth, the validated models are used to correct the bias for each pixel or polygon in which an animal location occurs.

What Is Biologically Significant?

An essential step in designing an experiment (or sample survey) is calculating the sample size necessary to meet the goals of the project. This step is rarely completed because to determine sample size we must specify the level of significance for the test (or confidence interval), power of the test (probability of inclusion in interval), population variance (estimated by sample variance from pilot study), and biological significance of the difference (bound on estimate) we wish to detect (Zar 1984). In the past, practitioners of radiotelemetry typically have focused their study design and interpretation of results on statistical significance rather than on the biological significance or a meaningful biological basis for specifying a bound on the error. Why have biologists not addressed biological significance? We propose that biologists have placed undue importance on statistical significance because that is much easier to define than biological significance.

Most biological studies seem to be designed primarily to obtain results that yield a desired level of statistical difference among treatment levels of interest. While statistical significance may be appropriate for manipulative studies that employ genuine treatment levels, experimental controls, and replication, its use in most observational studies is not appropriate (Burnham and Anderson 1998). Moreover, reliance on statistical significance, without consideration of biological effects, is inappropriate for all wildlife studies (Johnson 1999). A more germane question is whether the study is designed to detect differences that have biological significance (Tacha et al. 1982, Johnson 1999); that is, whether the study is designed to measure a response or address an issue that affects the species or ecological condition in a manner that has compelling biological effects or relations.

We suggest that users of radiotelemetry address biological significance by answering three related questions when planning a study. First, do the response variables directly or indirectly estimate demographic, social, or economic quantities of importance? Second, what magnitude of difference or response in these quantities is considered biologically important? Third, what is the broader research context (conceptual model or general research hypothesis; Ratti and Garton 1994) for identifying a particular difference or response as being important?

As an example, consider a radiotelemetry study designed to measure the difference in annual survival of brown-headed cowbirds (*Molothrus ater*) under two environmental conditions. Such a study obviously addresses an important demographic quantity (question 1). However, if the difference in annual survival under the two environmental conditions is 0.57 vs. 0.51, does this difference contribute to increasing vs. decreasing population growth of cowbirds in the two areas (question 2)? One must combine these estimates of survival rates with information from other demographic rates in a population model to assess their biological significance to the cowbird population. Moreover, one does not know whether the potential difference in population growth between the two areas is compelling in terms of its benefit to other bird species that are negatively affected by nest parasitism of cowbirds (question 3).

Granted, few radiotelemetry studies have the resources to fully address these questions. However, the process of attempting to answer such questions during planning helps identify key limitations of the study and allows for modification of the design to improve its capability to address issues of biological significance. For the cowbird example, designing a study to collect additional data on annual reproduction of cowbirds would allow researchers to calculate population growth for the species under each of the two environmental conditions (question 2). In turn, assessment of cowbird parasitism of avian nests in relation to changes in cowbird population growth would help determine whether certain environmental conditions, associated with lower cowbird population growth, ultimately benefit nest success of other bird species (question 3). Finally, a radiotelemetry study designed to assess annual survival and reproduction of the bird species affected by cowbird parasitism in the two environments would allow researchers to calculate differences in population growth of such species and assess the degree to which each environment may be associated with increasing or declining populations of such avian species.

ULTIMATE DESIGN: DEMOGRAPHIC RESPONSES TO LANDSCAPE CONDITIONS AND RESOURCE SELECTION

Most radiotelemetry studies of behavior, movement, space use, home range size, and resource selection have focused on documenting the choices that animals make within their environment. We believe that such studies could be improved substantially by linking these measures of animal choice with simultaneous measures of demographic performance and response. Unifying concepts for this approach are individual fitness and source-sink dynamics for metapopulations.

MEASURING INDIVIDUAL FITNESS

"In any given environment, the product of the relative survival value and relative reproductive capability of a given genotype constitutes its fitness or adaptive value" (Ehrlich and Holm 1963). By using radiotelemetry in the future it may be possible to actually measure fitness and relate it to individual decisions concerning habitats and behaviors. This possibility will open the road to great strides in understanding the mechanisms of action of key factors influencing survival and reproductive success of individuals and small groups of closely related animals occupying the same general landscape. To accomplish this ultimate goal in biology will require integrating information obtained from creative use of radiotelemetry with genetic measurements of relatedness, detailed observations of behavior, site-specific sampling of habitat, estimates of energy intake and expenditure, remote-sensed measures of landscape-scale environment, and interactions with predators and competitors. Clearly, this goal is not immediately within reach but radiotelemetry is a tool that will contribute to its eventual attainment.

METAPOPULATION DYNAMICS

The concept of a metapopulation, a "population of populations" (Levins 1969), has been used to describe a set of local populations in a region with differing rates of population change. One common pattern for such a metapopulation consists of a set of growing or stationary populations (sources) in favorable environments producing emigrants that maintain declining populations (sinks) in less favorable environments (Pulliam 1988, Pulliam and Danielson 1991).

The concept of mapping, predicting, and validating source and sink areas for a species, and understanding the environmental conditions that contribute to sources vs. sinks, is well within the technologies and methods that currently support the use of radiotelemetry (e.g., Powell et al. 2000a,b). That is, radiotelemetry can be used to collect data for simultaneous estimation of survival, reproduction, home range size, and resource selection of members of a population. In turn, these data can be used to calculate population growth and relate it to the local environment of that population. Ultimately, such an analysis would allow practitioners of radiotelemetry to calculate habitat-specific or spatially explicit measures of population growth or similar measures of performance, and to estimate the contribution of various environmental factors to variation in population performance on large landscapes (see Conroy 1993, Conroy et al. 1995, Dunning et al. 1995, Turner et al. 1995, for more discussion of these concepts). Such an approach, emphasizing habitat-specific measures of population performance, is radically different from conventional studies of resource selection, where habitats typically are identified as "selected" or "avoided" without information about whether such patterns have demographic consequences, and whether the consequences are of sufficient magnitude to be biologically significant (see Hobbs and Hanley 1990, for more discussion of this problem).

Although the technologies and methods are now available for conducting new studies that link animal choices with demographic consequences, development of experimental designs needed to conduct such studies has largely been ignored. New thinking about design of radiotelemetry studies is needed. Moreover, the analytical skills required to conduct and publish such studies is daunting, and many practitioners of radiotelemetry may need additional training to keep pace. Finally, the resources needed for such studies are more demanding, as such studies require higher sample sizes, longer time periods, larger spatial scales, automated tracking and mapping systems, increased data storage and management, and novel methods of analysis. Despite such obstacles, we believe the return on investment for studies that evaluate animal choices with demographic consequences is far greater than for more conventional radiotelemetry studies that estimate univariate patterns of resource selection, home range size, survival, or reproduction as "stand-alone" indices of animal or population performance. We urge users of radiotelemetry to pursue these new, integrated approaches as a central theme in future applications of this important technology.

SUMMARY

Designing studies that use radiotelemetry requires careful consideration of the goals of the project and resources available to meet those goals. Success in meeting the research goals depends on thoughtful planning of field methods and ancillary data collection, selection of telemetry equipment appropriate to the study animal and budget, careful execution of the field protocols, and creative analysis of the data. Some of the most important design factors include consideration of the study's purpose; degree of experimental manipulation, controls, and replication; selection of an efficient yet unbiased sampling scheme; definition of the sample unit; calculation of sample size requirements; identification and removal of sources of bias; and clear specification of biological significance. Notably, we summarized results that suggest that the sample sizes needed to obtain accurate estimates of home range size, resource selection, and survival with radiotelemetry may be much higher than is typical of past studies. Moreover, we emphasize the need to integrate univariate metrics of animal choice, such as estimates of home range size and resource selection, with metrics for the demographic consequences of these choices, all of which can be generated from radiotelemetry. We urge practitioners of radiotelemetry to pursue these integrated approaches as a central theme in future applications of this important technology.

CHAPTER 3

Effects of Tagging and Location Error in Wildlife Radiotelemetry Studies

JOHN C. WITHEY, THOMAS D. BLOXTON, AND JOHN M. MARZLUFF

Radio Tracking and Animal Populations
Copyright © 2001 by Academic Press. All rights of reproduction in any form reserved.

An important benefit of radio-tagging animals is the possible accuracy, precision, and completeness of resulting observations. When our study animal runs or flies out of view, the telemetry signal allows us to continue to collect location data. However, accurate inference in radiotelemetry depends on unbiased observations. Remotely gathered location estimates may differ in precision and accuracy from visual observations. Moreover, behavior may be influenced by the act of tagging or tracking. So, when we use telemetry locations to infer animal behavior and ecology we depend on a set of assumptions. Often these are tacitly dismissed as irrelevant to the current study. Rarely are they quantified, studied, and reported (Saltz 1994).

A fundamental assumption in wildlife studies reliant on radiotelemetry is that radio-tagged animals are "moving through the environment, responding to stimuli, and behaving in a manner similar to non-instrumented animals" (White and Garrott 1990). Despite the importance of this assumption, fewer than 100 published studies in leading journals address the effects of radio-tagging on study animals (Appendix 3.A). The lack of such studies may reflect difficulties in publishing methods studies or difficulties in studying effects. We suspect the latter to be most common because most animals studied with telemetry are inherently difficult to study without telemetry. Therefore, comparisons of behavior between tagged and untagged animals are difficult.

Another equally important assumption in radiotelemetry studies is that location estimates are accurate and free of bias. Few radiotelemetry studies report accuracy and bias (Saltz 1994) and those that do rarely distinguish precision from accuracy. Animal locations obtained remotely are *estimates* of the animal's actual location, and the probabilistic nature of that estimate defines the accuracy with which we can measure space use and resource selection (Nams 1989).

These two assumptions are not the only, or necessarily the most influential, sources of bias. Good experimental design and appropriate analysis are essential to make accurate inferences (see Chapter 2). Design considerations include the time of day and year observations are recorded, the frequency of tracking or recording locations, the number of animals under study, and the method of obtaining location estimates (Mech 1983, Kenward 1987, White and Garrott 1990, Beyer and Haufler 1994, Johnson et al. 1998, Otis and White 1999). Readers should consult Garton et al. (Chapter 2), White and Garrott (1990; Chapter 2), and Ratti and Garton (1994) for design considerations. Good experimental design allows researchers to obtain location estimates that represent their study animal's use of space in an unbiased way (Otis and White 1999; see also Chapter 5).

We begin this chapter assuming that proper experimental design is in place. However, even a well-designed study can produce biased and inaccurate results if the effects of transmitters on animals and the accuracy of location estimates obtained by radiotelemetry are not investigated. Understanding the magnitude of these two sources of bias and examining their influence on inference help researchers provide a complete and unbiased picture of how an animal behaves. In this chapter we review the scientific literature to (1) summarize the effects of transmitters on wildlife, (2) make recommendations for studying and minimizing effects, (3) investigate the magnitude and implications of location error resulting from conventional transmitters, and (4) discuss the importance of conducting site-specific beacon tests to better understand bias and error in specific studies. Our review does not include physiological biotelemetry or sonic telemetry used in aquatic environments, although biases related to effects on the animal and location error may be similar to those discussed below.

EFFECTS OF TRANSMITTERS ON ANIMALS

Documenting the effects of radiotransmitters on animals minimally requires direct observation of tagged animals and optimally requires a comparison between tagged animals and untagged controls. Because many animals studied with telemetry are difficult to observe, definitive tests of transmitter effects are rare. For example, we searched five leading journals that often publish telemetry-based articles (*The Journal of Wildlife Management, Wildlife Society Bulletin, Journal of Field Ornithology, Copeia,* and *Journal of Mammalogy*) from 1972 to April 2000 and found only 96 articles that assessed the effects of radiotransmitters on animals. Most studies compared tagged animals to untagged controls (78%, $n = 75$). Others assessed telemetry effects by comparison with published literature on untagged animals ($n = 5$), comparison among attachment techniques and/or transmitter weight ($n = 7$), or simple, direct evaluation of behavioral or physical changes ($n = 9$; e.g., skin abrasions, excessive feather wear, significant changes in body mass) following transmitter attachment. Studies that only made passing reference to possible effects were excluded from our review because they did not possess the necessary study design to evaluate transmitter effects. These articles generally contained statements such as, "transmitters did not appear to affect" or "there may have been effects on behavior x caused by the transmitters."

Studies of radio-tagging effects are not uniformly distributed among taxa. The majority of articles devoted to assessing the effects of transmitters were conducted on birds (79%); primarily waterfowl and upland game birds (Fig. 3.1). This is not surprising due to the obvious concerns over radiotransmitter attachment to flying animals. In our review, we did not find any studies that assessed effects of transmitters on amphibians or reptiles. Effects on mammals, especially large ones, are also poorly documented, presumably because

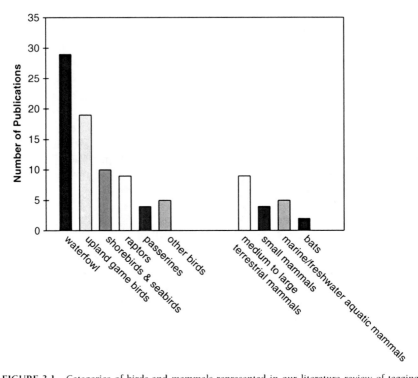

FIGURE 3.1   Categories of birds and mammals represented in our literature review of tagging effects. Ninety-six articles were reviewed from *The Journal of Wildlife Management, Wildlife Society Bulletin, Journal of Field Ornithology,* and *Journal of Mammalogy.*

researchers either assume effects to be negligible or fail to publish negative results.

A variety of effects from tagging have been documented. Tagging may alter survivorship, reproduction, behavior, or condition (White and Garrott 1990). However, for every study that shows an effect of tagging, another fails to show an effect (53% of all studies did not find effects and 47% did, regardless of type of effect or animal studied; Fig. 3.2). A review of studies concluding detrimental effects by attachment type reveals some surprising patterns. Backpack harness systems resulted in a relatively high frequency of effects (68% of studies reviewed), whereas implantation methods yielded a relatively low frequency of detrimental effects (24% of studies reviewed; Fig. 3.3). Although we hypothesized that studies published more recently might be less likely to discover effects, due to smaller radio packages and more refined attachment methods, 21 of 43 (49%) studies published before 1991, and 25 of 53 (47%)

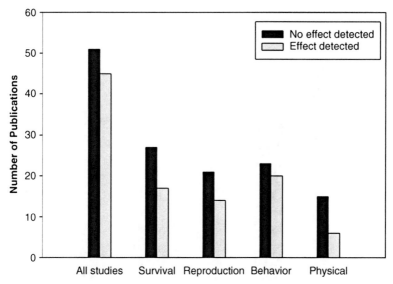

FIGURE 3.2 Number of publications (among 96 reviewed) with "no effect detected" and number with "effect detected" of radiotransmitters on survival, reproduction, behavior, or physical condition (includes skin abrasions, excessive feather wear, significant changes in body mass, and infection). "All studies" is a count of whether a publication reported a significant ($P < 0.05$) effect in *any* of the four categories. A paper that reported effects of radiotransmitters in more than one category would be included in each category. For example, Vekasy et al. (1996) reported no effect detected on reproduction of radio-tagged prairie falcons and a significant effect on hunting behavior in poor prey years.

studies published between 1991 and 1999, found no significant effects from tagging. Despite the relatively consistent proportion of studies finding specific effects, most tagging effects appear to be taxon- and technique-specific. Therefore, we discuss reported effects separately for each taxon-technique combination below. Our review of the available literature is not exhaustive; rather, it highlights some noteworthy findings.

WATERFOWL

Ducks and geese are sensitive to external transmitters, especially those attached using backpack harness systems (Dwyer 1972). Most (10 of 12) waterfowl researchers who tested backpack harnesses found negative effects (e.g., decreased survivorship). Diving ducks (*Aythya* spp.), in particular, appear very susceptible to any type of externally mounted transmitter (backpacks, Perry

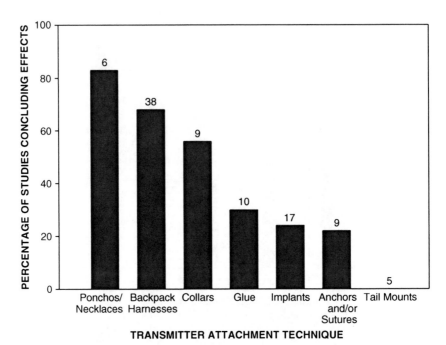

FIGURE 3.3 Percentage of studies for each attachment technique that found significant effects on survivorship, reproduction, physiological parameters, behavior, or physical condition due to radiotransmitters. Sample sizes (number of studies) are shown above bars. Attachment techniques with fewer than five reported studies are not shown.

1981; neck collars, Sorenson 1989). Effects from backpack harnesses ranged from short-term behavioral modifications (e.g., increased preening, stretching, or fluffing feathers; Greenwood and Sargeant 1973, Wooley and Owen 1978, Pietz et al. 1993, Blouin et al. 1999, Garrettson et al. 2000) to longer-term effects on survivorship (Ward and Flint 1995, Dzus and Clark 1996), and reproductive output (Rotella et al. 1993, Pietz et al. 1993, Gammonley and Kelley 1994). Backpack harnesses should not be used on canvasbacks (*Aythya valisineria*) because Perry (1981) reported 20% weight loss within 9 days, or on snow geese (*Chen caerulescens*) because Blouin et al. (1999) reported increased vulnerability to hunting during the post-release adjustment period.

Back-mounting techniques not reliant on harnesses, such as those utilizing anchors and/or sutures (Mauser and Jarvis 1991), resulted in fewer detrimental effects than backpack harnesses. Only 2 of 8 researchers detected effects on survivorship and/or reproduction from anchors and/or sutures. Implantation (abdominal or subcutaneous) appears to be the safest (i.e., fewest effects reported) transmitter attachment technique for waterfowl. Three studies compared implanted birds with those outfitted with external transmitters and found that implants had uniformly fewer significant effects (anchors, Paquette et al. 1997; backpacks, Dzus and Clark 1996, Rotella et al. 1993).

Schulz et al. (1998) developed and refined techniques to implant subcutaneous and intraabdominal transmitters with external antennas in mourning doves. (*Zenaida macroura*). Using a well-designed study, they monitored physiological and pathological responses to implanted transmitters. They concluded that subcutaneous implants with external antennas provide a better alternative to intraabdominal implants. Of interest, at 4–6 days postsurgery 153 of 160 doves had closed or healed surgical sites with no reported complications. Although further comparison of subcutaneous transmitters with conventional techniques is needed, these results are very promising.

Upland Game Birds

Backpack harnesses, breast harnesses (Erikstad 1979) and ponchos/necklaces (Small and Rusch 1985, Marks and Marks 1987, Pekins 1988, Marcstrom et al. 1989, Carroll 1990, Burger et al. 1991, Cotter and Gratto 1995, Bro et al. 1999) have been used extensively on upland game birds (i.e., galliformes). The impacts of backpack harnesses on upland game birds paralleled those found in waterfowl; 9 of 14 researchers using backpacks demonstrated negative effects as did 6 of 7 who utilized ponchos/necklaces or breast harnesses. Effects of backpack harnesses ranged from reduced takeoff angle, flight speed, and climbing rate in the gray partridge (*Perdix perdix*; Putaala et al. 1997) to longer term effects on survivorship and/or reproduction (Warner and Etter 1983, Small and Rusch 1985, Cotter and Gratto 1995). Reduced survivorship was the most common effect reported from poncho attachments on upland game birds. Because external harnesses have negative effects on upland game birds, further research on alternative attachment techniques, such as implants or adhesives, is necessary.

We found only two studies that assessed the effects of implanted transmitters on upland game birds (Hubbard et al. 1998, Schulz et al. 1998). Hubbard et al. (1998) noted reduced wing length growth in tagged birds but did not detect effects on body mass. They concluded that implants were better than backpack harnesses for wild turkey poults (*Meleagris gallopavo*).

Raptors

In contrast to waterfowl and upland game birds, raptors suffer few effects of externally mounted transmitters. All three studies that assessed tail-mounted transmitters found no negative effect (Taylor 1991, Sodhi et al. 1991, Hiraldo et al. 1994), and 2 of 6 studies found significant effects from backpack harnesses. Both studies documenting negative effects of backpack harnesses were on spotted owls (*Strix occidentalis*; Paton et al. 1991, Foster et al. 1992). Both groups of researchers advised against further use of backpacks on spotted owls because they found reduced survivorship and/or reproductive output due to backpacks.

Transmitter effects on raptors, and possibly other large birds, may depend on resource availability. Vekasy et al. (1996) concluded that prairie falcons (*Falco mexicanus*) wearing backpack harnesses captured different prey types than non-radio-tagged falcons during poor prey years, but not during years of normal prey abundance. Despite this behavioral difference, no effect of tagging on reproduction or survivorship was observed (Vekasy et al. 1996). Yearly variations in the effects of tagging were also noted in golden eagles (*Aquila chrysaetos*; Marzluff et al. 1997c). Tagged eagles bred less successfully in poor prey years than untagged eagles, but age differences in tagged vs. untagged eagles also may have contributed to differences. Thus it appears that tagging raptors is relatively benign, especially in years of normal to abundant prey availability.

Shorebirds and Seabirds

Adhesives, tail mounts, and anchor/suture methods have all met with success in shorebirds. Backpack harnesses have yielded mixed results in this diverse group, with detrimental effects noted on breeding behavior in the American woodcock (*Scolopax minor*, in the sandpiper family; Ramakka 1972, Horton and Causey 1984), whereas no effects from backpack harnesses were reported in a study on survivorship and reproduction in adult herring gulls (*Larus argentatus*; Morris et al. 1981).

Small Mammals

Few studies have been conducted on the effects of transmitters on small mammals. Mikesic and Drickamer (1992) found that neck collars reduced activity levels of wild house mice (*Mus musculus*) relative to uncollared mice in captivity. Berteaux et al. (1996) and Johannesen et al. (1997) did not detect an effect of collars on meadow vole (*Microtus pennsylvanicus*) behavior or root vole (*M. oeconomus*) survivorship, respectively. Reynolds (1992) implanted transmitters into the intraperitoneal cavities of captive deer mice (*Peromyscus maniculatus*) and prairie voles (*Microtus ochrogaster*) and monitored subsequent changes in body mass and white blood cell profiles. In this study, significant transmitter effects were documented in deer mice but not in prairie voles. No single technique is clearly more effective for small mammals.

Medium to Large Terrestrial Mammals

Neck collars rarely affect survivorship, reproduction, behavior, or condition in medium- to large-bodied terrestrial mammals. Effects on condition (body mass loss) were noted in San Joaquin kit foxes (*Vulpes macrotis mutica*) recaptured after 6–30 days (Cypher 1997) and in white-tailed deer (*Odocoileus virginianus*; Clute and Ozoga 1983). The latter case was reportedly due to excessive ice buildup on the transmitter during winter.

Other types of transmitters fare less well. Ear tag transmitters reduced moose (*Alces alces*) calf survivorship, but the researchers were unable to determine a mechanism because calves with plain ear tags survived similar to untagged controls (Swenson et al. 1999). Intraperitoneal implants did not affect yellow-bellied marmots (*Marmota flaviventris*; Van Vuren 1989), mink (*Mustela vison*; Eagle et al. 1984), and Franklin's ground squirrels (*Spermophilus franklinii*; Eagle et al. 1984). However, female armadillos (*Dasypus novemcinctus*) in poor condition at the time of capture suffered reduced survivorship and reproduction after implanting transmitters (Herbst 1991).

Aquatic Mammals

Various types of implants (intraperitoneal or subcutaneous) have been used without significant adverse effects on beaver (*Castor canadensis*; Davis et al. 1984, Guynn et al. 1987), river otter (*Lutra canadensis*; Reid et al. 1986), and sea otter (*Enhydra lutris*; Garshelis and Siniff 1983). Reid et al. (1986) noted that implants could affect river otter reproduction if the individual is in poor condition at the time of capture. Garshelis and Siniff (1983) compared implants (intraperitoneal and subcutaneous) to three external attachment methods (neck collars, flipper attachments, and ankle bracelets) and found that intraperitoneal implantation did not adversely affect sea otters during 1 year of tracking. Subcutaneous implants also fared well, but may have contributed to mortality in two individuals. All three external methods had some detrimental effect on sea otter behavior. As with waterfowl, implantation techniques appear preferable over external attachment techniques in aquatic mammals.

BATS

As with birds, transmitters may affect flight performance in bats. Flight maneuverability in the Yuma myotis (*Myotis yumanensis*) was affected by transmitters that weighed as little as 5% of the body mass in this species (Aldridge and Brigham 1988). Adhesive techniques (transmitters glued to backs) have been used with no apparent effect on the foraging success of hoary bats (*Lasiurus cinereus*; Hickey 1992). But further documentation is necessary to determine if other species are affected. Additional research should focus on identifying threshold weights, above which transmitters become too heavy to be carried by bats without biasing telemetry studies.

RECOMMENDATIONS

Although many studies document effects of radio-tagging, it is important to emphasize that many of the insights we have gained into the behavior of wide-ranging, rare, and secretive animals would not have been possible without the use of telemetry. The point of the review above is to minimize future effects by learning from past studies and anticipating how tagging might affect future telemetry studies.

Unrecognized effects of radiotransmitters on animals can bias inference made from radiotelemetry studies. The clearest example is if transmitters increase mortality and estimates of survivorship are calculated using radio-tagged animals. However, more subtle effects can also impact conclusions. If transmitter effects are greater during poor prey years, or when individuals are in poor condition, or when certain weather events occur, then studies that compare demographic variables across habitats may report important differences that are attributable to transmitters alone. Another bias may result from pooling observations from study sites when individuals from one area are more susceptible to transmitter effects than individuals in other areas. One must always be cautious when assuming transmitters have no effect.

Building on White and Garrott's recommendations (1990:37) we suggest the following for researchers planning a radio-tagging study:

(1) Use the smallest possible transmitter, especially when placing instruments on flying, aquatic, or small animals. A general rule of thumb for birds is that transmitter mass should not exceed 3% of the animal's total body mass.

(2) Transmitter attachments should be as inconspicuous as possible. Be especially cognizant of disrupting animal camouflage, status signals, and flight profiles.

(3) Test transmitter attachments in captivity before placing them on free-ranging animals.

(4) Avoid the use of external harness attachments, especially on waterfowl and upland game birds. Large passerines and raptors appear less influenced by harnesses.

(5) If harnesses are used, carefully consider their design and fit to minimize effects. Slippery materials (e.g., Teflon) should be used to reduce abrasion. The importance of fit cannot be overemphasized; harnesses that are too tight limit movements, but those that are too loose can easily entangle the animal.

(6) Anticipate growth and seasonal changes in physique and behavior when deciding on the type of transmitter to use and adjust the fit of the attachment technique accordingly.

(7) Anticipate how and when transmitters will be shed. Often it is more risky to have an attachment become partially disassociated, and entangle the animal, than it is to ensure the harness cannot be shed.

(8) Avoid placing transmitters on animals during times of stress (e.g., poor resource years, reproductive seasons). Also refrain from placing instruments on animals that appear to be in poor condition unless your study objectives or design specifically require it.

(9) Allow a period of several days to weeks for newly instrumented animals to acclimate to tagging before collecting data.

(10) Do not assume there will be no effect in your study, even if other similar studies failed to show an adverse effect from tagging.

(11) Report any results, both positive and negative, concerning transmitter effects.

(12) Consider several response variables, including behavioral and physiological factors, when evaluating the influence of a transmitter on an animal.

Consideration of these general recommendations is an important starting point in the study of radio-tagged animals. Understanding the types and magnitude of transmitter effects therefore should be an integral part of any radiotelemetry study. Valid population inferences cannot be drawn from a radio-tagged sample of individuals unless one concludes that transmitter effects (as well as other capture, marking, and tracking effects) are negligible or unrelated to the process under investigation. Therefore, our final recommendation is to design a rigorous study of transmitter effects whenever possible. When designing a telemetry effects study two questions should be considered:

(1) What should be measured? Each study should be designed to investigate how tagging affects the response variable(s) being measured. Studies of behavior must take into consideration how behavior is affected by tagging. Studies of demography must at least incorporate an understanding of how

survivorship and reproduction are affected by tagging. But such studies will also benefit from knowledge regarding how behavior and physiology are affected by tagging. Such knowledge may facilitate a mechanistic understanding of changed demographics.

(2) How should effects be measured? Ideally, captive studies will be coupled with a companion field study to determine what effects, if any, transmitters have on animals. Measuring changes in free-ranging, tagged animals before and after tagging compared to changes in untagged controls during the same time period is a powerful research design. However, resources and opportunities rarely allow such extensive research. Either comparison alone (pre- and post-tagging or tagged vs. untagged) would be the next most rigorous design. We recommend that researchers at least carry out this type of study. Detailed observation after tagging is the least rigorous design, but even this can reveal some insight into effects, especially severe ones. When making these kinds of observations it is important for researchers to challenge the observations by asking: *would an untagged animal do what that tagged animal just did?*

An important statistical issue must be considered when the effect of tagging is studied. Researchers are put in the awkward position of testing the null hypothesis that tagging has no effect. Therefore, at best we can only conclude that evidence is not sufficient to conclude that an effect exists; we cannot conclude that no difference exists (White and Garrott 1990). Our ability to reject the null hypothesis depends critically on statistical power. With enough power, we can conclude transmitters affect animals even if there is little evidence of a biological effect. Therefore, we recommend that tests of transmitter effects not rely on standard hypothesis testing but rather demonstrate biologically significant differences in central tendency between responses of tagged and untagged animals (or between before and after tagging) using techniques such as confidence interval comparisons (Johnson 1999, Chapter 2).

LOCATION ERROR

A broad range of factors may affect the accuracy of animal locations obtained by radiotelemetry. Techniques used in estimating locations include "homing in" on the animal, aerial tracking, satellite tracking, and triangulation using fixed and/or mobile receiving stations (White and Garrott 1990). Each technique has different sources and typical magnitudes of associated error. Along with greater reliance on radiotelemetry by wildlife managers has come, albeit slowly, greater concern for determining and reporting the quality of remotely sensed location estimates (Heezen and Tester 1967, Springer 1979, Hupp and Ratti 1983, White and Garrott 1990, Saltz 1994). Understanding the limitations of the technique selected is critical for a successful study. Although the necessary level of accuracy depends on the objectives of the study, researchers need to be conscious of assumptions they make about the accuracy of estimated animal locations.

Tests of radiotelemetry accuracy have shown signal reception and location accuracy to be affected by mapping error, signal bounce, vegetation cover, electromagnetic interference, animal movements, operator error, and distance to radio-tagged animals (Heezen and Tester 1967, Springer 1979, Deat et al. 1980, Hupp and Ratti 1983, Mech 1983, Lee et al. 1985, Kufeld et al. 1987, Saltz and Alkon 1985, Schmutz and White 1990, White and Garrott 1990, Nams and Boutin 1991, Parker et al. 1996). All radiotelemetry studies must consider the sources of error that most affect their location estimates. We will limit our discussion of error sources to conventional very high frequency (VHF) telemetry, as Rodgers discusses error associated with satellite and GPS tracking systems in Chapter 4. We conclude with strategies for testing and reporting location error, and suggest that researchers use a site-specific approach to assess accuracy in all radiotelemetry studies.

MAPPING ERROR

Placing the location of a radio-tagged animal or an observer on a map is not necessarily a simple task. With smaller scale maps (i.e., greater distances represented by the same distance on the map) even pencil-lead width errors can represent significant distances on the ground (Mech 1983, Samuel and Fuller 1994). Studies in habitats that lack landmarks (e.g., roads or topographic features) can easily suffer from mapping error. Studies using aerial tracking may encounter the same problem, depending on such factors as airspeed, altitude above ground, and observer experience or fatigue (White and Garrott 1990). Testing the accuracy associated with aerial tracking is especially important if animals are not relocated visually (White and Garrott 1990). Global positioning system (GPS) locations taken from aircraft have been reported accurate to within 100 m (mean linear error of uncorrected GPS estimates of 52 m, Leptich et al. 1994; 80 m, Carrel et al. 1997) and should be used when possible. Marzluff et al. (1994) recorded locations from an airplane using a GPS unit when directly over a radio-tagged prairie falcon, resulting in more precise location estimates than those obtained by conventional ground tracking (confidence ellipses 4.8 times smaller than ellipses obtained by ground-based triangulation). Experience in the study area will help with observers' navigation and mapping skills. In addition, hand-held

GPS units can be taken with the researcher and used to record specific locations and/or to quality-check locations marked on a map.

SIGNAL BOUNCE

Reflected signals ("signal bounce") are caused when the signal does not have a clear, direct path between transmitter and receiver. The term "line-of-sight" is used to convey whether such a clear path exists (e.g., a line-of-sight reference point had no visible obstructions between the point and receiving tower; Garrott et al. 1986). We generally think of topographic features or structures as providing obstacles between a radiotransmitter and a receiver, but Cochran (1980) described how radiotelemetric line-of-sight is not identical to visual line-of-sight. If the transmitter is less than half a wavelength from the earth (about 1 m at 150 MHz), signal propagation may be noticeably affected even if no other obstacles to the receiver exist (Cochran 1980). Hupp and Ratti (1983) noted that error was larger using hand-held antennas that were closer to the ground than other reception systems. However, if a direct signal path from transmitter to receiver is blocked by a topographic feature, the signals arriving at the antenna will be the result of bounced or reflected signal paths. The problem of signal bounce is greater when transmitting at higher frequencies (i.e., > 100 MHz; Macdonald and Amlaner 1980). Non-line-of-sight bearings may be highly inaccurate in comparison with line-of-sight bearings (absolute mean bearing error for 72 of 147 non-line-of-sight reference points was 21.2°, compared with less than 3° for line-of-sight points; Garrott et al. 1986). In one mountainous area non-line-of-sight signals were more likely to be weaker than unobstructed signals and were less precise than stronger signals (SD of bearing error = 9.6° for weak signals, 1.9° for strong; Kufeld et al. 1987).

The plane of polarization of the transmitted signal can also be affected by bounce. Because the orientation of the transmitter antenna relative to the earth depends on the position of the animal, we do not know if the signal is horizontally or vertically polarized as it leaves the transmitter. In addition, if the signal "bounces" off reflective surfaces, such as wet snow, canyon walls (Beaty and Tomkiewicz 1990), or dense vegetation (Samuel and Fuller 1994), the polarization of the signal can change. If receiving antennas are hand-held then their orientation can be changed manually to determine if a stronger signal is detected vertically or horizontally, but if fixed antennas are used then the optimal orientation (that which gives the greatest accuracy, not necessarily the greatest signal strength) should be determined before the study begins.

Because of the potential bias in recording reflected signals as accurate representations of the location of an animal, it is particularly important to assess the topography of a study area and determine what measures can be taken to decrease the effects of signal bounce. The routine availability of digital elevation maps and geographic information system (GIS) functions that calculate viewable space around any point in a study area can facilitate the determination of regions where bounce is likely to be problematic. To reduce problems in such areas, Lee et al. (1985) suggested placing receiver towers on ridges to increase the likelihood of line-of-sight signals. If reflected signals are frequently received (e.g., >10–15% of total; Garrott et al. 1986), taking three or more bearings also may be necessary (White and Garrott 1990). Then a robust estimator that differentially weights bearings based on perceived accuracy (e.g., Lenth 1981) may be used. Using one of Lenth's (1981) estimators to recognize signal bounce is better than common criteria used in the field (e.g., assessing signal strength or modulation, limits on acceptable angles of bearing intersection), as demonstrated by Garrott et al. (1986). In some areas it may be impossible to accurately estimate locations, and other techniques to relocate animals visually, such as aerial tracking or homing, should be considered.

Vegetation

Vegetation can affect signal transmission and reception. Signal strength may decrease if antenna elements are close to tree limbs (Hupp and Ratti 1983) or under large trees (Cottam 1988). Testing for error in three habitats with trees (flat, rolling, and mountainous terrain) resulted in error polygons too large to determine the location of test transmitters (Hupp and Ratti 1983). Chu et al. (1988) found that the presence of foliage in a hardwood study area increased the proportion of extreme bearing error. However, other studies have found no difference in accuracy in open vs. wooded habitats (Cottam 1988, Wallingford and Lancia 1991). These conflicting reports emphasize the importance of a beacon study in your particular study area. If a study is conducted at different times of the year, then researchers should test radiotelemetry accuracy in different seasons as well to determine whether growing vegetation affects accuracy.

Electromagnetic Effects

Bias in signal direction caused by power lines or other electromagnetic interference was discussed early in the radiotelemetry literature (Cochran and Lord 1963, Slade et al. 1965). In one of the few studies reporting tests of electromagnetic interference, Parker et al. (1996) found that bearing error measured at two receiving stations within 300 m of 200,000-V power lines in

New Zealand was significantly greater (negative bias) than error measured at stations more than 600 m away. They conducted the test only after discovering error in their tracking data and concluded that the stations close to power lines could not be used for radio-tracking despite clear lines-of-sight and unvegetated habitat. Diehl and Larkin (1998) described losing contact with radio-tagged birds due to electromagnetic interference from alarm systems or garage door openers but did not test the magnitude of the impact. Urban "noise" may decrease transmitter range and signal reception (Harris 1980, Cochran 1980). Cudak et al. (1991) recorded radio "noise" in the 148- to 412-MHz range from a plane flying over central Illinois. The largest city (Champaign-Urbana) had the highest noise readings (with a peak close to 9 dB over background levels in one example; Cudak et al. 1991), but all towns (including rural residential communities with populations under 2000) had identifiable noise signatures. Peak readings also varied by time of day and by month. However, they found no significant changes from similar measurements recorded in 1972 (Swenson and Cochran 1973), and the effect of measured noise on radiotelemetry was not tested. Although population growth and urban development over the last 30 years has increased the risk of electromagnetic signal interference for field researchers, how much of a problem interference is depends on the objectives and accuracy requirements of the study. To our knowledge no one has tested a tracking system for effects of cellular phone or pager transmissions interference specifically, but the proliferation of these devices and the use of spread spectrum technology, where signals spread over a range of frequencies, creates the potential for interference. During field studies researchers using hand-held or mobile antenna systems may be able to decrease audible interference somewhat by moving their position slightly or changing the plane of the antenna (Larkin et al. 1996).

ANIMAL MOVEMENT

When bearings are taken on a radio-tagged animal simultaneously by more than one receiving station, the animal's position relative to the receiver's position is fixed. Studies may also use a single person with a hand-held antenna or one or two people in a vehicle. If the animal moves between successive azimuths, then errors in the estimated location will result (Cochran 1980, White and Garrott 1990). A simulation of the location error caused by animal movement with a three-antenna triangulation system found up to 10-fold increases in error (Schmutz and White 1990). Linear error was significantly greater, compared with fixed test transmitter results, when estimating locations of hand-held transmitters moved around to simulate bird movements (Pyke and O'Connor 1990) and when tracking a person walking to simulate coyote movements (Laundré and Keller 1981). Simultaneous bearings are superior to sequential ones when tracking moving animals and should be used when possible (Schmutz and White 1990). However, the behavior of the study organism can help determine how significant animal movement may be to location error. In some cases, it may be assumed that animals are stationary during the tracking period (e.g., bearings taken during midday resting periods or on birds at their nests). Setting limits on the time between successive azimuths is another strategy used to minimize error caused by movements (Katnik et al. 1994, Chapin et al. 1997). Any such assumptions or procedures should be reported and justified.

DISTANCE EFFECTS

The distance effect, whereby location error increases with increased distance between the transmitter and receiving equipment, was first discussed by Slade et al. (1965). Since then, not all studies have confirmed this effect. Some have shown either poor correlation between distance and error (Kufeld et al. 1987), or no distance effect in some environments (Hupp and Ratti 1983). However, bearing error will magnify linear error at increased distance from the receiver (Hupp and Ratti 1983, Saltz and Alkon 1985), and for this reason Springer (1979) recommended minimizing the distance to the transmitter. The most precise estimates (i.e., the smallest error polygons) using two bearings are obtained adjacent to one of the receiving stations where the bearings intersect at 90°. Because there is a practical limit to signal reception in the field, if nonsimultaneous bearings are taken on moving animals, then shorter distances to the animal often mean shorter times between azimuths, which can increase overall accuracy. The locations from which fixes are obtained using a mobile tracking system, and therefore the distance to the radio-tagged animal, depend on particular characteristics of the study area, including road distribution, access to private land, topography, and sources of interference such as power lines (Sargeant 1980). These characteristics should be studied before the project is initiated to determine at what distances radio-tagged animals will typically be tracked. Then a beacon study can be designed to include the range of distances expected.

OBSERVER EFFECTS

Depending on the method used, observer error can contribute significant error to location estimates. For example, sighting a hand-held compass along the antenna (White and Garrott 1990) and establishing the direction of the vehicle in a truck-mounted system (Pace 1988) have been noted as sources of observer error. For this reason, when conducting a beacon study observers should take bearings on test transmitters using the same system and methods that will be used in the field.

Significant differences in bias among observers have not been detected often (Springer 1979, Lee et al. 1985, Kufeld et al. 1987, Zimmerman 1990). However, as with the detection of radiotransmitter effects, low sample sizes limit the power of statistical hypothesis tests to detect differences. For example, Springer (1979) reported mean bearing error (bias) of $-1.0°$, $0.2°$, and $2.2°$ for three observers ($n = 40$ in two cases and $n = 80$ in one case), but this difference was reported as not significant. The potential for interobserver variation in bearing error should be investigated and differences reported with confidence intervals, even if P values are not considered significant. In the case of aerial telemetry, use of different pilots may result in significantly different location error (Hoskinson 1976). Researchers should also be aware that observer error can increase when the personnel do not know they are being tested (Mills and Knowlton 1989).

Field personnel can affect animal movements, especially when homing methods are used (Cochran and Lord 1963). Bias may be introduced both during a single location effort (movement away from the researcher before the animal is relocated) and during the course of the study through continual harassment (White and Garrott 1990). Researchers should conduct tests using simultaneous tracking by triangulation and homing to determine how animals react to close approaches. If "remote" observers track the signal while another observer homes in on the animal, the frequency or timing of movement can be correlated with the approach of the observer. At a minimum, researchers should note how often animals move away (if it can be determined by the signal) before visual locations are recorded. Animals may also react to disturbance by seeking particular cover types and remaining there for longer periods (Lariviere and Messier 1998). Determining how animals react to moving or stationary vehicles may also be important (Ellis 1964). There may be times when approaching closely is of less concern, such as when homing to a bird at its roost site (as long as it is not flushed), in comparison with homing during breeding activity and potentially disrupting its nesting behavior.

TESTING AND REPORTING ERROR ESTIMATES

Researchers have emphasized testing and reporting error (Springer 1979, Hupp and Ratti 1983, Lee et al. 1985, White and Garrott 1990, Saltz 1994), but many studies neglect to report error altogether or provide incomplete data

(Saltz 1994). Given the variety of factors that may affect the accuracy of location estimates obtained by radiotelemetry, researchers should test the accuracy of their radio-tracking system before conducting their research. Only then will they be able to determine if their proposed methods will provide data precise and accurate enough to meet their objectives. Location estimates that are accurate to 1 km may be sufficient for studies of long-range migration or dispersal, whereas fine-scale resource selection studies must determine the accuracy required to detect use of the smallest patch defined as available. Location error is often greater when the animal is not resighted, but even studies using highly accurate GPS units to record visual relocations should report the error associated with their location data.

There is considerable confusion in the literature about how to measure and interpret telemetry error. Much of this is due to the fact that researchers often measure *precision* but then use it to indicate *accuracy*. Precision is a measure of the consistency of a tracking system. It is usually reported as the standard deviation of bearing errors (where bearing error is the difference between the estimated bearing and "true" bearing), or as the error polygon or confidence ellipse associated with a location estimate (Lee et al. 1985, Saltz and White 1990, Nams and Boutin 1991, Saltz 1994). Accuracy is a measure of how close an estimated location is to the actual location. It is usually reported as the mean difference between bearings to estimated and actual locations (bias), mean distance between actual and estimated locations (linear error), or the percentage of times that a confidence ellipse around an estimated location includes the actual location (coverage; Garrott et al. 1986, White and Garrott 1990).

Here we will discuss standard measures of precision and accuracy while arguing that although most studies reporting error estimates adequately consider precision, the accuracy of location estimates could be better estimated using measures of linear error. We also suggest testing for location error using a site-specific regression approach.

REPORTING ERROR ESTIMATES

We examined *The Journal of Wildlife Management* (JWM) from 1993 to 1999 (Volumes 57–63; Appendix 3.B) and found 43 papers that used triangulation to obtain location estimates. We did not include papers measuring survival that used triangulation but whose results depended on determining if animals were still alive, or papers that described how to test telemetry error. To compare our results to the reviews of Hupp and Ratti (1983) and Saltz (1994), we used the methods of Saltz to determine if "acceptable" error measures were reported (e.g., size of error polygon or confidence ellipse;

Saltz 1994). We also recorded how location accuracy was estimated and whether data censorship methods were reported.

Six years after the review by Saltz (1994), reporting on error in studies using triangulation to estimate animal locations remains unsatisfactory in JWM. We found the same proportion of papers with no reporting, inadequate reporting, and acceptable reporting on telemetry error as Saltz (1994) ($\chi^2_{(2)} = 3.69$, $P = 0.158$; Fig. 3.4). Combining our results with Saltz's, the percentage of papers from 1986 to 1999 without any error reporting (27%) has declined from the 75% reported by Hupp and Ratti (1983) in their review of papers from 1971 to 1980 ($\chi^2_{(1)} = 23.4$, $P < 0.001$). However, at least one study without mention of location error was found in each of the volumes of JWM reviewed for this chapter. In addition, although our analysis used Saltz's definition of acceptable reporting, we note that error polygon size is not a true measure of accuracy (see Error Polygons and Confidence Ellipses below)

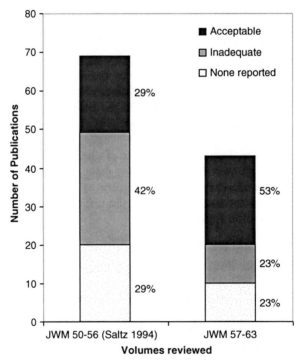

FIGURE 3.4 Number of studies using triangulation to obtain location estimates, with % shown at side, by category of "error reporting" (none, inadequate, acceptable) in two reviews (Saltz 1994, and this chapter). Volume numbers of *The Journal of Wildlife Management* included in each review are shown.

and that only 9 of the 23 papers with "acceptable" error reporting included measures of linear error (21% of the total). Eleven (26%) reported error polygon size, 2 reported confidence ellipse size, and 1 used visual follow-ups to estimate accuracy.

Ten studies we reviewed (23%) reported mean and standard deviation of bearing error, while 7 others reported either mean or standard deviation, but not both. When reporting bearing error the sample size of test transmitters and bearings should be reported, as well as both the mean and standard deviation of bearing error and whether bearings were found to be biased.

Reporting methods of data censorship are deficient as well: only 15 papers (35%) in our review reported methods of censoring data. Examples of censorship methods include using error polygon size or confidence ellipse size thresholds to exclude "unacceptable" locations, time limits between consecutive bearings, and upper or lower limits on the angle of bearing intersections. These procedures are designed to minimize location error, but the same authors often fail to address error adequately: only 10 of the articles (23% of the total) reporting censorship methods also reported an acceptable (*sensu* Saltz 1994) measure of accuracy. To evaluate published literature, peers need to know censorship procedures and the percentage of data actually used in the analysis (Garrott et al. 1986). Researchers should justify the specific time limits chosen for consecutive bearings through tests of accuracy with simulations of animal movements (see Beacon Tests below). If time periods chosen are too long, errors in location estimates will increase if animals move during the tracking period. If time periods are too short, then observers may be unable to obtain the necessary azimuths in the allowed time, and the study will be biased against location estimates in areas with fewer receiving points or difficult access. If poor signals are routinely ignored and tend to originate in certain parts of the study area, any resulting estimates of home range or habitat use could be unacceptably biased (Kufeld et al. 1987).

The only means to evaluate how researchers have censored data and measured the precision and accuracy of their location estimates is complete reporting in the literature. Editors should insist that their reviewers note whether location error was accounted for by authors and whether censorship procedures were adequately reported.

MEASURING BEARING ERROR

Beacon Tests

The most commonly used procedure to measure bearing error is a beacon study with test transmitters at known locations (White and Garrott 1990). The coordinates of these locations should be determined with the most accurate technique available, whether by survey, mapping on aerial photos, or GPS units. Test transmitters are often attached to wooden posts, although this may not accurately represent field conditions (White and Garrott 1990). Alternatives include attaching radiotransmitters to plastic bottles filled with saline (Hupp and Ratti 1983) or sugar (Larkin et al. 1996) solution, or to a road-killed carcass in order to simulate the absorption of radio signals by the study animals' bodies. If sequential bearings will be taken on moving animals, animal movements should also be simulated to test the effects of such movements on error. Otherwise the results of the beacon study will show less error than may actually occur in the field. If simultaneous bearings will be used, then animal movements are of less concern (Schmutz and White 1990). Cederlund and Lemnell (1980) described using a radio-tagged dog to simulate animal movements and thereby corrected bearings taken using a mobile tracking system. Passerine movements may be simulated by conducting tests on hand-held transmitters moved around in the air by a person (Pyke and O'Connor 1990).

With the selected beacon method, observers should take bearings on the test transmitters using the same system and methods that will be used in the field because the methods used may contribute a significant amount of error to the system (Pace 1988, White and Garrott 1990). Just as the test transmitter locations must be known, the locations at which bearings are taken should be known to the greatest accuracy available. For studies using mobile systems, this can be accomplished by establishing receiving points throughout the study area that can be mapped and used repeatedly. With both receiver and test transmitter locations known, the true bearing can be calculated (see Fig. 5.3 in White and Garrott 1990) and compared to the estimated bearing as recorded by the observer. Replicate bearings should be taken on the same test transmitter from the same receiving location, although some means to avoid observer expectancy bias should be taken. Examples include using a "dummy" compass rosette (White and Garrott 1990) or taking bearings from all different receiving locations before taking replicate bearings (Wallingford and Lancia 1991).

With the estimated and true bearings from each receiving location to all test transmitters, the mean error and standard deviation can be calculated for each receiver. Because the precision of a receiver is the amount of variation in the bearing error, the standard deviation of the measured error can be used as a measure of precision (Lee et al. 1985, White and Garrott 1990). If the mean error is significantly different from zero, then the receiver is biased. The cause of bias should be investigated before using the system for research. Typical problems include a misaligned compass or consistent loss of precision in a particular direction due to signal bounce or absorption, or electromagnetic interference (White and Garrott 1990).

Error Polygons and Confidence Ellipses

Using a beacon study to calculate the standard deviation of the bearing error allows one to construct confidence intervals around individual estimated bearings (White and Garrott 1990). Heezen and Tester (1967) introduced the error polygon method, which in the case of two receiving locations uses the error arcs formed by the intervals around estimated bearings to define a four-sided polygon (Heezen and Tester 1967, Springer 1979, Hupp and Ratti 1983, White and Garrott 1990). Given an actual or simulated transmitter location, the area of this polygon can be calculated (See Fig. 4.6 in White and Garrott 1990, Springer, 1979) and can be used to evaluate whether the telemetry system is precise enough for the objectives of the study (Hupp and Ratti 1983, Nams 1989).

For individual location estimates, if 95% confidence intervals are used to construct the error arc, then a polygon resulting from the intersection of two estimated bearings is expected to contain the actual location of the animal with a probability of (0.95^2) or 0.90 (Springer 1979). This calculation assumes that the angle errors are normally distributed and independent, which may be difficult to test conclusively in a limited study (Zimmerman 1990). Saltz and Alkon (1985) proposed using the largest diagonal of the error polygon as the most useful estimate of error but did not test how this measure corresponded with actual linear error. White and Garrott (1990) argued that because a given "longest diagonal" length may be found in error polygons with very different areas (nearly square vs. long and narrow), the area of the error polygon is preferable as a measure of precision.

Springer (1979) suggested extending the error polygon method when taking three estimated bearings on a radio-tagged animal. However, Nams and Boutin (1991) showed via simulations that the resulting polygon should, but does not, consistently contain the actual location with a probability of (0.95^3) or 0.86 (when a 95% confidence interval is used around each bearing). Both Nams and Boutin (1991) and White and Garrott (1990) recommended using one of Lenth's (1981) estimators (maximum likelihood, Huber, or Andrews) to determine the transmitter location and precision of the estimate. Pace and Weeks (1990) used their own estimator (nonlinear-weighted least squares estimator) to allow for nonhomogeneous error variances among the different signals used for a location estimate. In simulations, this technique showed improvement over Lenth's maximum likelihood and Andrews' estimator when error variances differed among receivers or were related to distance to transmitters. Using their estimator requires quantifying the error variance at a mix of distances, directions, and cover types for each receiving location, which may take more time than a typical beacon study (Pace and Weeks 1990).

Confidence ellipses are appropriate for measuring location precision, but most researchers extend their interpretation into a measure of location accuracy. Confidence ellipses may indicate something about accuracy (White and Garrott 1990, Rotella and Ratti 1992, Saltz 1994), but they are actually measures of precision that do not translate easily into measures of accuracy. Location estimates with small confidence ellipses are more precise than those with large ellipses. However, they are not necessarily more accurate. The probability that an estimated location is inside a 95% confidence ellipse is typically much lower than 0.95, depending on the ellipse size and which estimator is used: for example, the coverage of the best estimator for all ellipses of less than 0.6 ha was 78% (Andrews' estimator; Garrott et al. 1986). Zimmerman and Powell (1995) found only 70% of true locations within the maximum likelihood estimator 95% confidence ellipse, and suggested that a lack of bearing error independence may affect estimator performance. Linear error is not consistently correlated with ellipse size (Garrott et al. 1986), although a computer simulation indicated that it was positively correlated with the length of the ellipse's major axis (Saltz and White 1990). Assuming that the confidence ellipse is the area in which the true location is found 95% of the time (i.e., as a measure of accuracy) can lead to erroneous conclusions about what resources radio-tagged animals are using and through which areas they are moving.

IMPLICATIONS

Without conducting tests of telemetry system error, researchers cannot know how to properly interpret their observations of radio-tagged animals. Literature review and proper design can suggest ways to minimize effects, bias, and error, but field tests for each unique system are required to interpret results accurately. Reporting results of the field tests is essential for peer interpretation of a study's conclusions.

A clear bias from inaccurate location estimates in resource selection studies occurs when error areas include more than one resource type. Although the location estimate places the animal in one resource, it may actually be located in another resource. Researchers may then underestimate use of certain resources in some cases and overestimate it in others (Samuel and Kenow 1992). Nams (1989) reported that large telemetry errors bias measures of habitat selection (e.g., error areas >1.5 times habitat diameter) but also noted that increasing the sample size of location estimates (about 2.5-fold) can compensate if error diameter ≤ habitat diameter. However, as we noted earlier, traditional measures of "error area" (error polygons and confidence ellipses) are better measures of precision than of accuracy. With linear error estimates from a beacon study researchers can better evaluate at what scale appropriate selection analyses can be made.

Another important bias from inaccurate location estimates occurs when the likelihood of obtaining an acceptable location estimate, or the size of error associated with that estimate, varies by habitat. Increasing sample size or excluding data does not overcome this bias, and Rettie and McLoughlin (1999) suggested using GIS to analyze a circular area around each telemetry location. Our recommendation of using a regression model to predict linear error (see below) would provide the radius of that circle.

Bias in home range and other movement studies depends on the scale of movements that researchers are interested in detecting. It may be that coarse estimates of locations are acceptable, but that does not eliminate the need to test the radio-tracking system to be used.

RECOMMENDATION: A SITE-SPECIFIC APPROACH TO MEASURING ACCURACY

Rather than rely on measures of precision to indicate accuracy, we suggest that accuracy be directly measured by calculating the linear error of location estimates. This can be done by examining the relationship between accuracy and a suite of variables thought to influence it in a specific study area (Wallingford and Lancia 1991, Zimmerman and Powell 1995). Conducting an extensive beacon test where transmitters are placed throughout the study area and tracked under actual field conditions can be designed to accomplish this goal. The linear error of each estimate is determined, and multiple regression is used to relate linear error to variables such as distance from receiver to transmitter, temperature, and geometry of bearing intersections. The resulting predictive equation can then be used to estimate the linear error associated with location estimates on animals using the study area. For example, a regression model using the deviation of intersection angle from 90° (of two bearings) and the distance from receiver to estimated location explained 82% of the variance in linear error in a test of a truck-mounted tracking system (Wallingford and Lancia 1991). This approach has been used infrequently because it can be time consuming, expensive, and site-specific (Saltz 1994). However, with the advent of inexpensive, portable GPS units this technique is now much easier to implement. Moreover, a site-specific assessment of error is exactly what most researchers need to understand the accuracy of locations obtained by triangulation. Specific areas where signals are difficult to hear, or accurate estimates impossible to obtain, can be identified before extensive tracking takes place. Field procedures can be adjusted, for example to move closer to animals that may

be in such areas or to establish more receiving points from which to take
bearings.

In conjunction with a study of prairie falcons in southwestern Idaho
(Marzluff et al. 1997a), researchers investigated the accuracy of location
estimates obtained from simultaneous triangulation by four to six trackers.
Linear error was calculated directly in this study by estimating the location of
standard beacons (transmitters placed in a known location) and radio-tagged
falcons that were under direct observation at the time of triangulation (and
thus functioning as living beacons). We report the results of this study here
because it explicitly tests how confidence ellipses relate to accuracy as compared to a site-specific regression approach.

A field triangulation program calculated point estimates of transmitter
locations using three estimators: maximum likelihood estimator (MLE;
Lenth 1981), Andrews' estimator (Lenth 1981), and least squares estimator
(LSE; Pace 1988). These procedures also allowed for the calculation of a 95%
confidence ellipse around each point estimate. In addition, multiple regression
was used to predict the linear error associated with each point estimate.
Confidence circles (95%) were then created by buffering each point estimate
with a circle using the equation:

$$\text{Radius} = \text{predicted mean linear error} + (1.96)(\text{SE of mean linear error}) \tag{3.1}$$

Eight factors were suspected to correlate with linear error: (1) size of the 95%
confidence ellipse, (2) number of bearings used to obtain the location estimate,
(3) length of the major axis of the ellipse, (4) orientation of the major axis, (5)
length of the minor axis, (6) ratio of major to minor axis length, (7) tracking
zone within which the location was estimated, and (8) distance from the
location estimate to the center of the "tracking region" (Saltz and White 1990,
Wallingford and Lancia 1991, Zimmerman and Powell 1995). The tracking
region is a dynamic polygon with vertices formed by the positions of the
trackers involved in a particular location estimate.

The 95% confidence ellipses generated by Lenth's (1981) and Pace's (1988)
estimators were poor indicators of accuracy. Linear error increased slightly with
increasing ellipse size, but this relationship was significant only for Lenth's
MLE, and in all cases ellipse size explained a very small amount (1–4%) of the
variation in linear error (Andrews: $n = 126$, $R^2 = 0.03$, $F_{(1, 124)} = 3.61$,
$P = 0.06$; MLE: $n = 129$, $R^2 = 0.04$, $F_{(1, 127)} = 4.70$, $P = 0.03$; LSE: $n = 149$,
$R^2 = 0.01$, $F_{(1, 147)} = 1.28$, $P = 0.26$). The only variable that was significantly
related to linear error was the distance from the location estimate to the center
of the tracking region. Using the equation:

$$\text{Linear error} = -1071 + 0.53(\text{distance from center}) \tag{3.2}$$

to predict linear error around point estimates derived from the Andrews estimator we accounted for 54.2% of the variation in linear error ($F_{(1, 124)}$ = 148.8, $P < 0.001$).

Standard confidence ellipses covered the actual location of stationary beacons and falcons less frequently than the expected 95% (Table 3.1). Coverage was especially poor for the smallest (most precise) error ellipses. Confidence circles produced by regression models were significantly smaller than Lenth's error ellipses [median ellipse sizes (ha): 454, 502, 1930; median circle sizes (ha): 259, 276, 329, for Andrews', MLE, and LSE point estimates respectively] and produced significantly improved coverage that approached 95% (Table 3.1).

Use of a regression model can offer estimates of linear error and result in improved coverage and smaller ellipse sizes. In the process of gathering data for a site-specific model researchers have opportunity to test and analyze their telemetry system as a whole. The specific results of the study described may be most relevant to others studying wide-ranging birds, but the approach to understanding what factors affect telemetry accuracy in a specific study area is relevant to any study involving triangulation.

SUMMARY

Two important sources of bias and error in telemetry studies that are often dismissed are (1) the effects of capture, tagging, wearing transmitters, and

TABLE 3.1 Coverage Produced by Lenth's (1981) Andrew's and Maximum Likelihood Estimators and Pace's (1988) Least Squares Estimator, with Standard Confidence Ellipses ("Standard") and When Modified with Regression Model ("Regression"), for Beacon and Falcon Observation Data

| Source | Estimator | % Coverage | | $\chi^2_{(1)}$ | P |
		Standard	Regression		
Beacon	Andrews	65	89	21.07	<0.001
	MLE	60	89	30.35	<0.001
	LSE	66	76	3.49	0.06
Falcon observation	Andrews	70	100	10.59	0.001
	MLE	80	100	9.23	0.001
	LSE	73	90	1.18	0.28

Chi-square analysis compared the number of locations covered versus those missed by Lenth's standard confidence ellipses against regression model modified ellipses.
MLE, Maximum likelihood estimator; LSE, least squares estimate.

tracking, and (2) the inability to precisely and accurately estimate transmitter locations. These sources of error should be accounted for in all wildlife radiotelemetry studies.

Transmitters affect wildlife differently. Reported effects are most common in waterfowl and are most often associated with backpack harness attachments. Contrary to intuition, implant transmitters have the least effect. We offer 12 general guidelines to determine and minimize the effects of transmitters on wildlife, but note that each animal reacts differently and each new innovation in transmitter design should be evaluated. Therefore, generalizations are difficult to make and should serve only as a starting point for a careful investigation of potential effects and how to minimize them in each study.

A variety of factors can cause location error in radiotelemetry studies, and each source of error will have more or less influence on location estimates depending on the radio-tracking system used and the characteristics of the study area. The effect of location error is a loss of precision and accuracy of location estimates, which if unrecognized will bias inferences made from the study. The error associated with a telemetry system and criteria used to censor inaccurate locations are not always reported in studies, which hampers peer interpretation of conclusions. Traditional methods of evaluating error (e.g., error polygons and confidence ellipses) are useful as measures of precision, but not as direct measures of accuracy. Linear error can be measured and used to evaluate accuracy directly. The level of "acceptable" error in location estimates depends on the study objectives.

We encourage all researchers using radiotelemetry to conduct two studies *before beginning to collect data on actual study subjects*. First, study the effects of the transmitter and method of transmitter attachment on the study species. Ideally this would occur under captive settings initially, then continue in the field concurrently with the main study. When possible, the behavior, physiology, condition, reproduction, and survivorship of tagged animals should be related to those attributes of untagged, control animals. It is important to monitor the performance of tagged animals throughout the study because the severity of effects may be correlated with the abundance of food or other resources. Second, if animal locations are to be estimated remotely, conduct a thorough beacon study in the actual field site under expected field conditions. The goal of a beacon study is to produce a spatially explicit map of tracking accuracy for the study area and quantify the precision and accuracy of the telemetry system. Accuracy is especially important to understand and is best investigated using a regression-based approach to estimate linear error.

APPENDIX 3.A Journal Articles Reviewed for Transmitter Effects[a], Organized by Taxon and Type of Effects Studied[b]

Species	Effects studied	References
Birds		
Waterfowl		
Bean, Brent, and Barnacle Geese	Behavior	Giroux et al. 1990
Black duck	Behavior	Wooley and Owen 1978
Black brant	Survival	Ward and Flint 1995
	behavior	Ward and Flint 1995, Sedinger et al. 1990
Blue-winged teal	Behavior, physical	Greenwood and Sargeant 1973, Garrettson et al. 2000
Canvasback	Behavior	Perry 1981, Korschgen et al. 1996a (ducklings)
	Physical	Korschgen et al. 1996a (ducklings)
Common goldeneye	Behavior	Korschgen et al. 1984
Gadwall	Survival, reproduction	Pietz et al. 1995
Emperor goose	Survival, reproduction	Schmutz and Morse 2000
Eurasian wigeon	Behavior	Giroux et al. 1990
Harlequin duck	Survival	Esler et al. 2000
Mallard	Survival	Paquette et al. 1997, Mauser and Jarvis 1991 (ducklings), Mauser et al. 1994 (ducklings), Pietz et al. 1995, Dzus and Clark 1996
	Reproduction	Paquette et al. 1997, Pietz et al. 1993, 1995, Dzus and Clark 1996, Bergmann et al. 1994, Rotella et al. 1993, Orthmeyer and Ball 1990, Houston and Greenwood 1993, Ball et al. 1975
	Behavior	Greenwood and Sargeant 1973, Pietz et al. 1993
	Physical	Bakken et al. 1996 (ducklings), Greenwood and Sargeant 1973, Korschgen et al. 1996b, Houston and Greenwood 1993
Northern pintail	Reproduction	Guyn and Clark 1999
Redhead	Survival, reproduction, behavior	Sorenson 1989
Ring-necked duck	Behavior	Korschgen et al. 1984

(continues)

APPENDIX 3.A (*continued*)

Species	Effects studied	References
Snow goose	Behavior	Blouin et al. 1999
Wood duck	Survival, physical	Davis et al. 1999 (ducklings)
	Reproduction	Gammonley and Kelley 1994, Ball et al. 1975
Upland game birds		
Blue grouse	Survival	Pekins 1988, Hines and Zwickel 1985
	Reproduction, behavior	Hines and Zwickel 1985
Gray partridge	Survival	Carroll 1990, Bro et al. 1999
	Reproduction, physical	Bro et al. 1999
	Behavior	Putaala et al. 1997
Prairie chicken	Survival	Burger et al. 1991
Red grouse	Survival	Boag et al. 1973
	Reproduction	Boag et al. 1973, Lance and Watson 1977
	Behavior	Lance and Watson 1977, Boag 1972
Ring-necked pheasant	Survival	Johnson and Berner 1980, Marcstrom et al. 1989, Warner and Etter 1983
	Reproduction	Warner and Etter 1983
Rock ptarmigan	survival	Cotter and Gratto 1995
Ruffed grouse	Survival, Behavior	Small and Rusch 1985
Sharp-tailed grouse	Survival	Marks and Marks 1987
Spruce grouse	Survival, behavior	Herzog 1979
Wild turkey	Behavior	Nenno and Healy 1979
	Physical	Hubbard et al. 1998 (poults)
Willow grouse	Reproduction	Erikstad 1979
Shorebirds and seabirds		
American woodcock	Behavior	Horton and Causey 1984, Ramakka 1972
Black-legged kittiwake	Behavior	Wanless 1992
Chinstrap penguin	Reproduction, behavior	Croll et al. 1996
Common tern	Behavior, physical	Klaassen et al. 1992
European golden-plover	Behavior	Whittingham 1996
Great snipe	Survival, reproduction, behavior, physical	Kalas et al. 1989

(*continues*)

APPENDIX 3.A (*continued*)

Species	Effects studied	References
Herring gull	Survival, reproduction	Morris et al. 1981
Least tern	Reproduction, behavior	Hill and Talent 1990
Marbled murrelet	Survival	Newman et al. 1999
Snowy plover	Reproduction, behavior	Hill and Talent 1990
Xanthus' murrelet	Survival	Newman et al. 1999
Raptors		
Bald eagle	Survival	Buehler et al. 1995
Barn owl	Reproduction	Taylor 1991
Buteos (red-tailed, ferruginous, and Swainson's hawk)	Reproduction	Andersen 1994
Lesser kestrel	Survival, reproduction, behavior	Hiraldo et al. 1994
Merlin	Survival, reproduction	Sodhi et al. 1991
Prairie falcon	Reproduction, behavior	Vekasy et al. 1996
Snail kite	Survival	Snyder et al. 1989
Spotted owl	Survival, Reproduction	Paton et al. 1991, Foster et al. 1992
	Physical	Foster et al. 1992
Passerines		
Common yellowthroat	Behavior	Sykes et al. 1990
Hooded warbler	Behavior	Neudorf and Pitcher 1997
Wood thrush	Survival, physical	Powell et al. 1998 (migrating), Powell et al. 2000b
Other birds		
Acorn woodpecker	Behavior	Hooge 1991
Florida sandhill crane	Survival, behavior	Klugman and Fuller 1990
Mourning dove	Behavior	Sayre et al. 1981
	Physical	Schulz et al. 1998
Sandhill and whooping crane	Behavior	Melvin et al. 1983
Mammals		
Medium to large terrestrial mammals		
Armadillo	Survival, reproduction	Herbst 1991
Franklin's ground squirrel	Survival, physical	Eagle et al. 1984

(*continues*)

APPENDIX 3.A (*continued*)

Species	Effects studied	References
Mink	Survival, physical	Eagle et al. 1984
Moose	Survival	Swenson et al. 1999 (calves)
Mountain goat	Survival, reproduction, behavior	Côté et al. 1998
Mule deer	Survival	Garrott et al. 1985 (fawns)
Pronghorn	Survival	Keister et al. 1988
San Joaquin kit fox	Survival, reproduction, physical	Cypher 1997
White-tailed deer	Physical	Clute and Ozoga 1983
Yellow-bellied marmot	Survival, reproduction, physical	Van Vuren 1989
Small mammals		
Deer mouse	Physical	Reynolds 1992
House mouse	Behavior	Mikesic and Drickamer 1992
Meadow vole	Behavior	Berteaux et al. 1996
Prairie vole	Physical	Reynolds 1992
Root vole	Survival	Johannesen et al. 1997
Aquatic mammals		
Humpback and right whales	Behavior	Goodyear 1993
North American beaver	Behavior	Davis et al. 1984
	Physical	Davis et al. 1984, Guynn et al. 1987
River otter	Reproduction	Reid et al. 1986
Sea otter	Behavior	Garshelis and Siniff 1983
Bats		
Hoary bat	Behavior	Hickey 1992
Yuma myotis	Behavior	Aldridge and Brigham 1988

[a]Transmitter effects studied: behavior, survival, physical and reproduction. [b]Journals searched for this review were as follows: *The Journal of Wildlife Management, Wildlife Society Bulletin, Journal of Field Ornithology, Copeia,* and *Journal of Mammalogy* (1972–April 2000).

APPENDIX 3.B Page Numbers of Articles That Used Triangulation to Estimate Animal Locations, by Volume in *The Journal of Wildlife Management* (1993–1999), That Were Reviewed for Error Reporting[a]

			Volume			
57	58	59	60	61	62	63
85–91	88–94	98–103	154–164	112–122	171–178	116–125
258–264	280–288	228–237	422–430	140–150	205–213	210–222
282–290	462–469	238–245	777–787	343–350	280–285	261–269
475–494	536–545	392–400	962–969	435–443	1020–1035	291–297
861–867	600–607			497–505	1264–1275	546–552
868–874	608–618			634–644		593–605
				707–717		711–722
				736–746		853–860
				1115–1126		
				1213–1221		

[a]This appendix can be combined with the appendix in Saltz 1994.

PART **III**

Equipment and Technology

Recent Telemetry Technology

ARTHUR R. RODGERS

Power Supplies
Microcontrollers
Coded Transmitters
Sensors
Archival Tags
Satellite Telemetry Systems
    Argos-Based Systems
    GPS-Based Systems
Hyperbolic Telemetry Systems
Implications for Data Analysis
Implications for Researchers
Future Directions

When used in a broad biological sense, "telemetry," or "biotelemetry," refers to the remote determination of an animal's status (Priede 1992). This may include the individual's current level of activity, some physiological measurement (e.g., body temperature, heart rate, etc.), and its physical location. Thus, telemetry is used in an extensive array of agricultural, physiological, and medical research activities to monitor the status of individuals. In these studies, measurements are usually made at close range, typically within a few hundred meters. Ecologists, field biologists, and wildlife researchers, on the other hand, generally use "telemetry" to estimate a series of geographic locations for individual animals, often from distances exceeding several hundred meters. Furthermore, since radio signals emanating from a device attached to each animal are commonly used to estimate locations, the technique is often referred to as "radiotelemetry." This description of the method is not inconsistent with applications in other fields of study because many of the devices attached to free-ranging animals include sensors that provide information about the individual's physiology or behavior. However, in those studies where sensors are not used or the primary purpose is to estimate animal locations, some researchers prefer to call the technique "radio-tracking" (Kenward 1987, White and Garrott 1990).

For almost 40 years wildlife biologists have used radiotelemetry to study free-ranging animals in their natural environment. Throughout that time researchers have been quick to adapt new telemetry technology to their particular study needs. Consequently, this technology has provided abundant knowledge about the physiology, behavior, daily movements, migrations, survival, habitat use, and population densities of many wildlife species. Miniaturization of electronic components has allowed radiotelemetry to be used on virtually all orders of mammals, as well as many species of fish, amphibians, reptiles, and birds.

The conventional approach to radio-tracking involves the attachment of a very high frequency (VHF) radiotransmitter to an animal in the form of a "collar" or "tag," which also includes a power supply and antenna. Pulsed radio signals received by a radio receiver are then used to estimate the animal's location. Individual animals are identified by the unique frequency and pulse rate of the transmitter they carry. Location estimates can be obtained by "homing-in" on the animal using a hand-held directional antenna and receiver (Mech 1983), or by triangulation of at least two directional bearings recorded at known locations remote from the animal (White and Garrott 1990). Among other problems, a major limitation of these techniques is the signal range of the transmitters. Whereas VHF transmitters may have a range of 3–20 km on the ground (Kenward 1987), many large animals and migratory species commonly move over large areas of several hundred or even thousands of square kilometers. Consequently, aircraft are often used to track species that move long distances. Aerial tracking can improve the signal range of VHF transmitters to 35–100 km (Samuel and Fuller 1996), but there are still a number of limitations to this approach. In particular, costs escalate quickly as the size of the area to be searched, the number of animals to be located, and the frequency of location estimates increase.

To overcome some of the limitations of conventional VHF tracking and to reduce costs, several satellite tracking systems were developed during the 1980s. The best known of these uses the CLS/Service Argos Data Collection and Location System (Toulouse, France) carried on Tiros-N weather satellites operated by the U.S. National Oceanic and Atmospheric Administration (NOAA). For brevity, this system is commonly referred to as simply the Argos system. The Argos system uses the Doppler shift of ultra-high frequency (UHF) radio signals arriving at a satellite in successive intervals during an orbital pass to estimate the location of the transmitter on the earth's surface. Location estimates are then relayed from the satellite to regional receiving stations strategically located around the world, then to global processing centers in Landover (United States) or Toulouse (France) for redistribution to individual researchers. The Argos system has been very successful in tracking the long-distance movements of highly mobile animals in remote areas (for examples, see White and Garrott 1990).

Advances in VHF and satellite telemetry systems up until the late 1980s are well documented. Most of the early technological developments of these systems are described in a number of general references to radiotelemetry, including Brander and Cochran (1971), Amlaner and Macdonald (1980), Cochran (1980), Cheeseman and Mitson (1982), Mech (1983), Kenward (1987), White and Garrott (1990), Priede and Swift (1992), and Samuel and Fuller (1996). Fancy et al. (1988) and R. Harris et al. (1990) have provided reviews of the Argos system. In addition to scientific journals such as *The Journal of Wildlife Management*, *Wildlife Society Bulletin*, and *Journal of Fish Biology*, annual proceedings of conferences convened by the International Society on Biotelemetry have also documented many of the early developments and provide an ongoing source of information about the most recent technological advances. Many of these references go well beyond descriptions of the technology and provide essential information about the principles of radio-tracking, considerations in study design, effects of telemetry devices on animals, methods of attachment, data analysis, and even the details of circuit construction for ambitious researchers who want to build a system. In spite of technological change, much of the information in these publications is still relevant and provides the necessary background for most wildlife telemetry studies.

Telemetry technology has continued to progress throughout the 1990s. Advances in power management, electronic components, and microcomputers, as well as the emergence of several innovative systems based on new technology, have all contributed to an ever-increasing variety of telemetry options available to field researchers. The purpose of this chapter is to provide an update on the most significant technological advances of the last decade and an overview of the most recent telemetry systems. As will become evident, the trends toward reduced size, improved accuracy, and greater automation of telemetry systems that have been the driving forces behind the development of telemetry technology since the early 1960s have continued in the 1990s. Automated systems in particular present a challenge to data management and analytical procedures because of their capacity to generate enormous quantities of data. The implications for analysis of location data resulting from these recent advances in technology are considered and some speculation on the future directions of telemetry technology is offered.

POWER SUPPLIES

Most telemetry devices are powered by batteries, which convert chemical energy to electrical energy. Batteries suitable for telemetry applications come in a wide range of shapes, sizes, and capacities that use a variety of chemical and physical configurations to generate the required electrical output. Two fundamentally different types of batteries are available for use in telemetry units; primary cells that use their chemical energy just once, and secondary cells that can be recharged and reused several times (Takeuchi 1995). From a telemetry system design standpoint, the most important characteristics of batteries are voltage, energy density, and the response to load current. Whereas voltage is determined almost entirely by chemistry, energy density and the response to load current can also be affected by the physical design of a battery and the external environment (Takeuchi 1995).

The output power of telemetry devices, and their resulting signal range, is directly related to the voltage of the power supply. Signal range generally increases with battery voltage, although the relationship is not strictly linear (Kenward 1987). Batteries can be connected in series to increase system voltage and extend signal range, but there are limits because this strategy will obviously add to both the weight and volume of collars or tags attached to animals.

Battery capacity determines the length of time a telemetry device will operate and is specified as ampere-hours (Ah) for larger cells or milliampere-hours (mAh) for smaller cells. So, for example, a battery with a 300-mAh capacity is capable of supplying a load current of 1 mA for 300 h, 10 mA for 30 hs, etc. In general, capacity increases with cell size. Dividing the battery capacity by the current requirement of the telemetry device provides an estimate of operational life expectancy. Such estimates are usually optimistic, however, because of a number of factors that might affect capacity. For example, capacity may be lost at high temperatures through "parasitic" chemical reactions that occur in some types of cells (Takeuchi 1995). The external environment also affects the capacity retained by batteries during storage. Less capacity is lost by storing batteries at lower temperatures ($\sim 5°C$) than higher temperatures (Takeuchi 1995). This is an important consideration because telemetry collars or tags are often kept in storage for a period of time before deployment on study animals, thereby reducing capacity and operating life in the field. By connecting batteries in parallel rather than in series, the capacity of a telemetry device can be increased. As before, this strategy has limits imposed by the weight and volume of units that can be attached to animals.

A particularly useful measure for comparison of batteries is energy density, which combines voltage and capacity with the volume and weight of a cell.

Energy density is typically expressed volumetrically as watt-hours per milli-liter $(Whml^{-1})$ or gravimetrically as watt-hours per gram (Whg^{-1}). Both measures are used because batteries with the highest volumetric energy density do not necessarily have the highest gravimetric energy density (Takeuchi 1995). In either case, energy density standardizes the trade-off between energy and size in cells with different chemical and physical configurations, thereby permitting direct comparison. Batteries with the highest energy density are preferred for most telemetry applications.

The primary-cell batteries most commonly used in telemetry devices manu-factured during the 1990s have been based on chemical reactions involving silver oxide or lithium. Silver oxide cells contain aqueous electrolytes and have an output of 1.5 V, whereas lithium-based cells vary from 1.5 to 3.6 V and have nonaqueous electrolytes because water reacts with lithium (Kenward 1987, Takeuchi 1995). Both cell types are available in a range of capacities, but lithium batteries come in larger sizes, have higher energy densities, and lose less of their capacity during storage than silver oxide cells (Kenward 1987, Takeuchi 1995). The different type of electrolyte in these cells is an important determinant of their response to load current. Cells with aqueous electrolytes usually have higher conductivity than nonaqueous cells, so there is less voltage drop when a load current is applied (Takeuchi 1995). However, aqueous cells lose their conductivity faster than nonaqueous cells as tempera-ture declines and the voltage drop in silver oxide batteries may prevent telemetry units from operating much below $0°C$ (Kenward 1987). Thus, lithium batteries are often best suited for telemetry devices used in cold environments.

The choice between silver oxide or lithium batteries in a telemetry device is based primarily on size. Silver oxide batteries are commonly used in the smallest telemetry units suitable for body or stomach implants and external attachment to birds, small mammals, fish, amphibians, or reptiles (Kenward 1987). However, the smaller size of silver oxide cells usually translates to lower capacity and reduced operating life of the telemetry unit. Consequently, if small size is not a primary concern, lithium-based batteries are used whenever possible because of their higher voltages and capacities, which provide longer life and greater signal range than silver oxide batteries. In addition, lithium batteries have a long shelf life and tolerate the wide temperature variations experienced by animals in many research programs.

Secondary-cell batteries have generally been used as a "back-up" power supply in telemetry devices that use photocells (i.e., photovoltaic cells that convert sunlight to electrical energy) as a primary energy source (Patton et al. 1973). Although very small telemetry units can be powered entirely with photocells, they cease to function at low light intensities caused by nightfall and heavy shade (Kenward 1987). If an animal can tolerate a larger device, then a better strategy is to use photocells to recharge secondary-cell batteries that can provide power at low light intensities. The most common secondary-cell batteries used in these applications are nickel-cadmium (NiCad) with aqueous electrolytes. NiCad batteries have an output of 1.5 V but have substantially less energy density than primary cells (Takeuchi 1995). However, because they can be recharged, the total amount of energy produced by secondary cells over their entire operational life is potentially very high. Unfortunately, this potential is rarely achieved because most secondary cells, such as NiCad batteries, have shown substantial "memory" effects. These effects result from repeated partial discharging of NiCad cells that prevent them from being completely discharged the next time they are used. This is a common occurrence in telemetry devices that use photocells to recharge NiCad batteries because the secondary cells are partially discharged for short periods when an animal moves under shade or cover, or at night. The end result is a reduction in the operational life of the telemetry collar or tag in which these cells are used. In fact, some telemetry manufacturers consider this problem so serious that instead of attempting to recharge secondary cells they simply use a lithium-based primary cell as a back-up power supply under low-light conditions. Recently developed lithium ion secondary cells may alleviate the problems associated with recharging. Lithium ion batteries with nonaqueous electrolytes can output up to 4.0 V and do not show substantial memory effects (Takeuchi 1995). Whereas NiCad batteries can be recharged up to 500 times under ideal conditions, lithium ion cells can be recharged as many as 2000 times (Takeuchi 1995).

Regardless of the energy source, the power supply makes up more than half of the total volume and weight of most telemetry collars or tags. Therefore, reducing the size of the power supply can have the greatest impact on overall weight and volume of telemetry units attached to animals. This can sometimes be achieved through trade-offs between system parameters such as operating life and signal range (i.e., using a smaller, lower voltage battery that reduces signal range but maintains the operational life of the device). More often, size limits force researchers to simply use smaller power supplies that diminish both signal range and the operational life of telemetry units. In spite of substantial advances during the 1970s and 1980s, little progress was made in the reduction of volume and weight of power supplies during the 1990s. Instead, the manufacturers of telemetry equipment have concentrated their efforts on the development of power management strategies that conserve energy and use it more efficiently to meet the requirements of transmitters and other device components. Additional energy savings have been achieved through the continuing development of more efficient electronic components (i.e., circuit boards, microchips, etc.). The net result is that a smaller power supply can often be used to meet the energy needs of electronic components that required larger power sources in the past, thereby reducing the overall size of telemetry collars and tags. Out of concern for the potential effects of telemetry devices on individuals, considerable emphasis continues to be placed on reducing the volume and weight of units attached to animals.

MICROCONTROLLERS

The single most important component of telemetry equipment to undergo further development during the 1990s has been the microcontroller. Often referred to as "microprocessors," these electronic components originally did little more than execute a preprogrammed series of instructions. Although these simple microcontrollers were available in the early 1980s and used in some receiving equipment, their size, power consumption rates, and voltage requirements were too high for most telemetry devices attached to animals. Instead, complex circuitry employing "discrete logic" was used to control device functions such as the pulse rate and pulse duration of signals from very high frequency (VHF) transmitters. In earlier motion-sensing units, for example, an integrated circuit would be used to change the signal pulse rate after a "preset" time interval during which a mercury tilt-switch had not moved (Kenward 1987). The length of the time interval was determined (i.e., "preset") by the operating characteristics of a complementary metal oxide semiconductor (CMOS) field effect transistor (FET) in the integrated circuit. Discrete logic systems were effective in expanding the telemetry options available to researchers but were not always reliable, and there were limits imposed by size on the number of components that could be incorporated into collars or tags attached to animals.

In the late 1980s and early 1990s, single-chip microcontrollers became available that weighed about 4 g and had both low voltage (2.0–3.0 V) and low current (0.015–0.020 mA) requirements. These devices were far more sophisticated than their predecessors and incorporated erasable programmable read-only memory (EPROM), random access memory (RAM), input/output (I/O) communication ports, and, perhaps most importantly, quartz crystal timing. In effect, these microcontrollers had all of the capabilities of a basic microcomputer. Their size and energy requirements had been reduced to the point where they could replace the discrete logic components found in many telemetry collars or tags, with the prospect of providing researchers with far more options and control over transmission and data collection than ever before. Combined with a power supply in suitable packaging, microcontroller tags weighing as little as 20 g were available in the mid-1990s. Soon after, ongoing development reduced current requirements of some microcontrollers to 0.010 mA and their size had shrunk to the point where complete micro-controller transmitters can weigh as little as 1.5 g.

Incorporation of quartz crystal timing in microcontroller transmitters represents a particularly significant advance that deserves special attention. The highly accurate and stable digital time base of quartz crystal timing provides more precise and consistent control of signal pulse rates and pulse widths than could be achieved with discrete logic circuits of the past. These improvements to signal quality greatly enhance the ability of receiving systems to detect transmissions in "noisy" environments. Perhaps the greatest advan-tage of the digital time base, however, is the ability to control the on/off duty cycles of various functions in telemetry units. Thus, it is possible to slow down the pulse rate, reduce the length of the duty cycle, or turn off transmitters for long periods and restart them later. This feature can be advantageous, for example, in studies where aerial tracking is used to determine animal loca-tions. If tracking is conducted during particular intervals (e.g., the last week of each month), microcontroller transmitters can be programmed to turn on only during those time periods and remain off at other times. The transmitters can also be programmed to turn on during daylight hours when aerial tracking takes place and to turn off at night. Transmitters might also be turned off during periods of animal hibernation. Studies of migratory species can benefit by allowing animals to be captured and fitted with collars or tags at times and places that are most efficient while delaying the startup of various telemetry functions for days, weeks, or even months (Lacroix and McCurdy 1996). Because duty cycle schedules and other options are set by software program-ming rather than specific hardware components, changes and upgrades to microcontroller tags can be made efficiently and economically. The greatest advantage of scheduling strategies such as these is that the capacity of the power supply can be conserved. Although the microcontroller will continue to draw some current while other functions are turned off, it may only be about 10% of that required to power a typical transmitter. In the end, the energy saved can be used to extend the operational life of the telemetry device in the field by as much as 80–90%. Alternatively, the ability to schedule on/off duty cycles can be viewed as a power management strategy that makes it possible to reduce the size or number of batteries in a telemetry unit by distributing the capacity available from fewer or smaller batteries more effectively over the duration of a study. In many cases these power management strategies make it possible to deploy a smaller telemetry device on animals than in the past, which can help address concerns about the size of units attached to animals while still meeting research objectives.

The ability to schedule on/off duty cycles of transmitters can also be useful in situations where the number of available frequencies is limited and it is necessary to monitor a large number of animals. Several individuals could be assigned the same frequency, but each transmitter could be programmed to turn on and off at different times to avoid conflicts.

Although microcontrollers have had their greatest impact on devices attached to animals, they have also become an integral part of other system equipment and have facilitated the development of several new telemetry systems. They are an essential component of automated telemetry systems. Although available in the 1980s (Kenward 1987), programmable VHF telemetry receivers with scanning capability were not widely used, but with the further development of microcontrollers during the 1990s they are now commonplace. These receivers allow the user to automatically step through a set of programmed frequencies, listening to each one for a short, user-defined time interval. If a signal is detected, scanning can be stopped while the animal is located, then resumed until another signal is received, and so on. Many of these receivers also have built-in data logging components or can be connected to external devices for automatic data recording. This feature makes it possible to configure one of these receivers for remote, self-contained operation with on-board storage of data, or control and retrieval of data via cellular phone, radio modem, satellite, etc. (Voegeli 1989, Eiler 1995). Additional options available to researchers include the ability to identify and monitor more than one individual on a single frequency, designate transmitters as "sensors" (i.e., record temperature, activity, or mortality, or process signals from electromyogram transmitters), monitor both ultrasonic and VHF signals with the same receiver, and control switching among multiple antennas. With these capabilities, a remote, solar-powered installation can automatically scan for signals received at up to eight antennas from a large number of transmitters. When a signal is received, the system can pause automatically and record data such as the date and time, frequency, pulse rate, signal strength, antenna identifier, and sensor information, then resume operation. During the 1990s, remote receiving/data logging systems such as these were used extensively in fisheries research (e.g., Eiler 1995, Moser and Ross 1995, Hinch et al. 1996, Lacroix and McCurdy 1996, Weatherley et al. 1996, Smith and Smith 1997). Wildlife studies have also benefited from these automated systems, for example, to determine the responses of cottontail rabbits to microclimate (Althoff et al. 1989); to monitor the migration of canvasback ducks (Korschgen et al. 1995) or peregrine falcons (Howey et al. 1989); to determine activity of mule deer (Relyea et al. 1994), presence of female Florida panthers at dens (Land et al. 1998), and attendance of female American marten at natal dens (Ruggiero et al. 1998); and to measure the thermal microclimate of tree cavities used by big brown bats (Kalcounis and Brigham 1998).

CODED TRANSMITTERS

Coded transmitters, or "tags" as they are often called, have been available since the early 1980s (Lotimer 1980). These devices permit the unique identification of more than one individual using a single frequency. They were originally developed to solve the problem of having to monitor a large number of individuals in situations where the number of available frequencies was limited. By coding the transmitted signals, more than one individual could be assigned to each frequency and the same codes could be used on multiple frequencies. In the earliest coded tags, the identity of each transmitter was determined by the repetition rate of a series of signal pulses where the time between each pulse (i.e., pulse rate) remained constant for each tag but varied among tags (Fig. 4.1). Development of CMOS pulse control circuitry in the

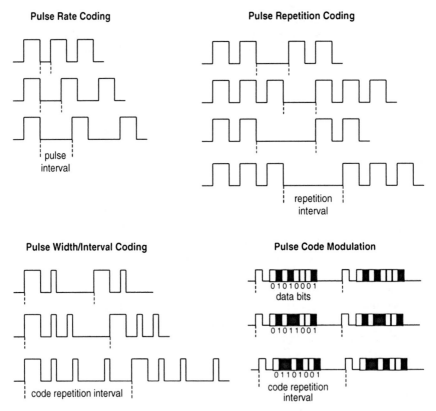

FIGURE 4.1   Signal pulse coding formats commonly used with telemetry transmitters.

1980s allowed the single pulse to be replaced with a group of two or more equally spaced pulses in which the number of pulses or the time between pulses, as well as the repetition rate, could be used to identify individual transmitters. In both of these coding schemes, the signal pulses are of constant duration (i.e., pulse width). Prior to the development of more sophisticated telemetry receivers in the late 1980s, the duration of each pulse had to be long enough to be detected by the human ear. The energy requirements of these audible signals limited the maximum number of pulses in a group to four. In addition, older receivers often had difficulty separating groups of pulses from background noise. Advances in receiver technology allowed the development of new coding schemes in which the duration of each pulse was no longer required to be audible, thereby permitting greater variation in the duration of pulses in a group. Since pulses could be of shorter duration, the number of pulses in a group could exceed the previous limit of four without a significant increase in power consumption. Pulse width coding together with new receiver technology provided significant improvements in the numbers of transmitters that could be assigned to a single frequency as well as greater immunity to background noise, but there were still a few problems.

Most of these early pulse repetition coding schemes relied heavily on the ability of the transmitter to produce precisely timed signal pulses (i.e., pulse rates and pulse widths). Consequently, any internal or external factors that might cause even minor shifts in timing (e.g., temperature or low battery voltage) could affect the performance of these systems. Even without timing shifts, the presence of a large number of coded tags in a small area could increase the risk of "code collisions" (i.e., signals arriving at the same time from two or more transmitters with similar codes that could not be distinguished by a receiver because of background noise) and subsequent errors in identification. Not surprisingly, then, coded transmitters benefited tremendously from the incorporation of microcontrollers in the early 1990s. As suggested earlier, microcontrollers with quartz crystal timing provide precise and consistent control of signal pulse rates and pulse widths. Moreover, microcontrollers allowed more reliable coding schemes to be introduced that could further increase the numbers of tags using a single frequency and reduce the risk of code collisions. For example, phase-modulated Manchester code uses a series of positive (i.e., in phase) and negative (i.e., out of phase) pulses to produce a binary sequence that can be translated by an appropriate receiver. Thus, in a 24-bit sequence, the first 8 bits are used to "wake up" and synchronize the receiver, the next 8 bits are used to provide up to 256 unique numerical identification codes, and the last 8 bits can be used to transfer sensor data (Korschgen et al. 1995). This simple on/off coding scheme, as well as several others that produce numerical codes (Cupal and Weeks 1989), improves the accuracy of tag identification by removing the need to discriminate signal pulses of different widths, which also lowers the probability of code collisions. Compared to pulse repetition coding, these digitally coded tags also use less energy. These energy savings can be used to extend the operating life of the transmitters, to extend the range of the system by increasing signal strength, or to reduce the overall size of the device deployed on study animals.

By allowing the unique identification of more than one individual using a single frequency, coded tags optimize the use of available frequency space and make it possible to monitor a large number of individuals in a minimum amount of time. Since many animals can be accommodated on each frequency, less time is spent scanning through a set of frequencies and there is less chance of missing an individual on each pass through the set. Using a multichannel receiver that can simultaneously monitor several frequencies can further enhance this advantage of coded systems. Receiver-coprocessors capable of monitoring up to 25 frequencies simultaneously have been available since the early 1990s. Combined with a coding scheme that assigns 50 individuals to each frequency, for example, these systems can simultaneously monitor 1250 transmitters. Because the potential for code collisions is obviously very high, especially in noisy environments, receivers used with coded tags are programmed with special algorithms to discriminate valid signal pulses from extraneous noise and ensure unique code identification. Some systems incorporate digital spectrum analyzers to further improve code discrimination capabilities. As a result, receivers used with coded tags effectively have greater sensitivity than conventional receivers do and, depending on the coding scheme, may actually provide better audible detection of signals.

With their enhanced abilities, coded telemetry systems are well suited for remote stand-alone installations (Howey et al. 1989). In addition to automated data recording, remote receivers/data loggers used with coded tags can also provide multiple antenna switching and two-way communication via cellular phone, radio modem, or satellite. These features make coded systems useful for monitoring the migratory movements of large numbers of individuals, such as fish moving along rivers (Eiler 1995, Moser and Ross 1995) or waterfowl moving over broad geographic areas (Korschgen et al. 1995). These systems might also be used in situations where animals are in view for very short periods of time, as is the case with whales, seals, or sea turtles at the water surface, or fish moving rapidly through spillways around hydroelectric dams. Although coded systems are amenable to automatic monitoring, they may also prove useful in aerial tracking of wide-ranging animals, such as wolves, or migratory ungulates because of the advanced code discrimination capabilities of receivers used with these systems that can extract distinct identification codes from high levels of background noise.

SENSORS

Much of the recent technology that has benefited coded telemetry systems has also enhanced the performance of sensors incorporated in devices attached to animals. Sensors for monitoring temperature and movement, which have been available since the early days of field telemetry, have undergone numerous changes, and many new types of sensors have been developed. Moreover, microcontrollers have allowed sensors to be linked to other device functions, resulting in many new options that can be exploited by researchers.

Temperature sensors have been used to monitor the body temperatures of animals in relation to ambient temperature, activity, reproductive state, and general health (Samuel and Fuller 1996). In some cases, temperature changes at specific body sites have been examined (e.g., stomach). Telemetry sensors have also been used to determine temperatures in specific microhabitats, such as nests, dens, or hibernacula. Temperature sensors use a thermistor to alter the pulse rate and duration of signals from VHF transmitters. Thermistors are semiconductors that function as heat-sensitive resistors. As temperature increases, resistance declines, allowing more current to pass through the circuit. In the earliest and simplest systems, the current would flow through a thermistor to a pulse capacitor that would increase the signal pulse rate with increasing temperature (Kenward 1987). Changes in pulse rate at each temperature were determined by the characteristics of the particular thermistor and other circuit components. Since current flow was regulated by the thermistor, the operating life of these devices was dependent on the sensitivity of the thermistor and the operational temperature range; greater sensitivity and prolonged use at high temperatures could dramatically reduce operating life (Kenward 1987). This problem was solved through advances in CMOS pulse control circuitry in the 1980s, which allowed the current available for transmitting signals to become essentially independent of the thermistor. Since temperature measurements ultimately relied on changes in pulse rate, however, factors that might affect components other than the thermistor, such as ambient temperature, low battery voltage, and aging components, made it necessary to periodically recalibrate these devices. A strategy developed in the late 1980s removed the need for repeated recalibration by incorporating a reference circuit into the transmitter. In these units the pulse rate set by the thermistor could be compared with a more stable pulse rate determined by a more accurately known resistance. Independent measurement of the two pulse rates with a precision timer then allowed corrections of temperature measurements to be calculated in the field based on initial calibrations. With the advent of microcontrollers in the 1990s, corrective strategies such as these were incorporated directly into devices attached to animals. Automatic temperature correction and more precise control of pulse rates permits current microcontroller transmitters to have a resolution of ± 0.1 to $\pm 0.2°C$ over a wide range of operating temperatures (-40 to $+50°C$). Some devices offer a temperature resolution of $0.03°C$ over a slightly narrower operating range (-20 to $+35°C$). Since transmission of temperature measurements through modification of signal pulse rates is essentially a form of coding, improvements in receiving–coprocessing systems that have benefited coded telemetry systems have also enhanced the performance of temperature sensors. Indeed, temperature measurements are readily amenable to various digital coding schemes and can be stored or transmitted as part of the data stream in coded telemetry systems. This makes it possible to accurately monitor temperature changes through remote stand-alone installations.

Motion sensors are used to monitor animal activity, posture, and mortality. These sensors are based on a mercury tilt-switch containing a drop of mercury that moves through a small tube to open and close a circuit. When the circuit is open, current flows through a single pulse capacitor that sets a specified pulse rate and duration of VHF signals from the transmitter (Kenward 1987). When the circuit is closed, a second parallel capacitor increases pulse duration and decreases pulse rate. Typically, the tilt-switch is oriented such that the circuit is closed (i.e., slow pulse rate) when the animal is motionless and open (i.e., faster pulse rate) when it is moving. Tilt-switches may also be strategically mounted to indicate a particular posture associated with specific behaviors (e.g., to indicate feeding activity by monitoring whether an animal's head is "up" or "down"). Deaths of animals might also be determined by listening for signal changes induced by an activity sensor at times when individuals are expected to be active. However, it was recognized early on that such inferences could be difficult because animals might remain inactive for significant periods of time for other reasons. As suggested by Cochran (1980:511), the "critical measurement is the length of time since the subject last moved." Thus, discrete logic systems of the 1980s incorporated a time delay mechanism into their integrated circuits, as outlined previously. For mortality studies, the delay time was usually set at several hours, after which the signal pulse rate might increase or decrease to indicate that the status of the tilt-switch had not changed. If movement occurred before the time period had elapsed, the delay mechanism was reset. Unfortunately, this created a different problem: the delay mechanism could be reset by the actions of predators or scavengers feeding on a dead animal, so accurate detection and determination of the time and cause of death were often belated or difficult. Microcontroller tags of the 1990s mitigate this problem by providing the ability to define sensor levels that might distinguish mortality from low-level activity, as well as more precise and consistent measurement of time-delay periods. For example, a unit might be programmed to transmit a mortality signal only if the tilt-switch does not change its status more than twice in a 4-h period followed by a 6-h period in which the switch status does not change more than 10 times. The tag might also be programmed to increase the pulse rate by a set amount (e.g., 5 beats/min) following each successive interval (e.g., 4 h) in which these conditions are met. By measuring the pulse rate when the tag is recovered, it is then possible to determine the approximate time of death. Alternatively, the microcontroller tag might store the activity counts (i.e., the number of changes in tilt-switch status), an average of these counts, or an index based on preset values over a predefined period, along with time and date information to be transmitted or recovered later on. Similar to temperature data, activity measurements are readily transmitted as part of the data stream in coded telemetry systems, making it possible to monitor activity and mortality through remote stand-alone installations. Thus, another option is to prevent tags from sending out a signal until the mortality criteria are satisfied. Then, the unit transmits a numeric identification code that can be received and relayed by cellular phone, radio modem, satellite, and so forth to notify the researcher that mortality has occurred, as well as identifying the affected animal. This strategy of designating transmitters as sensors in remote situations takes advantage of many of the features of recent coded systems and has the added benefit of conserving battery power.

Numerous other types of sensors have benefited from recent advances in radiotelemetry, together with the miniaturization and development of other biosensitive electronic devices. Implantable electrocardiogram (ECG) and electromyogram (EMG) transmitters, which measure the voltage of electrical impulses produced by the heart or other muscle activity through two electrodes, have benefited from the incorporation of automatic gain control circuitry and the development of new polymers for sealing implants, as well as Teflon coating for internal wires and electrodes. Measurement of light intensity, typically monitored with photovoltaic panels or phototransistors (Kenward 1987, Althoff et al. 1989), has been improved and is now possible with implanted devices through advances in fiber-optic or "light pipe" collecting technology. Largely as a result of progress made in medical and agricultural applications, an ever-widening variety of electrochemical sensing transmitters have become available. These "biosensors" can be used to measure chemical characteristics in minute samples of various substrates such as soil, water, gas, blood, or urine and translate the results into a measurable radio signal (Rogers and Gerlach 1996). Some of these devices respond to changes in conductivity (e.g., salinity sensors), but many of the more recent biosensors use enzymes or immunoassays (i.e., antibodies) to detect specific compounds (Van Emon and Lopez-Avila 1992, Rogers and Gerlach 1996). A particularly intriguing approach has been the development and incorporation of genetically manipulated microorganisms that produce optical signals when they detect specific DNA sequences, into a microchip that converts the optical signals to radio signals for telemetry (Vo-Dinh et al. 1994). It seems that the possibilities are endless and only limited by the imagination of the researchers seeking new knowledge.

The miniaturization of sensors and other hardware components along with the incorporation of microcontrollers into devices attached to animals allows researchers to combine several options into a single unit. Moreover, microcontrollers allow various device functions to be mediated by sensors. For example, a photosensor can be used to turn off a transmitter during the night or day, depending on when animals are expected to be active. In the past, this option might have been criticized because signals (and indeed the tag itself) may be lost if an animal died with the tag covered by vegetation or debris (Kenward 1987). However, it is now relatively easy to also include a much improved mortality sensor in the device, as outlined above, along with the photosensing capability. Another possibility is to turn the transmitter off when an animal enters torpor or hibernation based on a threshold temperature programmed into the tag. Time-depth recorders, used to study diving behavior of marine birds and mammals, can be programmed to record data only when activated by a salinity sensor (e.g., Bowen et al. 1999). These and many other similar options have significant potential for conserving battery life. As before, these energy savings can be used to extend the operating life of the transmitters or to lower the overall size of the device attached to study animals by reducing the size of the power supply.

Sensor control of transmitters has also provided a solution to a particular problem in studies of diadromous fish and some marine mammals and birds that spend time in both freshwater and salt water. Whereas VHF radio frequencies (typically 100–200 MHz) readily penetrate freshwater, acoustic transmitters (typically 30–100 kHz) must be used in salt water (Priede 1992). In addition, VHF signals are attenuated more than acoustic signals with increasing water depth and are almost undetectable below 30 m. On the other hand, acoustic transmitters do not perform well in fast-moving water, and their signals can be severely attenuated in shallow water or by aquatic vegetation and turbid conditions (Priede 1992). The simplest solution to these problems is to use a device that transmits both VHF and acoustic signals (Smith and Smith 1997). In some freshwater situations, this strategy has the added advantage of providing range and bearing to a location by measuring the time delay between the arrival of VHF and acoustic signals at a dual receiver (Armstrong et al. 1988). However, the energy required to power both transmitter subsystems will clearly reduce the expected operating life of these devices. Because acoustic transmitters use almost 10 times more power than VHF transmitters (Priede 1992), a better strategy might be to deactivate the acoustic subsystem after a programmed time interval when individuals are expected to have entered freshwater (Solomon and Potter

1988). Alternatively, the VHF and acoustic transmitters could be activated or deactivated by a salinity (i.e., salt water) sensor programmed with specific conductivity thresholds. As well, the subsystems might be linked to a pressure sensor that switches between VHF and acoustic signals depending on depth. This latter strategy might also be employed in studies of entirely freshwater species in deep water where acoustic signals could be more useful than VHF.

ARCHIVAL TAGS

Archival tags, also referred to as "data loggers," were originally proposed by Holter (1961) as a way to record heart rates of patients in medical applications where continuous monitoring may not be possible. Thus, their use is sometimes referred to as "Holter monitoring" (Priede 1992). Conceptually these devices are quite simple: rather than transmitting data as they are collected, information is stored until it can be sent to a receiver or retrieved upon recovery of the unit. Although a variety of relatively simple devices have been available since the early 1960s (e.g., Cresswell 1960), it was not until low-power microcontrollers were introduced in the early 1990s that these tags began to realize their potential.

The development of archival tags for wildlife applications has been driven largely by studies of diving behavior in marine mammals and birds. Lacking the more sophisticated signal processing technology of the 1990s, early researchers found that data could only be transmitted effectively when an animal was at the water's surface. Consequently, they needed a device that could store data such as dive depth, duration, and physiology (ECG or EMG) until it could be retrieved. Units developed in the 1980s used a pressure sensor or transponder to initiate data transmission when an animal surfaced (Kenward 1987). Small amounts of data were then transmitted repeatedly until the animal's next dive. Although transmission redundancy may ensure data transfer, it is clearly an inefficient use of the power supply. To compensate for these energy requirements, devices were necessarily large and had short operating lives.

Archival tags developed in the 1990s incorporate many of the features conferred by microcontrollers. To begin with, archival tags used in marine studies are much smaller than their predecessors, weighing as little as 16–30 g with dimensions as small as 18 × 57 mm. Similar to other microcontroller tags, quartz crystal timing has greatly improved the accuracy of time-based functions, such as the measurement of dive duration. As well, archival tags can be programmed to delay startup for up to 1 year and the duty cycle schedule can be set to record data at intervals from 1 s to 99 days. Sensors incorporated into present day archival tags are also much improved. Internal and external temperature sensors can provide 0.06°C accuracy with a resolution of 0.03°C over an operating range of −20 to +35°C. Pressure sensors for measuring dive depth may have an accuracy of 0.15–7.5 psi with a resolution of 0.03–0.5 psi to depths of 1000 m. Ambient light sensors have also improved through the development of light-pipe collecting technology that allows internal or external tag attachment. Other optional sensors (e.g., EMG, paddlewheel velocity meter, etc.) can also be incorporated into archival tags. As outlined in the preceding section, various sensors can be used to control device functions. For example, a depth threshold might be used to trigger measurements of temperature and light above or below a programmed value. Depending on the sampling schedule, current archival tags may collect data for more than 10 years and retain up to 2 megabytes of nonvolatile memory for more than 20 years (Block et al. 1998). With such large data storage capacity, redundant data transmission is no longer necessary. A VHF or acoustic transmitter in the device can be used to track and recapture animals for data retrieval. Alternatively, external tags might also have a remote release mechanism that is triggered by a command signal or at a preset time. The unit itself can then be tracked and recovered for data retrieval. However, if preferred, data might still be transferred to a receiver or other external device with data logging capability, then relayed by cellular phone, radio modem, satellite, etc. In this case, the archival tag functions much like other microcontroller tags and can be further enhanced by taking advantage of features associated with coded systems, as described in previous sections.

In addition to sensor data, archival tags can be configured to determine and store animal locations. Some units provide highly accurate (< 10 m) position estimates derived from the global positioning system (GPS), as described in a later section. Others calculate relatively coarse locations based on ambient light levels. These latter position estimates are sometimes referred to as "geolocation" or "geoposition" data. To make these calculations for marine species, archival tags record depths and corresponding ambient light levels throughout each day and use these data to estimate surface light intensity (Hill 1994). More advanced units provide improved estimates of surface light intensity by using light sensors that respond only to blue–green light and by applying correction factors to account for variation of water clarity with depth. Estimates of surface light intensity are then used to determine times of sunrise and sunset. By knowing these times and the date on which they were determined, comparisons can be made with Greenwich mean time (GMT) to estimate geographic location. Longitude is established from midnight or local noon based on the average of recorded sunrise and sunset times, while latitude is determined from the estimated day length. Longitudinal errors associated with these methods are typically ±0.5° (Delong et al. 1992, Welch and Eveson 1999). Errors in latitude, on the other hand, are usually on the order of $\pm 1.0°$ (111 km). At the equinoxes it is not possible to estimate latitude with this approach because day length is independent of latitude (Welch and Eveson 1999). This limitation of the method can be overcome if the archival tag also stores temperature data that can be compared to sea surface temperatures at the known longitude (Block et al. 1998). Surface temperatures measured by remote sensing are widely available from a variety of sources (e.g., NOAA, RADARSAT International, etc.). Although geolocation data are obviously crude in comparison with more sophisticated methods, they can be useful in tracking the gross movements of individuals over extremely wide geographic areas. To date, these types of archival tags have been used primarily for studies of teleost fish under high exploitation pressure (Block et al. 1998) and some marine mammals (Stewart and Delong 1991). However, there is great potential for using this recent technology for further studies of marine mammals and sea birds, as well as adapting it for use on many terrestrial species that may be too difficult or costly to study by other methods.

SATELLITE TELEMETRY SYSTEMS

ARGOS-BASED SYSTEMS

Satellite-linked telemetry systems were originally developed to monitor environmental data, such as atmospheric pressure and temperature, as well as providing geographic locations, for meteorologists and oceanographers. The first receivers for these systems were thus carried on board Nimbus weather satellites (Kenward 1987, White and Garrott 1990). Similarly, when the Argos system became fully operational in 1978, receivers were carried on Tiros-N weather satellites operated by NOAA under the Polar Orbiting Environmental Satellite (POES) program. The original sensors and data transmission components weighed 4.5–16 kg and were designed to be carried on various "platforms," such as meteorological balloons and moored or drifting buoys (Buechner et al. 1971). As such, these transmitters were referred to as "platform transmitter terminals," or "PTTs" for short, and this terminology continues to be used today.

Size and Performance

The first satellite transmitters for wildlife studies were tested on elk (*Cervus elaphus*; Buechner et al. 1971) and used to record ambient temperature and light levels in a black bear (*Ursus americanus*) den (Craighead et al. 1971). These prototype PTTs were extremely large, weighing 5–11 kg (Kenward 1987,

Fancy et al. 1988, White and Garrott 1990), similar to devices deployed on buoys and weather balloons. The size of devices attached to animals did not decrease significantly until the second generation of satellite transmitters was produced in the mid-1980s in conjunction with the Argos system.

Second generation devices weighed 1.6–2.0 kg (Fancy et al. 1988), functioned for 12–18 months, and facilitated numerous studies on a wide variety of both marine and terrestrial wildlife species (Fancy et al. 1988, White and Garrott 1990). Along with their smaller size, these PTTs provided improved reliability and performance (Fancy et al. 1988) and incorporated a variety of sensors relevant to wildlife applications (Fancy et al. 1988, Garner et al. 1988, R. Harris et al. 1990, Taillade 1992). In spite of significant advances, PTTs were still too large to be suitable for studies of small or even medium-sized animals.

The availability of smaller, low-power microcontrollers in the late 1980s facilitated the development of the next generation of satellite transmitters. In addition to a reduction in the size and power requirements of electronic components, the power output of the transmitter was reduced from 1 W to 0.25–0.4 W, allowing smaller batteries to be used. The first of these PTTs weighed about 175 g. NiCad batteries that were recharged by solar panels provided the power source. However, problems with rechargeable NiCad batteries, as outlined in a previous section, quickly led to their replacement with lithium cells. Removal of the recharging components and the switch to lithium batteries further reduced the weight of PTTs to 110–150 g. These smaller PTTs were used for only a few years to study marine animals and a few of the largest species of birds (Jouventin and Weimerskirch 1990, Priede and French 1991) before the next significant size reduction occurred.

Although satellite transmitters weighing as little as 50 g were predicted from the beginning (Buechner et al. 1971), and some suggested that they might eventually weigh as little as 10 g or less (Cochran 1980), PTTs approaching these sizes were not achieved until the early to mid-1990s. Size reductions were again made possible primarily through advances in microcontroller technology that resulted in smaller electronic components with lower voltage requirements. In fact, voltage requirements were reduced by almost half, declining from the 7–15 V of earlier devices to 4–7 V. This meant that the power supply could also be cut in half. For example, instead of requiring four batteries connected in series, only two were required. The 50% reduction in battery weight and volume resulted in miniature PTTs weighing 45–120 g with dimensions as small as 5.6 × 3.3 × 1.8 cm, suitable for a variety of large-bird species (Meyburg and Lobkov 1994, Brodeur et al. 1996, Higuchi et al. 1996, Ely et al. 1997, Brothers et al. 1998, Ueta et al. 1998). By the late 1990s, miniaturized PTTs weighing 30 g or less were readily available and used to study medium-sized birds weighing 700–1500 g, such as osprey (*Pandion haliaetus*; Kjellen et al. 1997), white-faced whistling ducks (*Dendrocygna viduata*; Petrie and Rogers 1997), and peregrine falcons (*Falco peregrinus anatum*; Britten et al. 1999).

Where size is not a critical issue larger PTTs have the distinct advantage of a potentially longer operational life. Whether or not that potential is realized depends on both the on/off duty cycles of the transmitter and the repetition period of transmitted signals (i.e., time between transmissions from the PTT, which is typically 60–90 s, when it is turned on). In one study, for example, PTTs weighing 1.1–1.2 kg were deployed on wolves (*Canis lupus*) and programmed to transmit for 6 h then turn off for 42 h, resulting in an average operational life of 181 days (Ballard et al. 1995). By contrast, 30 g PTTs on white-faced whistling ducks that were programmed to transmit for 8 h and then turn off for 158 h lasted an average of 239 days (Petrie and Rogers 1997). Differences in the duty cycles specified in these two studies were presumably related to the sample size requirements and duration of each project. Thus, on/off duty cycles must be carefully planned to maximize opportunities for successful signal transmission to satellites while meeting study objectives with respect to sampling intensity and project duration (see Chapter 2).

Accuracy

Regardless of PTT size, the accuracy of Argos location estimates depends on the number of Doppler measurements obtained (i.e., signals received) during each satellite overpass, stability of the transmitter's frequency oscillator (which is affected by temperature), altitude of the transmitter (which is assumed to be at sea level in position calculations), ionospheric propagation errors (which may be affected by sun spot activity), and errors in satellite orbital data (Fancy et al. 1988, R. Harris et al. 1990, White and Garrott 1990, Keating et al. 1991, Taillade 1992). Second-generation PTTs used in the mid-1980s had average location errors ($\pm$ SE) of 829 ± 26 m, with 90% of estimated locations within 1.7 km of the true location, and a maximum error of 8.8 km (Fancy et al. 1988). Beginning in April 1987, Service Argos established location quality classes (Table 4.1) ranging from 1 (poorest) to 3 (highest), to provide users with estimates of location precision (R. Harris et al. 1990, Keating et al. 1991, Taillade 1992). These classes are based primarily on the number of signals received during each satellite overpass, the temporal distribution of received signals, and the stability of the transmitter's frequency oscillator. Tests of these Argos location quality classes have shown that actual errors often exceed specifications (R. Harris et al. 1990, Keating et al. 1991, Ballard et al. 1995, Mate et al. 1997, Brothers et al. 1998). As a result, some researchers pool these location quality classes and specify the accuracy of

TABLE 4.1   Argos Location Quality Classes, Specifications, and Estimated Accuracy for 68% of Locations.

Class	Specifications	Accuracy
3	> 4 messages received by satellite; pass duration > 420 s; internal consistency < 0.15 Hz; quality control of oscillator drift; unambiguous solution	150 m
2	> 4 messages received by satellite; pass duration > 420 s; internal consistency < 1.5 Hz; quality control of oscillator drift; unambiguous solution	350 m
1	4 messages received by satellite; pass duration < 420 s; internal consistency < 1.5 Hz; lacking quality control of oscillator drift	1 km
0	2 messages received by satellite; pass duration < 240 s; poor internal consistency; unstable oscillator	> 1 km
A	3 messages	> 4 km[a]
B	2 messages	> 10 km[a]
Z	Invalid locations	

[a]Based on data of Brothers et al. (1998) and Britten et al. (1999).
Note: 68% of locations are expected to fall within this distance.

position estimates calculated by Service Argos simply as 1 km or less (Meyburg and Lobkov 1994, Higuchi et al. 1996, Kjellen et al. 1997).

A fourth "zero" location quality class with unspecified accuracy was added in January 1988. This class was intended specifically for wildlife researchers primarily interested in keeping track of their study animals under difficult field conditions with the likelihood of few successful signal transmissions. Class 0 positions are calculated from as few as two Doppler measurements and consequently are of lower quality than previously established classes. However, each class 0 record also includes a location index that can be used to determine why normal processing failed (R. Harris et al. 1990, Taillade 1992). Evaluations of class 0 position estimates have revealed average errors ($\pm$ SD) ranging from 6.4 ± 5.1 km (Britten et al. 1999) to 9.2 ± 6.4 km (Brodeur et al. 1996), with 68% of locations within 7.5 (Mate et al. 1997) to 14.3 km (Keating et al. 1991) of true locations, and maximum errors ranging from 15.8 (Britten et al. 1999) to 396.2 km (Keating et al. 1991). In June 1994, following the revision of algorithms used to calculate positions, Service Argos defined class 0 accuracy as greater than 1 km and added three more classes (A, B, and Z) with unspecified accuracies (Brothers et al. 1998, Britten et al. 1999). Brothers et al. (1998) recently tested 85- and 120-g PTTs and reported average ($\pm$ SD) latitude and longitude errors of 5.0 ± 8.6 km and 4.8 ± 6.2 km, respectively, for class A locations, and 10.2 ± 10.7 km and 13.2 ± 14.9 km for class B locations. In a more recent test of 30-g PTTs, Britten et al. (1999) found average errors ($\pm$ SD) of 4.1 ± 2.7 km and 35.4 ± 63.1 km with maximum errors of 9.7 and 285 km for location classes A and B, respectively. They concluded that locations they obtained with miniature PTTs were sufficient to identify movements over broad spatial scales but were not suitable for studies of habitat use. However, the accuracy of locations obtained from satellite transmitters will improve dramatically through the recent incorporation of GPS receivers into PTTs, as described in a later section. In this configuration, position estimates determined by GPS are treated much like sensor data and relayed to users via the Argos system. Since most GPS receivers can recalculate locations as often as once per second, higher temporal resolution than is currently available through the Argos system will also be possible.

Future Enhancements

The overall performance of the Argos system is expected to improve substantially over the next few years, and enhancements to the system may lead to further reductions in the size of PTTs. These improvements are expected to result from the recent launch of the NOAA-K satellite in May 1998, which is carrying the first in the next generation of Argos instruments (Argos-2). Together with previous NOAA satellites (H and J), the NOAA-K satellite will provide greater coverage of the Earth's surface. As well, the Argos-2 instrument features a wider frequency bandwidth (80 kHz) than previous Argos units (24 kHz). By spreading the signals across a wider bandwidth, the ability to discriminate transmissions arriving simultaneously at the satellite will be enhanced and more messages will be received intact for processing. The Argos-2 instrument also has twice as many (i.e., eight) data recovery units as its predecessor, allowing more messages arriving at the satellite at the same time with similar frequencies to be successfully processed. To accommodate these additional data recovery units, the communication link with the Tiros information processor on board the satellite has been increased from 960 (NOAA-H) or 1200 bits/s (NOAA-J) to 2560 bits/s. These enhancements will nearly quadruple the capacity of the Argos system by permitting the number of PTTs visible to the satellite that can be processed at a given time to increase from about 500 to approximately 1750. In addition to greater capacity, the sensitivity of the Argos-2 receiving system has been increased from -128 dB to -131 dB. Although this may not appear to be a big change in sensitivity (-3 dB), the implications are significant. To begin with, low-power signals will have a better chance of being detected so more messages and locations will be processed. This should also result in an increase in the proportion of estimates in the higher location quality classes (i.e., 1–3). Because the receiver is capable of detecting weaker signals than previous units, the output requirement of PTTs may again be reduced by almost 50%. Thus, a 0.4-W transmitter may be reduced to 0.2 W. Similar to previous reductions in output requirements, this change might allow smaller or fewer batteries to be used, leading to an additional 50% reduction in the size of PTTs. Alternatively, the operational life of existing PTTs could be almost doubled.

Further improvements to the Argos system are expected in the near future with the launch of the NOAA-L satellite in September 2000 and the planned launch of NOAA-M in 2001, which will also carry Argos-2 instruments. In addition, an enhanced Argos-2 instrument is planned for the ADEOS-II satellite, which is scheduled for launch by the Japanese Space Agency (NASDA) in 2001. This enhanced instrument will provide downlink messaging to PTTs, enabling transmitters and sensors to be turned on and off, recalibrated, and so forth. This "two-way" communication will provide greater system flexibility, improved performance, and more effective and economical data collection. The next generation of Argos instruments, the Argos-3, with even greater capability, is planned for satellite systems such as the European METOP series in the near future.

GPS-Based Systems

Although satellite-based telemetry systems such as Argos are well suited for tracking movements of animals over broad spatial scales, particularly in remote areas, location accuracy is generally insufficient for detailed resource selection studies. This limitation prompted the development of a wildlife telemetry system based on the NAVSTAR Global Positioning System (GPS) in 1992 (Rodgers et al. 1995, 1996, 1997). Since then, a variety of configurations have been designed and tested, and a number of improvements to these systems have been made.

The GPS Network

Initiated in 1973 by the U.S. Department of Defense (Wells 1986), the GPS network became fully operational in 1993, making it possible for users to obtain position estimates (i.e., "fixes") 24 h/day (Rodgers et al. 1996). The system has three main components: (1) a space segment of 24 satellites (plus several backups) that continuously broadcast spread-spectrum radio signals; (2) a network of ground monitoring and control stations that maintain the system time standard and calculate exact orbital information (i.e., ephemerides) for all satellites; and (3) GPS signal receivers carried by individuals. Thus, GPS reverses the roles of satellites and units carried by individuals. Rather than signals being broadcast by ground-based devices and received by satellites, a GPS receiver on the ground determines its position by simultaneously synchronizing on the spread spectrum signals from three satellites to obtain a two-dimensional fix (i.e., latitude and longitude) or four satellites to obtain a three-dimensional fix (i.e., latitude, longitude, and altitude). The receiver estimates its ground position from measurements of the time it takes for signals to travel from the satellites, as well as from their orbital details. These location estimates can be updated as often as once per second.

The accuracy of GPS locations is determined primarily by the synchronization of the receiver and satellite clocks. For reasons of national security, the U.S. Department of Defense has the ability to degrade the accuracy of civilian GPS receivers by incorporating unpredictable satellite clock and ephemeris errors into transmissions. This "selective availability" policy can limit the accuracy of horizontal position estimates to within 100 m of their true location 95% of the time (Wells 1986, Rempel et al. 1995). Selective availability is readily overcome, however, by placing a reference GPS receiver (often called a "base station") at a surveyed location and recording deviations from the known coordinates of the site (Rempel et al. 1995, Moen et al. 1997). Correction factors can then be computed and applied to fixes obtained from GPS units carried by individuals within 300 km of the base station. This process, referred to as "differential correction," can reduce location errors to less than 10 m (Moen et al. 1997, Rempel and Rodgers 1997).

On May 2, 2000, the U.S. Department of Defense discontinued its selective availability policy 6 years ahead of schedule (Lawler 2000). Consequently, most civilian GPS receivers should obtain horizontal position estimates within 20 m of their true location 95% of the time (Wells 1986). In fact, shortly after selective availability was "turned off," several fixed-station GPS receivers recorded 95% of locations within a radius of less than 10 m, similar to differentially corrected GPS data (Moen et al. 1997, Rempel and Rodgers 1997). Thus, for many GPS applications, differential correction of GPS data may no longer be necessary. However, because differential correction can also remove other sources of error, such as satellite orbit and clock errors, as well as ionospheric and tropospheric errors, it may still improve the accuracy of location estimates. Without selective availability, the accuracy of differentially corrected GPS data is less than 10 m, 95% of the time, and median errors of 1–3 m are possible (Wells 1986). Whether or not this level of accuracy is required depends on the objectives of specific research projects. Potential users of GPS-based telemetry systems should also be aware that even though the U.S. Department of Defense has discontinued global degradation of GPS, they have retained the capability to selectively deny GPS signals on a regional basis when national security is threatened. Thus, researchers in some parts of the world may be advised to collect data amenable to differential correction, even if they do not need the greater precision, in the event that selective availability is reinstituted in their study region.

GPS units designed for wildlife research function much like archival tags described in a previous section. Data such as geographic coordinates or GPS measurements (i.e., "pseudoranges" and satellite identification numbers) from which coordinates can be calculated following differential correction, dates and times of location estimates, and optional sensor information are stored in devices attached to animals until they can be retrieved. During the development of the first GPS units for wildlife in the early 1990s, three data retrieval options were considered: (1) store data on board until the unit can be remotely released by a radio-activated "break-away" mechanism or recovered by recapture of the animal; (2) transmit stored data for retrieval by secondary low earth orbit (LEO) satellite link; or (3) transmit stored data through a local, user-controlled communication link (Rodgers et al. 1996, 1998).

Local Communication Link

The first GPS-based animal location system used a local point-to-point communication link for data retrieval (Rodgers et al. 1995, 1996). This option was selected primarily because it provides two-way communication, allowing retrieval of data and real-time status information (e.g., battery condition, memory utilization, current position), as well as reconfiguration of the system (e.g., change the number of fixes attempted per day, change the ratio of day to night fixes, control data storage strategies, etc.) at the user's convenience. The system consists of remote GPS units carried by study animals and a "command" unit controlled through a laptop computer operating from a vehicle or aircraft (Fig. 4.2). UHF radio modems in both the animal units and the command unit provide the communication link within a range of 15 km (ground-to-air). The modem receiver in each animal unit is turned on at specified intervals of a unique schedule in anticipation of a signal from the command unit. When a signal is detected, the animal unit sends a response in its unique time slot and a communication link may be established. Precise timing of events is possible because the animal and command units have identical GPS receivers that provide an absolute time reference obtained from the GPS satellites. Once a communication link has been established, the animal unit transmits its last recorded position, providing a general reference point for navigation during the communication session. The laptop computer controls the data link protocols and supervises the transmission of data and diagnostic information, such as remaining battery life. After all of the information has been received and stored by the computer, the random access memory of the animal unit can be cleared and the communication link closed.

FIGURE 4.2   A GPS-based animal location system with a local communication link. The animal unit uses signals received from at least 3 GPS satellites to estimate locations that are stored until they can be transmitted by UHF radio modem to a command unit carried on board an aircraft or other vehicle.

In addition to a GPS receiver, the animal units are equipped with a standard VHF radio beacon that can be used to locate animals with conventional direction-finding techniques (Rodgers et al. 1995, 1996). With an operating range of 15 km (ground-to-air), the VHF beacon provides further navigation support during communication sessions. If the animal unit's main operating system fails due to battery exhaustion or some other cause, an independent emergency power source for the VHF beacon allows it to serve as an emergency locator for about 90 days. A wide-range temperature transducer and a dual-axis motion sensor are also incorporated into animal units. Thus, data stored by each unit include geographic coordinates, date and time of each fix attempt, GPS fix type (i.e., two- or three-dimensional) and associated horizontal dilution of precision (an indicator of fix quality), as well as sensor information. The first units had a data storage capacity of 3640 records. However, because these units did not store GPS measurements necessary for differential correction of position estimates (i.e., pseudoranges and satellite identification numbers) and selective availability was operational, the median location error of three-dimensional fixes they recorded was 45.5 m with 95% of estimates within 120 m of their true location (Rempel et al. 1995). A second version of these animal units, capable of storing the information necessary for differential correction, was developed shortly after and has a data storage capacity of 1640 records. With selective availability, these new units provided location estimates following differential correction with a median error of 4.2 m, and 95% of the positions were within 21 m of their true location (Fig. 4.3; Rempel and Rodgers 1997).

Similar to other telemetry devices, the operating life of GPS units is determined primarily by the duty cycle and the size of the battery pack used. However, the first animal units developed offer a variety of additional power management strategies to extend operating life. To begin with, the modem and VHF beacon can be programmed to turn off during periods when communication sessions are unlikely (e.g., at night). Clearing the

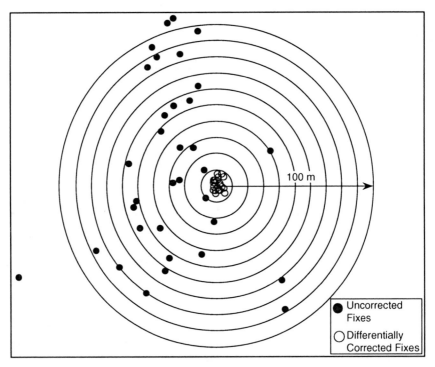

FIGURE 4.3   Location estimates with and without differential correction obtained by a GPS animal collar placed in an open field. (Data from Rempel and Rodgers 1997.)

animal unit's memory after data have been transferred and stored removes redundancy in subsequent communication sessions and also increases life expectancy. Scheduling GPS fixes less than 4 h apart, whether required or not, alleviates the need for the receiver to acquire new satellite ephemeris data (i.e., an "almanac"), which can take up to 12 min and consume substantial energy. Another extremely important parameter for a GPS unit is the amount of time the receiver is allowed to attempt a fix. Conceivably, a GPS receiver could be in a situation where it has difficulty simultaneously synchronizing on signals from the minimum requirement of three satellites. Without limitation, the receiver would keep trying to synchronize signals until it was successful or the power supply was exhausted. The first animal units were thus allowed 170 s to synchronize signals from three satellites for a two-dimensional fix and 10 additional seconds to lock onto signals from a fourth satellite to estimate a three-dimensional position (Rempel et al. 1995, Rodgers et al. 1995, Moen et al. 1996). The second version of these animal units allowed the GPS receiver a maximum of 185 s, but more time was allotted to acquiring signals from a fourth satellite, which substantially increased the proportions of three-dimensional fixes as well as the overall success rate of fix attempts (Moen et al. 1997, Rempel and Rodgers 1997). Employing these strategies, the first GPS animal units were fitted with a 400-g lithium battery pack that was expected to provide 1 year of operation at 8 fix attempts per day (i.e., every 3 h) with VHF beacon and modem functions available 16 h/day (Rodgers et al. 1995, 1996). A larger (700-g) battery pack, expected to provide 2 years of operation, was also available. Based on early field trials (Rodgers et al. 1995, 1998), the life expectancies of these units was downgraded, and the small battery pack is now expected to provide 268 days of service at 6 fixes per day (i.e., every 4 h), whereas the large battery pack may provide 575 days with the same fix schedule. The total weight of these GPS animal units, which were originally designed for caribou and moose, is 1.8 kg with the small battery pack and 2.2 kg with the large battery pack.

The next GPS-based telemetry system to be developed also provided a local communication link to retrieve data stored on animal units. In this case, animal units continuously broadcast stored data through individual pulse-coded VHF signals to a receiver and data logger during preset time periods. Thus, the VHF transmitter functions as both a direction-finding beacon and a radio link for data transfer. The data transmission range when operating on the ground is about 1–5 km and increases to 15–25 km when operated from an aircraft. Stored data include geographic coordinates, dates and times of location estimates, sensor information (activity and temperature), and battery status. Since communication is unidirectional, the GPS fix schedule and VHF transmission periods must be programmed before units are attached to animals and cannot be changed without recovery of individual units. Likewise, stored data cannot be cleared from the memory of animal units after they have been transferred. Rather than specifying an operational life based on time, these units are rated to provide different numbers of records based on battery capacity. The GPS fix schedule and VHF transmission periods then determine how long a unit will function in the field. Depending on battery size, animal units are available in a wide range of sizes from 475 g, capable of recording 400 geographic coordinates (i.e., non-differential data), to 1.8 kg, capable of recording 4600 geographic coordinates.

Satellite Link

GPS-based telemetry systems that use a local communication link for data retrieval may be limited by transmission range or constraints on personnel time, logistics, economics, weather, and daylight requirements for data recovery by vehicle or aircraft. Thus, in the mid-1990s a system was developed that transfers GPS data via the Argos satellite system (Rodgers et al. 1997, 1998, Arthur and Schwartz 1999, Schwartz and Arthur 1999, Bennett et al. 2001, Biggs et al. 2001). Devices attached to animals include a standard PTT and a GPS receiver, as well as a VHF beacon and microcontroller circuitry for managing duty cycles and data storage. GPS fixes with associated dates and times are stored and incorporated into the Argos data stream transmitted from the PTT to NOAA satellites and then to Service Argos ground stations as previously described. Although GPS location data are provided by the system, Argos positioning remains available and can serve as a backup. Since communication is unidirectional, the GPS fix schedule, VHF beacon, and PTT uplink periods must be programmed before units are attached to animals and cannot be changed without recovery of individual units. As well, because the longest message that can be transferred by the Argos system is 32 bytes, GPS data required for differential correction cannot be transmitted and only five to seven geographic coordinates can be sent in each Argos message. GPS data required for differential correction are actually stored on board the animal unit and can be used when the unit is recovered. In fact, animal units can store up to 2750 differentially correctable records or, optionally, 5500 geographic coordinates. Since the PTT continuously transmits the most recent five to seven geographic coordinates when turned on, multiple uplinks of the same information can be obtained, ensuring data integrity. On the other hand, since the oldest GPS location stored for transmission is replaced with each new position estimate, some data may not be immediately available if the PTT fails to successfully transmit recently stored data before they are overwritten. Once again, all is not lost because these data are stored in the animal unit and can be downloaded when the unit is recovered. So this limitation may only be important to studies requiring up-to-date, "real time" location estimates. This concern can also be addressed by using an Argos uplink receiver operated locally from a vehicle or aircraft, or configured for automated remote data logging, to intercept transmissions (within 1 km) from PTTs (Bennett et al. 2001). In this case, the system has the same limitations as others that use a local communication link for data retrieval.

Similar to other GPS animal units, the operating life of combined GPS-Argos devices is largely determined by the GPS fix rate and the size of the battery pack used. However, the operating life of combined GPS-Argos units also depends on the PTT uplink schedule. For example, a large GPS-Argos unit having a total weight of 2.2 kg, allowed 90 s to attempt one GPS location/day, and with a 6-h Argos uplink (900 ms every 200s) every 5 days (i.e., 6 h on, 114 h off), has an expected operating life of 1084 days. At 5 GPS location attempts per day and a 6-h Argos uplink each day, the expected operating life decreases to 259 days. All else being equal, a smaller GPS-Argos unit having a total weight of 1.7 kg, attempting 1 GPS location/day with a 6-h Argos uplink every 5 days, has an expected operating life of 481 days. At 5 GPS location attempts per day and a 6-h Argos uplink each day, the expected operating life of the smaller unit decreases to 115 days.

Store-On-Board

Ironically, the last GPS-based telemetry system to be developed is conceptually the simplest and is the most similar to an archival tag. Differentially correctable GPS data, or geographic coordinates, along with dates and times of location estimates and optional sensor information are stored on board devices attached to animals until the unit can be remotely released by a radio-activated "break-away" mechanism or recovered by recapture of the animal (Merrill et al. 1998, Arthur and Schwartz 1999, Schwartz and Arthur 1999, Bowman et al. 2000). These units are now available in a variety of configurations from several manufacturers of telemetry equipment. The primary reason for their development was to reduce the size of GPS units deployed on animals, so that a wider range of species might benefit. By removing data transmission components, such as UHF modems or PTTs, the weight of animal units can be reduced by almost 50%, from the 1.7 to 2.2-kg devices used with elk, bears, caribou, and moose, down to 0.9 to 1.1-kg units that can be used on wolves and white-tailed deer. Similar to other telemetry devices, the total package weight of store-on-board GPS units is determined primarily by the size of the battery pack, which in turn affects the expected operating life. Since the weight of additional memory microchips is negligible relative to the weight of communication components removed, store-on-board GPS units have large data storage capacities ranging from 2750–4400 differentially correctable GPS records to 5400–8800 nondifferential geographic coordinates. Hence, even though manufacturers might specify the operating life of store-on-board units as, for example, 10.5 months or 450 days at 6 GPS fix attempts per day, a quick calculation with respect to a particular device will show that seldom, if ever, will the entire data storage capacity be used. Therefore, store-on-board units are better rated on the basis of the numbers of expected GPS fix attempts for a given battery capacity rather than specifying an operational life based on time. When rated this way, small store-on-board GPS units are capable of 990–1260 GPS fix attempts, and larger units are capable of 1890–2700 GPS fix attempts. The GPS fix schedule and VHF beacon availability will then determine how long a store-on-board GPS unit will actually function in the field.

Because there is no communication link with store-on-board units, VHF beacon availability and the GPS fix schedule must be programmed before units are attached to animals and cannot be changed without recovery of individual units. The primary role of the VHF beacon is to aid recovery of individual units and to serve as an emergency locator if a unit's main operating system fails due to battery exhaustion or some other cause. However, to compensate for the lack of data transfer while in the field and to reassure users that units are functioning properly, most devices have several VHF beacon modes that indicate the status of individual units. In some units this might consist of a simple mortality mode, much like other telemetry devices, that decreases the VHF pulse rate after a specified time delay if an animal does not move. However, much more sophisticated schemes are available that might, for example, indicate that the last GPS fix attempt was successful by emitting a particular sequence of single and double "beeps" (Merrill et al. 1998).

To reduce the risk of injury and stress to study animals, as well as lowering capture costs, some store-on-board GPS units are equipped with radio-activated release mechanisms. One such unit (Merrill et al. 1998, Bowman et al. 2000) incorporates a commercially manufactured mechanism that was originally developed as part of a radio-activated drug delivery system (Mech et al. 1984) used to immobilize white-tailed deer (*Odocoileus virginianus*) (Delgiudice et al. 1990, Mech et al. 1990) and wolves (Mech and Gese 1992). This release mechanism, weighing less than 20 g, adds little to the total weight of the GPS animal unit and can be triggered from distances of 1–2 km. The unit can also be programmed to drop off automatically as the main system battery nears exhaustion (Merrill et al. 1998). Following release, the VHF beacon pulse rate decreases and continues transmitting for about 25 days to facilitate recovery of the device.

Cost

In spite of a growing number of studies suggesting that GPS telemetry systems can be cost effective (Rodgers et al. 1996, 1997, Merrill et al. 1998, Schwartz and Arthur 1999, Bennett et al. 2001, Biggs et al. 2001), there continue to be concerns about the price of using this new technology (Moen et al. 1996, Arthur and Schwartz 1999, Otis and White 1999). However, the development of store-on-board GPS units has also produced an almost 50% reduction in cost. Whereas the first commercially available GPS units with local communication capability were priced at about US$5000 and current GPS-Argos devices cost about US$4000 (Arthur and Schwartz 1999), store-on-board units are currently available for about US$2500. This is about the same price as a large animal PTT (Merrill et al. 1998, Schwartz and Arthur 1999), but without the additional costs of data handling that must be paid to Service Argos. Similar to other microelectronics, prices are likely to decline further with future availability of cheaper components. Prices may also drop in the near future as competition increases and manufacturers of telemetry equipment recoup their initial investments in the development and production of GPS devices.

Observational Bias

One of the most important concerns with GPS-based telemetry systems is the potential for observational bias (i.e., the possibility that GPS fix attempts might be more or less successful in some habitats than others depending on the characteristics of the forest canopy) (Rempel et al. 1995, Moen et al. 1997, Johnson et al. 1998, Rettie and McLoughlin 1999). Interestingly, this potentially significant source of error for habitat use-availability studies does not seem to have been given serious consideration until GPS-based telemetry systems were developed (Johnson et al. 1998). Certainly other automated telemetry systems, such as Argos PTTs, have the same potential for observational bias because transmitted radio signals would have similar difficulty reaching receivers through forest canopy or might be affected by terrain, either of which may result in animals going undetected in some habitats (North and Reynolds 1996). Even manual tracking of VHF radio signals might be biased since this can usually be done only during daylight hours and under appropriate weather conditions (Beyer and Haufler 1994, Rodgers et al. 1996, Arthur and Schwartz 1999, Schwartz and Arthur 1999). As well, VHF telemetry may bias results if persistent harassment of study animals causes them to change their normal movement or behavior patterns (White and Garrott 1990, Schwartz and Arthur 1999). In fact, it could be argued that GPS-based telemetry systems have less potential for observational bias than conventional

VHF methods because GPS fix attempts can be made 24 h/day and under all weather conditions without direct disturbance of study animals (Arthur and Schwartz 1999, Schwartz and Arthur 1999). Because GPS fix attempts are made far more frequently (typically <4 h apart) than with other systems (sometimes once a day but usually 1 or 2 times per week), the likelihood of obtaining at least a few fixes in seldom-used habitats is much greater. Nonetheless, observational bias is a potentially serious source of error that must be given due consideration in habitat use–availability studies.

Under controlled conditions, observation rates of the first GPS units were 97% in the open and ranged from 10% to 92% in pure stands of various boreal forest tree species (Rempel et al. 1995). Similar results (95% in the open and 60–70% under canopy) were obtained in a separate study by Moen et al. (1996). Differences in observation rates among forest types were attributed to differences in canopy closure and, more specifically, the density of thick, woody material that can interfere with signal transmission (Rempel et al. 1995). With the next version of these devices, observation rates improved to 100% in the open and ranged from 71% to 97% in the same forest stands (Rempel and Rodgers 1997). Observation rates of the first units deployed on free-ranging black bears (*Ursus americanus*) in the boreal forest of northern Ontario averaged only 46% (Obbard et al. 1998). On free-ranging moose, observation rates improved marginally from an average of 66% (Rodgers et al. 1995) to 69% (Rodgers et al. 1998) with the second version of the device. Using the same model of GPS collars on free-ranging moose, Dussault et al. (1999) found almost identical observation rates (68.9%) in Quebec, and Edenius (1997) recorded slightly higher rates (75%) in Sweden. These performance improvements were primarily attributable to better search algorithms in GPS receivers that allowed them to acquire satellite signals more rapidly than earlier versions. Remarkably similar mean observation rates have been recorded for other devices, such as GPS-Argos units deployed on elk (69%; Biggs et al. 2001) or store-on-board units deployed on bears (67%; Schwartz and Arthur 1999), wolves, and white-tailed deer (70%; Merrill et al. 1998), even though these units incorporate different GPS receivers. The observation rates of the first units on moose in northwestern Ontario subsequently diminished from 69% to 62%, then dramatically increased to 75% (Fig. 4.4) with yet another enhanced GPS receiver. Most recently, GPS units on moose recorded average observation rates of 93% during their first 5 months of deployment in 1999. Again, better satellite search algorithms in the latest GPS receivers are primarily responsible for the improved performance. Although the possibility of observation bias may never be entirely eliminated from GPS-based telemetry systems, it appears that this source of error may be effectively mitigated and could be less significant than with other systems.

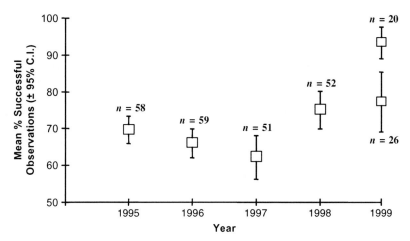

FIGURE 4.4   Mean (± 95% C I) observation rates (i.e., % of successful fix attempts) of GPS collars deployed on moose in northwestern Ontario during 1995–1998 and the first 5 months of 1999. New GPS receivers were incorporated into collars in 1998 and again in 1999. In 1999 some collars were deployed with the latest GPS receiver boards (*n* = 20) and others had the previous version (*n* = 26). The most recent GPS boards have attained the highest observation rates.

Future Development

Within their first 5 years, GPS-based telemetry systems developed rapidly to address concerns about their size, cost, and performance. Not surprisingly, these are the same apprehensions that were expressed as each new telemetry system (e.g., Argos PTTs) was introduced over the last 40 years. Nonetheless, these concerns may have at least some short-term validity. With respect to size, there has been an almost 50% reduction in weight with the development of store-on-board units, as outlined above. However, GPS units are still too large to be used on animals weighing less than approximately 20 kg. Further weight reductions are imminent, as telemetry manufacturers have recently acquired new GPS receivers that have less than half the current requirements of present devices. This should allow another weight reduction of at least 50% through a decrease in the size of the required power supply. Alternatively, the number of GPS position estimates obtained or the operational life of current GPS units could be more than doubled.

Much like PTTs, enhancements to the Argos system are expected to improve the performance of GPS-Argos devices attached to animals during the next few years. In the short term, an increase in the number of satellites and the improved capacity and sensitivity of the Argos-2 instrument will provide better coverage and an increased number of messages to be successfully processed. The next generation of Argos instruments, capable of two-way communication, will enable rescheduling of PTT uplink periods and GPS fix rates, thereby providing greater system flexibility.

With the recent termination of selective availability by the U.S. Department of Defense, the accuracy and precision of locations has improved 10-fold since the first GPS-based telemetry systems were developed for wildlife research. Future plans include the addition of a second satellite broadcast frequency in 2003 that will compensate for ionospheric errors, and a third frequency in 2006 that will provide even greater accuracy (Lawler 2000). Ultimately, 1–3 m accuracy, or better, may be available to all users.

GPS-based telemetry systems will continue to evolve. Like their predecessors, GPS units will continue to decrease in size and cost while improving in performance. Although new systems may be developed to transfer GPS data using current strategies, the next generation of these devices will probably make use of local cellular phone networks for data transmission.

HYPERBOLIC TELEMETRY SYSTEMS

The idea of applying hyperbolic navigation principles to wildlife telemetry has been around for almost 30 years, but early systems were quite expensive and met with limited success because of technological restrictions (Patric and Serenbetz 1971, Lemnell 1980, Yerbury 1980, Lemnell et al. 1983). Hyperbolic telemetry, sometimes called "reverse LORAN" (White and Garrott 1990), involves the measurement of differences in the time it takes transmitter signals to reach three or more receiving stations (Fig. 4.5). Hyperbolic functions are formed by measuring differences in the arrival times of signals at any two stations. The intersection point of these hyperbolas is used to determine the location of the transmitter. Although conceptually simple and quite similar to directional triangulation methods, the technology required for hyperbolic telemetry is highly sophisticated. Like so many other devices described in this chapter, advances in the development of microcontrollers and other electronic components during the 1990s have increased the feasibility of hyperbolic telemetry and rekindled an interest in further development of these systems.

Hyperbolic telemetry has several potential advantages. Although not suitable for migratory species, hyperbolic telemetry can provide GPS-level location accuracy for species that do not cover large distances. For example, one of the first systems tested on wildlife was deployed on moose in a 3000-ha forest and provided location accuracies of ± 40 m with three receiver stations 6 km apart (Lemnell et al. 1983). A system developed in the mid-1990s for tracking fruit bats in a tropical rainforest had a potential spatial resolution of ± 3 m and

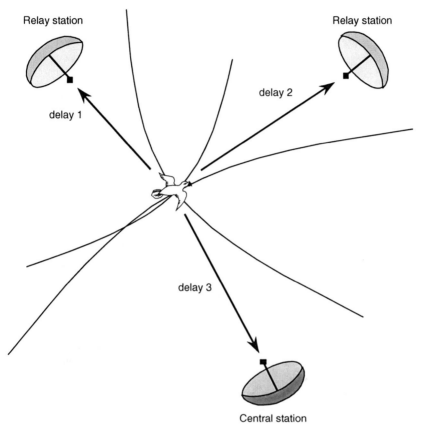

FIGURE 4.5 Components of a hyperbolic telemetry system. Signal arrival times are recorded at two relay stations and a central station. Data are retransmitted from the relay stations to the central station where differences in time-of-arrival (i.e., delay times) between pairs of stations are calculated to form hyperbolic curves. The transmitter is located at the intersection of the curves.

achieved accuracies of ±25 m in initial field tests (Spencer and Savaglio 1995). Another system currently under development (TICOM, Inc., Austin, Texas, USA) has also achieved accuracies of ±25 m in initial field tests. Most recently, a system developed for aquatic studies (Lotek Engineering Inc., Newmarket, Ontario, Canada) provided three-dimensional positions within 1m^3 of the true transmitter location (Fig. 4.6). As well as providing highly accurate position estimates, hyperbolic systems use relatively lightweight, pulse-coded transmitters that have undergone significant development since the 1980s, as described in an earlier section. Thus, numerous transmitters can

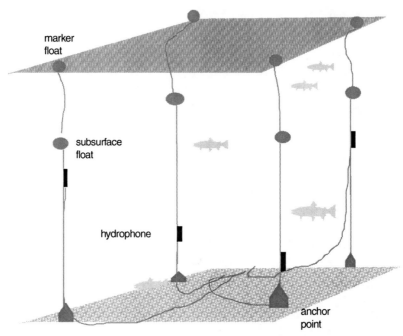

FIGURE 4.6 An aquatic hyperbolic telemetry system that uses four receiving stations to provide three-dimensional positions that are within 1 m^3 of the true transmitter location. (Courtesy of Lotek Engineering Inc., Newmarket, Ontario, Canada.)

be accommodated on a single carrier frequency, distinguished by their unique codes. Whereas earlier transmitters deployed on moose weighed approximately 800 g, units developed for fruit bats weighed only 25 g. Hence, hyperbolic systems can provide GPS-level location accuracy for small and medium-sized animals, as well as larger species.

In a typical hyperbolic telemetry system, one receiving station functions as a "base" or "central" station and the others as "relay" or "repeater" stations (Lemnell et al. 1983, Kenward 1987, Spencer and Savaglio 1995). Transmitter signals received at the relay stations are retransmitted on a new frequency to the central station for processing and storage of location estimates. Thus, signals received at the central station are delayed only by the travel time between the transmitter and the receiver. Retransmitted signals, on the other hand, are also delayed by travel times between the relay stations and the central station. To calculate the time delay between the relay stations and the central station, receivers must be synchronized and the relative positions of each must be known precisely. Whereas early hyperbolic systems required traditional survey methods to establish the locations of receiving stations and CMOS clocks had to be updated regularly to counter drift (Lemnell et al. 1983), current systems can take advantage of GPS to provide position estimates and an absolute time reference that is updated automatically to synchronize receivers. Following the removal of fixed delays between relay stations and the central station, differences in the arrival times of signals can be calculated to generate hyperbolic functions and determine the location of the transmitter.

The main limitation of hyperbolic systems is available signal strength, which can be affected by signal interference, Rayleigh fading (i.e., loss of signal strength with distance due to scattering of electromagnetic radiation by particles smaller than the radiation wavelength), and the effects of multipath (i.e., signal distortion resulting from individual radio signals taking two or more paths to a receiving antenna; the signals usually have different amplitudes and phases, which causes signal distortion) (Spencer and Savaglio 1995). These factors limit the ability of receivers to distinguish differences in the time of arrival of signals, which is the underlying principle of hyperbolic systems. Since electromagnetic waves travel at close to the speed of light (300,000 km/s), each 10-ns delay represents 3 m distance on the ground. Therefore, a system capable of 1 m resolution must be able to discriminate time intervals with an accuracy of 3 ns. To achieve this level of resolution, and to counter the effects of signal interference, fading, and multipath, the most recent systems use postdetection digital-signal processing techniques to remove unwanted noise and retain desired signals. Spread-spectrum signal processors at each receiving station determine time-of-arrival using cross-correlation analysis. Cross-correlation matches patterns of incoming signals with expected signal patterns stored in the processor's memory. Thus, even very weak signals from low-power animal transmitters (10–50 mW) can be detected in extremely noisy environments and reflected or multipath signals, which will be delayed behind direct signals, can be identified and ignored. Subsequently, the accuracy of time-of-arrival measurements is determined by the rate at which incoming signals are processed. If signals are processed frequently enough to discriminate time-of-arrival measurements at 10-ns intervals, then a spatial accuracy of 3 m on the ground can be achieved.

White and Garrott (1990:45) suggested that LORAN-C was the "state of the art" in determining animal locations in the field. Ten years later, reverse LORAN may well represent the current state of the art, at least for some applications. These hyperbolic systems combine much of the recent technology that has advanced many other telemetry systems. As a result, hyperbolic systems are able to provide a very desirable combination of high accuracy and small package size that can be deployed on a wide range of species. Since currently available pulse-coded transmitters can be used, the only "specialty" costs of hyperbolic systems are related to receiving stations. Although current hyperbolic systems are estimated to cost about US$45,000 per receiving station, these costs are expected to fall below US$15,000 per receiving station over the next few years. Similar to other systems, prices are likely to decline further with future availability of cheaper components and competition among manufacturers of hyperbolic telemetry systems.

IMPLICATIONS FOR DATA ANALYSIS

Recent advances in telemetry technology have resulted in systems capable of recording enormous amounts of highly accurate animal location data that may present a serious challenge to data management and analytical procedures. Until the mid-1990s, most computer software programs commonly used to analyze animal location data could accept a maximum of 500 position estimates, and only a few could handle 3000 sets of points (Larkin and Halkin 1994, Lawson and Rodgers 1997). These limits are readily exceeded by recent automated systems, such as GPS-based telemetry, that may record several thousand locations per animal per year (Rodgers et al. 1996). Since the mid-1990s, some of the earlier software programs have been upgraded to deal with larger datasets and to interface with geographic information systems (GIS) that are designed to cope with large amounts of geographic location data. There have also been a few new analysis programs developed entirely within a GIS framework (Hooge and Eichenlaub 1997, Rodgers and Carr 1998). Although GIS has been around for almost 20 years and has been commercially available for at least 10 years, wildlife researchers have been reticent to use it, in part because of the misconception that a GIS is simply a utility for creating and viewing maps. Until recently, there has also been a lack of inexpensive, accessible, "user-friendly" GIS software. Moreover, researchers have lacked significant quantity and quality of spatial data that could benefit from a GIS. Thus, very little research has been done to explore the potential application of powerful temporal and spatial statistical algorithms, now available in most GIS software, to analyses of animal location data. Advances in computer technology, with reduced processing time, improved user interfaces, and better statistical and analytical algorithms, have removed some of the restrictions and now make it attractive for researchers to consider using a GIS to develop new analytical methods (e.g., Ostro et al. 1999, Rettie and McLoughlin 1999).

In the late-1990s, several general articles were written about statistical testing of hypotheses in wildlife research and many more were specifically concerned with the problems of using telemetry data. Several authors suggested that estimation with confidence intervals might be a more appropriate method of comparing use-availability data than statistical tests of hypotheses

(Steidl et al. 1997, Cherry 1998, Johnson 1999). Due to the large sample sizes that can be obtained with the latest telemetry systems, the standard errors associated with these confidence intervals may be minimized, thereby providing better estimates and facilitating comparisons of use-availability data. In the end, the quality and quantity of data available through recent advances in telemetry systems may lead to greater consideration of biological rather than statistical significance in wildlife research and perhaps more testing of scientific as opposed to statistical hypotheses (Cherry 1998, Gerard et al. 1998, Johnson 1999).

Although the latest telemetry systems are capable of recording large amounts of location data, they cannot solve the problem of replication requirements in statistical analyses. Simply increasing the number of location estimates for individual animals does not solve replication problems because most studies are concerned with making inferences about populations of animals. Thus, the appropriate sampling unit in most hypothesis tests of habitat selection should be the animal, not the individual locations (Otis and White 1999). Consequently, the number of individual animals involved in a particular study has more effect on inferences that may be drawn than the number of locations obtained per animal. Some might even suggest that obtaining too many locations for individuals could be problematic because autocorrelation (i.e., locations taken too closely in time that may not be statistically independent) may cause negatively biased estimates of home range size (Swihart and Slade 1985b). However, a growing number of studies suggest that adequate sample size is more important than independence of location estimates (Reynolds and Laundré 1990, Minta 1992, McNay et al. 1994, Swihart and Slade 1997, Otis and White 1999, Seaman et al. 1999). Otis and White (1999) contend that concerns about autocorrelation may not be important if an appropriate study design for collecting a representative sample of locations during a specific time frame is applied and inferences are constrained accordingly (see Chapter 5).

Some of the more recently developed telemetry systems are also capable of providing highly accurate position estimates for animals that may increase discrimination among habitats used and not used (Rempel et al. 1995, Rempel and Rodgers 1997). Data obtained from hyperbolic or GPS-based telemetry, for example, have the advantage of maintaining the same spatial resolution as habitat mapping data derived from satellite imagery (e.g., the 30-m resolution of Landsat TM data). Consequently, there is less chance of assigning animal locations to incorrect habitat categories when these data are plotted on a satellite image in a GIS. This represents a significant advance over previous VHF or Argos telemetry, but even these new systems have inherent sources of error (Rettie and McLoughlin 1999). In most cases, these errors may be small relative to habitat patch size, but there is still potential for error polygons to overlap more than one habitat category. As with all telemetry location data, this problem can be resolved by recognizing that animals may select mosaics of habitat rather than specific points because nearby habitats may also play an important role in the habitat selection process (Rettie and McLoughlin 1999). Thus, circular buffers with a radius greater than or equal to the known telemetry error can be placed around point locations and the composition of habitats within buffers can be used to determine selection. This approach is consistent with many of the methods previously used in habitat selection analyses (Aebischer et al. 1993, Manly et al. 1993, Arthur et al. 1996), as well as the latest techniques based on discrete choice models (Cooper and Millspaugh 1999; see also Chapter 9). However, circular buffers may be considered an arbitrary estimate of habitat available to individuals unless substantial data per individual are obtained (McClean et al. 1998), so care must be taken to establish an appropriate buffer size for a particular study (Rettie and McLoughlin 1999; see also Chapters 8 and 9). On the other hand, examination of habitat composition within circular buffers of varying size may be used to determine the spatial scale at which individuals select habitats, thereby providing essential information for resource management (Lehmkuhl and Raphael 1993, Thome et al. 1999).

In spite of attempts by White and Garrott (1990) and Kenward (1992) to outline the desirable features of a data analysis system for animal location data, such a system has yet to be developed. Perhaps the task is too daunting for anyone to take on. More likely it is because wildlife researchers cannot agree on which analytical methods are most appropriate and because new methods are being developed at an increasing rate in response to new technology. Unfortunately, this has produced such a wide variety of analytical techniques and computer software that comparisons among, and even within, research studies may be invalid or misleading (Lawson and Rodgers 1997). The development of a comprehensive data analysis system for animal location data thus remains a highly desirable goal for wildlife researchers.

IMPLICATIONS FOR RESEARCHERS

There is likely to be an ever-increasing variety of telemetry systems available to researchers in the future. As always, researchers will have to carefully select an appropriate system for a particular project. They must recognize that new technology is often fraught with problems, and they must plan accordingly and be prepared to accept the consequences of premature system failures. Researchers must avoid the lure of technology that is currently "in vogue" or "prestigious" when study objectives can be met more efficiently with proven systems (Ballard et al. 1995). The technology employed must be matched against study objectives. Researchers must also be prepared to undertake the higher costs associated with new systems and resist the temptation to reduce replication or sampling intensity to save money. Economic concerns rather than firm scientific and statistical principles too often determine the quality of wildlife studies and, consequently, their value (Rodgers et al. 1996). Above all else, researchers must develop and adhere to an experimental design appropriate to well-defined research objectives. Only then can they select a suitable telemetry system, determine sampling schedules, estimate costs, and identify analytical procedures.

FUTURE DIRECTIONS

As technology progresses, improvements to current telemetry systems can be expected. Improved power supplies may not be imminent, but greater efficiency of microelectronic components and application of advanced power management strategies may lead to further reductions in the size of devices attached to animals, allowing studies on a wider variety of species than presently possible with some systems. Hyperbolic systems and data transfer by cellular phone networks or LEO satellites may well represent the next wave of technology available to wildlife researchers, but emergent technology will undoubtedly lead to the development of new telemetry systems. The basis of future systems is almost impossible to predict because most telemetry systems use technology developed for other purposes. For example, most electronic components were developed to reduce the size and increase the capabilities of personal and laptop computers (i.e., microcontrollers). Smaller batteries are the result of efforts to power watches, cameras, and other small electronic devices. Light pipe technology used in archival tags has been adapted from developments in fiber optics. The Argos system was developed to monitor environmental data for meteorologists and oceanographers. GPS was originally a U.S. Department of Defense initiative. Recent interest in hyperbolic telemetry systems stems from military interests and requirements to provide Emergency 911 service to cell phone users throughout the United States. So where will radiotelemetry find new technology to adapt in the future? It is difficult to speculate. Often, the most exciting aspect of a new technology is not its successful application to the problem it was designed to solve but rather its application to problems never envisaged during its development. Such applications are limited only by the imagination.

PART IV

Animal Movements

Analysis of Animal Space Use and Movements

Brian J. Kernohan, Robert A. Gitzen, and Joshua J. Millspaugh

Using Home Range Estimators to Analyze Animal
 Space Use
 Sampling Considerations
 Evaluation of Home Range Estimators
 Description of Kernel-Based Estimators
Analysis of Site Fidelity
 Distance and Direction Moved
 Distance between Home Range Centroids
 Multiresponse Permutation Procedures
 Home Range Overlap
Analysis of Animal Interactions
 Static Interaction Analyses
 Dynamic Interaction Analyses
The Future: Modeling the Movement Process
 Descriptive Approaches
 General Movement Models
 Biological Models
Summary

Understanding the organization of animals in space and time is a central question of ecology. The dynamics of a population are directly linked to the spatial arrangement and movements of individuals caused by internal or external pressures on the population. Animal space use and movements are best understood through direct observation. Although this method may allow locations to be precisely recorded, it is very labor intensive and the potential to influence the animal's behavior is high, often resulting in a biased estimate of distribution patterns. With animals that are wide-ranging or difficult to find, direct observation may be impractical. In the absence of direct observation,

Radio Tracking and Animal Populations
Copyright © 2001 by Academic Press. All rights of reproduction in any form reserved.

radiotelemetry has filled an important gap. Whether radiotelemetry is used on animals that are wide ranging and difficult to study, or simply for convenience, the data generated are well suited for investigating space use and movement patterns of animals.

At the core of many space use analyses is the estimation of an animal's home range. Other analytical techniques allow investigators to determine changes in an animal's position or fidelity to an area with respect to both space use and movements. This can be done for a single animal or among animals in the form of animal interaction analyses. The integrated use of tools such as global positioning systems (GPS), geographic information systems (GIS), and computer packages that compute movement statistics have helped advance the analysis of radiotelemetry data. In this chapter, we will describe a variety of techniques used to analyze animal space use, including home range estimation, analysis of site fidelity, and animal interactions. Limited attention will focus on "traditional" techniques (e.g., most home range estimators) as these are sufficiently covered in other reviews such as White and Garrott (1990). Instead, we focus our attention on recently applied techniques such as kernel estimators. Finally, considerations for increased emphasis on modeling movement processes in radiotelemetry studies will be discussed.

USING HOME RANGE ESTIMATORS TO ANALYZE ANIMAL SPACE USE

Since Burt (1943:351) first defined "home range" as "that area traversed by the individual in its normal activities of food gathering, mating and caring for young," animal space use data have been used to quantify home ranges. Several authors have redefined or criticized the concept of home range since this definition was proposed, (Mohr 1947, Jewell 1966, Baker 1978, Hansteen et al. 1997). Inherent problems with the original definition, including use of the ambiguous term "normal" and the inadequacy of not specifying a temporal component, have been discussed by White and Garrott (1990). Morris (1988) and Hansteen et al. (1997) recognized the need to specify a time frame and have done so with revised definitions of home range. We accept the central ideas of these discussions. However, to date no one has combined all of the suggested modifications into a succinct statement. Although we recognize that "home range" is a concept, not an entity (Morris 1988), we suggest that home range be defined as the extent of area with a defined probability of occurrence of an animal during a specified time period.

Hayne (1949) introduced the concept of "center of activity." Inherent in the center of activity concept is that a biological understanding of an animal's home range must include some information about the intensity of use in various parts of the home range (Hayne 1949). Subsequently, the identification and estimation of such "activity areas" within the home range have received much attention (Samuel et al. 1985, Samuel and Green 1988, Hodder et al. 1998). Most recently, the term "core area" has replaced Hayne's (1949) original center of activity terminology (Hodder et al. 1998). Both terms refer to the ability to distinguish between nonuniform use patterns within a home range. For purposes of this chapter, we will use the terms center of activity (singular), multiple centers of activity (plural), and core area interchangeably.

Given these definitions, it is not surprising that a variety of estimators have been developed that quantify home range. The various analytical techniques have been reviewed by Kenward (1987), White and Garrott (1990), and S. Harris et al. (1990), among others. Similarly, several authors have compared and contrasted the more common techniques, including Van Winkle (1975), Macdonald et al. (1980), Jaremovic and Croft (1987), Worton (1987, 1989, 1995), Boulanger and White (1990), Kenward (1992), Seaman and Powell (1996), and Seaman et al. (1999).

Home range estimators are used to address a variety of research questions, including estimation of home range size as first described by Burt (1943), home range shape and structure (e.g., Kenward 1992), movement or site fidelity analyses as defined by temporal change in home range position (e.g., Phillips et al. 1998), establishment of management boundaries (e.g., Edge et al. 1986), resource availability estimates (e.g., Johnson 1980), and analysis of animal interactions (e.g., dynamic interactions; Macdonald et al. 1980, Doncaster 1990). Rather than provide an exhaustive review of these analytical techniques, we will focus on three areas of analyzing spatial distributions of animals with home range estimators. First, we will provide a brief description of factors to consider when designing or comparing home range studies. Second, we will provide a brief evaluation of home range estimators using several important criteria. Third, we will describe kernel-based home range estimators.

SAMPLING CONSIDERATIONS

The accuracy of home range estimates is affected by several artifacts of sampling, including the time between consecutive locations (Swihart and Slade 1985a, b), the number of observations used to obtain the estimate (Stickel 1954, Jennrich and Turner 1969, Bekoff and Mech 1984, Seaman et al. 1999), and the technique used to collect location data (e.g., live trapping vs. radiotelemetry; Adams and Davis 1967). When designing a radiotelemetry study to estimate home ranges, a researcher must carefully consider how best to allocate resources (Otis and White 1999). That is, how much time should elapse between successive locations, what is the minimum number of locations that must be recorded to accurately estimate an animal's home range, and what is an appropriate sampling unit (e.g., individual location points or individual animals; see Chapter 2). Inevitably, trade-offs will have to be made between the number of animals monitored and number of locations per animal (Otis and White 1999; see also Chapter 2).

Effects of Autocorrelation: Determining the Appropriate Sampling Interval

During the past 15 years, much attention has been paid to the effect of autocorrelation in location data on home range estimates. An inherent assumption in most existing probabilistic estimators of home range is that location data used to estimate the home range are independent (Dunn and Gipson 1977). Swihart and Slade (1985a) suggested that autocorrelation exists when locations at time t are dependent on the animal's location at time $t - 1$. Assuming a fixed time between relocations, autocorrelated datasets are created when (1) the animal has too little time to move away, (2) the animal simply does not move between consecutive observations, or (3) the animal periodically returns to a previously used portion of its range (Hansteen et al. 1997). There is a misconception that nonprobabilistic techniques, such as the minimum convex polygon, are less likely to be affected by autocorrelated data because they do not assume independence (Swihart and Slade 1985b). Using computer simulations, Swihart and Slade (1997) found that convex polygon home range estimates were actually more biased by autocorrelated data than kernel estimators.

The main assumption related to the effects of autocorrelation on home range estimates is that n observations that are autocorrelated yield less information than do n independent observations (Swihart and Slade 1985b). That is, when data are autocorrelated, the distance moved between consecutive observations decreases, resulting in a lower proportion of the home range traversed (see figure 1 in Swihart and Slade 1985a) and an underestimation of home range size. In contrast, independent data allow the animal enough time to move throughout its home range, thus yielding more information about an animal's movement patterns (Swihart and Slade 1985a, b).

Although a few researchers have noted the assumption of observation independence (e.g., Dunn and Gipson 1977), Swihart and Slade (1985a) were the first to present a formal framework for the testing of autocorrelation in radiotelemetry data. Swihart and Slade (1985a) proposed Schoener's (1981) ratio statistic as an analytical tool to determine the time interval necessary to achieve statistical independence. This interval, called the time to independence (TTI), is the time at which autocorrelation is nonexistent in successive observations. This technique is based on the t^2/r^2 ratio (Schoener 1981), which is the ratio of the mean squared distance between successive observations (t^2) expressed as:

$$t^2 = \sum (X_{i+1} - X_i)^2/m + \sum (Y_{i+1} - Y_i)^2/m \tag{5.1}$$

and the mean squared distance from the center of activity (r^2) (Swihart and Slade 1997) given by:

$$r^2 = \sum (X_i - \bar{X})^2/(n - 1) + \sum (Y_i - \bar{Y})^2/(n - 1), \tag{5.2}$$

where (X_i, Y_i) are location data coordinates and m is the number of pairs of successive observations ($m = n - 1$ when all pairs are used). Swihart and Slade (1985a) defined TTI as the smallest time lag when nonsignificant t^2/r^2 ratios exist, followed consecutively by two or more nonsignificant ratios. Significance is evaluated by comparing the observed t^2/r^2 ratio to empirically derived values of the t^2/r^2 distribution calculated using successive pairs of observations (Swihart and Slade 1985a).

Despite the remarks of Swihart and Slade (1985a, b), the use of Schoener's (1981) ratio to assess the degree of autocorrelation in radiotelemetry data may be problematic. McNay et al. (1994) noted that when animals exhibit migratory movement patterns or make sporadic moves to uncommonly used portions of the home range the TTI was unrealistically long. If McNay et al. (1994) relied on their computed TTI, about 90% of their data would have been autocorrelated and the resulting independent data would likely not have been enough to compute an accurate estimate of home range. McNay et al. (1994) attributed this lengthy TTI to a skewed data distribution, which led to an apparent lack of independence (McNay et al. 1994:426). However, in a recent reevaluation of TTI, Swihart and Slade (1997) discussed that the t^2/r^2 ratio test is robust when data are not normally distributed, as stated in the original Swihart and Slade (1985a) paper. Despite the confusion about normally distributed data, the analysis of McNay et al. (1994) demonstrated one important feature about Schoener's t^2/r^2 ratio test: it is based solely on the distance traveled between successive locations in relation to the average distance from the center. Therefore, animals exhibiting movements that are not centered around one focal use area will exhibit an unrealistically long TTI, as observed by McNay et al. (1994). When using Schoener's (1981) test, home ranges are assumed to be stationary during the period of analysis (Swihart et al. 1988). When ranges are nonstationary, t^2/r^2 values decrease, resulting from an increase in r^2 (Swihart and Slade 1997). Therefore, before the t^2/r^2 ratio test is employed, it is wise to test for constancy in the activity center. If this assumption is violated, the TTI test presented by Swihart and Slade (1985a) has little value.

Other concerns have arisen with the TTI test, as outlined by Swihart and Slade (1985a). Some researchers have found no difference in home range estimates based on independent vs. sequential locations (e.g., Andersen and Rongstad 1989, Gese et al. 1990). Reynolds and Laundré (1990) reported that home range size for coyotes (Canis latrans) and pronghorns (Antilocupra americana) were underestimated when data were collected at long enough intervals to be considered independent. In their study, autocorrelated data provided a more accurate estimate of home range size than estimates obtained from statistically independent data (Reynolds and Laundré 1990).

Given these shortcomings in the TTI concept, what is an appropriate time interval between successive observations to accurately estimate an animal's home range? When home range is defined as the extent of area with a defined probability of occurrence during a specified time period, the answer is dependent on study objectives and therefore differs among species and studies. If a researcher is interested in daily distances moved and other short-time-frame measures of animal behavior, then statistical independence is irrelevant (Reynolds and Laundré 1990). Reynolds and Laundré (1990) reported that daily distances moved derived from independent observations (more than a 12-hour sampling interval) resulted in movement estimates that were not representative of true daily movements. Restricting sampling intervals to times that were statistically independent resulted in the loss of important biological information (Reynolds and Laundré 1990). Therefore, strict adherence to the TTI concept in all movement studies is not advised.

If the primary research focus is on home range estimates, as opposed to other movement analyses (e.g., migration), we recommend a systematic sampling scheme or other designs discussed by Garton et al. (Chapter 2) that most accurately estimate animal spatial distributions during the sampling period to which inferences are drawn, while using the individual animal as the sampling unit (White and Garrott 1990, McNay et al. 1994, Otis and White 1999). Adequately sampling animal locations throughout the duration of the study is more important than determining a time interval between sampling that is statistically independent. This requires that researchers incorporate important biological traits of the animal under study into the sampling schedule. For instance, for potentially nocturnal animals, an accurate estimate of home range should include both day and night locations if the study objective is to characterize 24-h home range characteristics (Smith et al. 1981, Beyer and Haufler 1994). Ignoring nocturnal movements may bias location estimates toward movements that are restricted to one particular behavior or activity (e.g., bedding during the day) (Smith et al. 1981). Swihart and Slade (1997:60) indicated that "selection of a sampling interval < TTI (i.e., an interval yielding autocorrelated observations) generally will not invalidate several common estimates and indexes of home range size provided that the time frame of the study is adequate."

Sample Size Requirements

To maximize the efficiency of home range studies, researchers often attempt to collect the minimum number of locations necessary to accurately portray the home range (Heezen and Tester 1967). As previously discussed, a researcher assumes that data are representative of the movements and activities exhibited over the period to which inferences are drawn. We consider this the "biological sample size," which inherently requires the researcher to consider the biology of the animal under investigation. To adequately delineate this sample size, the researcher must also explicitly specify the time period over which inferences are drawn (Otis and White 1999). However, there is also a statistical sample size, whereby home range estimators perform better with increasing number of data points (Seaman et al. 1999). Increasingly accurate estimates of home range are generally obtained with more data points as home range size increases to an asymptote (Swihart and Slade 1985b, Seaman et al. 1999).

The effect of sample size (n) on the accuracy of home range estimates is further related to the duration between observations. Increasing n will improve accuracy; but merely increasing n with smaller sampling intervals will do little to improve the estimate (Swihart and Slade 1985b). That is, it will generally require more autocorrelated data to achieve greater accuracy in estimates of home range than if data were independent. Therefore, there is an unavoidable trade-off between sampling interval and sample size (Hansteen et al. 1997). As the time between locations becomes longer and data presumably become more useful for home range estimation, the number of locations that may be collected is inherently reduced.

In theory, home range size estimates reach an asymptote when an adequate sample size is reached (S. Harris et al. 1990; but see also Gautestad and Mysterud 1995). The sample size required for unbiased estimates can be determined by plotting the area estimate against sample size (i.e., area-observation curve) (e.g., Gese et al. 1990). The minimum sample size is attained when the area estimates do not increase as more locations are added. To produce area-observation curves, fixes can be removed or added randomly or sequentially from the entire set of data points. S. Harris et al. (1990) suggested random deletion or addition for discontinuous data and sequential deletion or addition for continuous data. In order for an area-observation curve to be valid, there must be a constant center of activity over the period of analysis (Gautestad and Mysterud 1995). A difficulty in using area-observation curves is that they do not account for influences of either sample size or sampling interval (Hansteen et al. 1997, Otis and White 1999).

Several authors have used field data and computer simulations to determine how home range estimators perform under varying sample sizes (Boulanger and White 1990, Worton 1995, Seaman and Powell 1996, Seaman et al.

1999) and the results are contradictory. Gese et al. (1990) found that the minimum sample size needed to construct coyote home ranges ranged from 23 to 36, based on the asymptote of the area-observation curve. For minimum convex polygons, 100–200 locations are often needed (Bekoff and Mech 1984, Laundré and Keller 1984, S. Harris et al. 1990). Doncaster and Macdonald (1991) and Gautestad and Mysterud (1993) demonstrated that home range estimates might still increase beyond sample sizes of 100–200 (although this may be further complicated by autocorrelation in these data). Unfortunately, these contradictory results do not provide much guidance on necessary sample sizes to achieve a given level of accuracy or precision in home range studies. A necessary analysis step after data collection is to examine each dataset individually by plotting the estimated size of the home range against the number of locations used to generate the estimate, whereby appropriate conclusions can be drawn regarding levels of accuracy and precision.

While this *post hoc* analysis step is necessary, proper study design requires *a priori* consideration of the number of locations needed per animal to give a high probability that each home range will be estimated with desired accuracy. Recent work by Seaman et al. (1999) and Garton et al. (Chapter 2) may help provide some guidance for minimum sample sizes for the currently popular kernel home range estimators (Worton 1989, 1995). Seaman et al. (1999) used combinations of bivariate normal distributions to create regularly and irregularly shaped home ranges, intended to mimic real animal home ranges. Bias and variance approached an asymptote at about 50 locations per home range, and Seaman et al. (1999) recommended that researchers obtain a minimum of 30, and preferably more than 50, observations for kernel home range estimates. In contrast to other simulation work (Boulanger and White 1990), Seaman et al. (1999) found limited support for the hypothesis that the accuracy of the estimate was dependent on the distribution of the data (i.e., simple vs. complex). Another interesting finding in these simulations was that at small sample sizes, kernel home range estimates overestimated home range size (Seaman et al. 1999). This is in direct contrast to other home range estimates in which home range size is underestimated at small sample sizes. Other empirical research supports the hypothesis that kernel home ranges overestimate with fewer than 30 observations (Millspaugh 1995). Garton et al. (Chapter 2), using different evaluation techniques, suggests that up to 200 locations per animal may be necessary for adaptive kernel estimates.

EVALUATION OF HOME RANGE ESTIMATORS

Even when care is given to appropriate sampling, the accuracy of home range estimators varies widely. Home range estimators have been evaluated by use of a variety of techniques from comparison of methodology and assumptions (White and Garrott 1990, S. Harris et al. 1990) to computer simulations with known home range areas (Boulanger and White 1990, Seaman et al. 1999). Rather than reiterate the methodologies and underlying assumptions for each home range estimator, we will evaluate twelve of the most commonly used home range estimators using seven criteria (Table 5.1) and discuss advantages and disadvantages of each estimator as they relate to these criteria. The purpose of this evaluation is not to determine which one estimator is best because no one approach is best in all cases; rather, the purpose of our evaluation is to highlight important criteria to consider when designing and implementing an animal home range study and to rank estimators based on those properties.

The criteria used included (1) required sample size, (2) robustness with respect to autocorrelated data, (3) utilization distribution calculation, (4) parametric or nonparametric estimation, (5) multiple centers of activity calculation, (6) sensitivity to outliers, and (7) comparability to other estimators. Each estimator was assigned a score of 0 or 1, according to adequacy guidelines for each criterion. Therefore, a single estimator could score a maximum of 7 points (i.e., one for each criterion). Individual scores were used to identify benefits and shortcomings within and between estimators, and the overall score was used to rate estimators relative to one another. The evaluation was conducted by reviewing pertinent literature and through personal experience.

Our evaluation encompassed three basic groups of home range estimators: (1) polygon methods, (2) grid cell methods, and (3) probabilistic methods. The minimum convex polygon method is the oldest and most common home range estimator (Mohr 1947, Seaman et al. 1999). This and other polygon techniques (e.g., cluster analysis) construct home ranges by connecting a series of locations to form either convex or concave polygons. The grid cell method proposed by Siniff and Tester (1965) uses a set grid, which is overlaid on a series of location points. The cumulative set of occupied grids allows for simple three-dimensional contouring of ranges but does not calculate home range area as effectively as "outlining techniques" (S. Harris et al. 1990). Outlining techniques refer to those methods that contour around different intensities of use, most often resulting in an irregular, smoothed outer boundary and frequently multiple centers of activity (depending on the distribution of location points). Throughout this chapter we refer to these estimators as contouring methods or utilization distribution techniques (e.g., harmonic mean, kernel methods). Probabilistic models attempt to assess an animal's utilization distribution either by assuming a particular probability distribution (e.g., bivariate normal) or by attempting to characterize a variety of distributions accurately (e.g., harmonic mean and kernel methods) (S. Harris et al. 1990). Bivariate normal models best represent probabilistic models that

TABLE 5.1 Evaluation of 12 Home Range Estimators Relative to 7 Criteria, Including Sample Size Requirement, Robustness with Respect to Autocorrelated Data, Ability to Compute Utilization Distributions, Parametric or Nonparametric Estimation, Ability to Compute Multiple Centers of Activity, Sensitivity to Outliers, and Comparability between Estimators [a]

Home range estimator	Sample size [b]	Auto-correlation [c]	Utilization distribution [d]	Non-parametric [e]	Center of activity [f]	Outliers [g]	Com-parability [h]	Score
Minimum convex polygon	0	1	0	1	0	0	1	3
Peeled polygon	1	1	0	1	0	0	1	4
Concave polygon	0	1	0	1	0	0	1	3
Cluster analysis	1	1	0	1	1	0	1	5
Grid cell count	0	1	0	1	1	0	1	4
Jennrich–Turner	0	0	1	0	0	0	1	2
Weighted bivariate normal	0	0	1	0	0	0	1	2
Dunn estimator	0	0	1	0	0	0	1	2
Fourier series smoothing	0	0	1	1	1	1	0	4
Harmonic mean	0	0	1	1	1	1	0	4
Fixed kernel	1	1	1	1	1	1	0	6
Adaptive kernel	1	1	1	1	1	1	0	6

[a]Each criteron could receive one (1) point, and score represents the sum of those points.
[b]Calculated home range extent often stabilizes with ≤ 50 location points.
[c]Estimator is less sensitive to autocorrelated data.
[d]Estimator calculates home range boundary based on the complete utilization distribution.
[e]Estimator is nonparametric.
[f]Estimator calculates multiple centers of activity.
[g]Estimator is less sensitive to outliers.
[h]Estimator is comparable to other estimators when using the same dataset.

assume a bivariate normal distribution. White and Garrott (1990) provided a good analytical overview of several bivariate normal models, including the Jennrich-Turner estimator (Jennrich and Turner 1969), the weighted bivariate normal estimator (Samuel and Garton 1985), and the Dunn estimator (Dunn and Gipson 1977). Probabilistic models that do not require a known probability distribution include the harmonic mean method developed by Dixon and Chapman (1980), the Fourier series transformation method (Anderson 1982), and kernel methods (Worton 1987, 1989).

Required Sample Size

As previously discussed in this chapter, sample size requirements are a critical consideration when computing home ranges. For this criterion, we were most interested in the statistical sample size requirement, i.e., the point at which home range size reaches an asymptote (Swihart and Slade 1985b, Seaman et al. 1999). For our evaluation, estimators that stabilized at ≤ 50 data points were considered optimal.

Robustness with Respect to Autocorrelated Data

Despite the inherent assumption with most current probabilistic estimators that data be independent (Dunn and Gipson 1977), previous remarks highlighted the need to consider the intended use of home range estimates and to design radiotelemetry studies accordingly (Otis and White 1999). As discussed, a systematic strategy that adequately samples animal movements throughout the duration of the study is more important than determining a time interval between sampling that is statistically independent (McNay et al. 1994). Therefore, the accurate estimation of home range should not be constrained by an assumption of independence between location points. We rated each home range estimator on its ability to perform well with correlated data. Home range estimators that do not require independent data or are robust to some autocorrelation were considered better than those sensitive to autocorrelated data.

Utilization Distribution Calculation

The utilization distribution (UD) describes the relative frequency distribution for the location data over a specific time period (Van Winkle 1975). Recently, this concept has become increasingly important in home range studies because it assesses an animal's probability of occurrence at each point in space (Millspaugh et al. 2000; see also Chapter 12). UD techniques (i.e., methods based on densities of locations) have been contrasted with methods that estimate home ranges based on minimizing distance between locations (e.g., minimum convex polygon, cluster analysis) (Kenward et al. in review). Some parametric (Jennrich and Turner 1969) and most nonparametric (Dixon and Chapman 1980, Anderson 1982, Worton 1989) estimators base their calculations on the UD. An advantage of UD methods is that a home range boundary is calculated based on the complete distribution of an animal's location points, rather than simply characterizing the outermost set of points. Despite criticism that even UD methods cannot precisely describe home range size (Gautestad and Mysterud 1993), estimators were rated favorably in our evaluation if their computations were based on the UD.

Parametric or Nonparametric Estimation

Estimators that do not assume that locations come from a particular statistical distribution are nonparametric. Intuitively, the lack of assumptions implies a more robust method and allows estimators to conform to more irregular location distributions. Techniques that require a particular distribution (e.g., bivariate normal) for calculation are examples of parametric methods. Inaccurate calculations can result with these methods if the stated assumptions are not met. For our evaluation, nonparametric estimators were considered better than those with assumptions regarding the underlying location distribution.

Calculation of Multiple Centers of Activity

Following Hayne's (1949) description of center of activity, it is evident that multiple centers of activity can exist within a single home range area (Samuel et al. 1985). Several methods have been developed for identifying areas of more intense use, or multiple centers of activity, within a home range, including grid cell approaches (Voight and Tinline 1980, Samuel et al. 1985, Samuel and Green 1988), contouring approaches (Dixon and Chapman 1980, Worton 1989), and the incremental cluster approach (Kenward 1987). Given that animals generally live in spatially heterogeneous environments, it is important to understand the internal structure of their home range (Hodder et al. 1998). Therefore, those estimators that could compute multiple centers of activity scored higher than other estimators.

Sensitivity to Outliers

Animals may occasionally use areas outside their "normal" activity areas, thereby producing extreme locations (outliers) that can dramatically affect home range estimates (Ackerman et al. 1990). Depending on the estimator used, these effects can lead to an exaggerated outer boundary or a misrepresented UD. A common technique of dealing with outliers is to remove them from the location distribution prior to final analysis. However, a preferred method of dealing with outliers is through the use of home range estimators that are less sensitive to aberrant locations while still considering them as part of the overall location distribution. Therefore, a home range estimator that is minimally affected by outliers scored favorably.

Comparability to Other Estimators

The question of home range comparability continues to be elusive. Several authors have conducted studies that compare home range techniques (Van Winkle 1975, Macdonald et al. 1980, Boulanger and White 1990). S. Harris et al. (1990) and Lawson and Rodgers (1997) demonstrated that different methods could produce different results, even when analyzing identical data. Furthermore, Worton (1989) discussed the difficulties in transforming home range estimates made with different methods to a common basis for comparison. There are several reasons that these two situations arise: (1) differing analytical techniques, (2) same technique but different software (Lawson and Rodgers 1997), (3) different number of locations (Seaman et al. 1999), and (4) different user-defined options, such as cell size (Hansteen et al. 1997) or bandwidth (Worton 1989). We reviewed each home range estimator to determine its relative comparability to other estimators when using the same dataset and scored those that were more comparable 1 point.

Summary of Home Range Evaluation

We summarized the seven criteria as they related to 12 home range estimators, including minimum convex polygon (Mohr 1947), peeled polygon (Kenward 1985), concave polygon (Kenward 1987), incremental cluster (Kenward 1987), grid cell count (Siniff and Tester 1965), Jennrich–Turner estimator (Jennrich and Turner 1969), weighted bivariate normal (Samuel and Garton 1985), Dunn estimator (Dunn and Gipson 1977), Fourier series smoothing (Anderson 1982), harmonic mean (Dixon and Chapman 1980), and fixed and adaptive kernels (Worton 1989).

Evaluation scores ranged from 2 to 6 points with 42% of the estimators ($n = 5$) receiving 3 points or less (Table 5.1). The adaptive and fixed kernel estimators received the highest score (i.e., 6 points each), indicating better performance relative to the other home range estimators evaluated. The cluster analysis technique received the second highest score of 5 points. The importance of required sample size was demonstrated previously (see Sampling Considerations). There are practical limitations to this issue. Several methods, including the grid cell count and minimum convex polygon, require more than 100 fixes, often 200–300, to obtain reliable estimates of home range size (Bekoff and Mech 1984, Laundré and Keller 1984, S. Harris et al. 1990, Doncaster and Macdonald 1991). Sampling intensity of that magnitude is difficult to achieve in most non-GPS telemetry studies. The ability of a home range estimator to accurately estimate home range extent with less than 50 location points offers more flexibility to end-users. Otis and White (1999) discussed optimizing sampling strategies in home range studies as a function of the number of radiotelemetry locations needed for a particular analysis. The number of radio-collared animals is conditional on the required number of locations and the resources available to collect locational data. The appropriate number of locations that should be recorded will vary with study objectives. Seaman et al. (1999) recommended obtaining a representative sample of 30 or more locations, preferably 50 or more locations within a given time frame, when using kernel estimators. We extend this recommendation to applications involving other estimators (e.g., harmonic mean) as well (Table 5.1). This recommendation applies to each subset of the study period for which home ranges will be computed. For example, if the researcher is computing both seasonal and yearly home ranges, at least 30–50 locations per season should be collected. Using 50 location points as a goal, a researcher can determine, within the constraints of the study, how many animals can be monitored a minimum number of times. This approach ensures that researchers equip only as many animals as can be adequately monitored.

Estimators capable of computations based on the UD are generally affected by autocorrelated data (Table 5.1). For example, contouring and bivariate estimators use the complete UD when estimating home ranges and are sensitive to autocorrelated data. Conversely, both polygon and grid cell estimators are generally not affected by autocorrelated data; whereas their calculations are not based on complete UDs. The relationship between these criteria has been discussed previously in this chapter (see Effects of Autocorrelation: Determining the Appropriate Sampling Interval). Despite this conflict, most applications will benefit from estimators that compute home range boundaries based on UDs.

Parametric-based estimators offer greater comparability between studies (Table 5.1). However, radiotelemetry datasets rarely conform to regular distributions. Comparability between noncontouring estimators is attributed to these regular distributions and parametric assumptions. A disadvantage of nonparametric estimators is the lack of variance expressions suitable for *a priori* sample size calculation, in contrast to parametric estimators such as bivariate normal models (White and Garrott 1990). Despite these benefits of parametric estimators, nonparametric estimators are preferred, particularly given the generally irregular distribution patterns exhibited in animal distribution datasets.

Due to nonuniform use of areas contained within a home range, behavioral differences among animals in their movement patterns, or potential restrictions to animal movements, the ability of estimators to calculate multiple centers of activity and compute estimates based on the animal's complete UD has become critical in analyses of radiotelemetry data. However, if a simple outer boundary were required, then noncontouring estimators (e.g., minimum convex polygon) would be an appropriate choice. For example, if a researcher were interested in establishing the outer boundary of an animal's home range where a barrier was known to restrict movement (e.g., fence), a contouring method would likely establish the boundary beyond the barrier, whereas an estimator that did not consider the UD would conform to the barrier. Along with contouring methods, both the cluster analysis technique (Kenward 1987) and grid cell count method (Siniff and Tester 1965) are capable of calculating multiple centers of activity (Table 5.1). The ability of these estimators to differentiate disproportionate use within the home range sets them apart from the other lower ranked estimators in this evaluation.

Despite the care given to collecting location points, aberrant locations will often be recorded. Generally, contouring methods of home range estimation can accommodate outliers better than parametric and other noncontouring methods (e.g., minimum convex polygon) (Table 5.1). There are a variety of tests designed to identify outliers. Ackerman et al. (1990) used three procedures for identifying outliers: (1) a binomial test of observation density, (2) a weighted bivariate normal technique, and (3) examining a list of points with large harmonic values. At a minimum, outliers should be considered for removal using tests similar to these for estimators identified as being sensitive to outliers (e.g., minimum convex polygon). Other approaches such as the peeled polygon and cluster analysis treat outliers by "peeling them off" and do not include them when calculating home range boundaries. However, significant reduction of sample size can be detrimental to home range estimation. Kernel estimators, harmonic mean estimators, and Fourier series smoothing techniques that are less sensitive to outliers are preferred.

S. Harris et al. (1990) suggested using at least two home range estimators with any dataset, one of which should be the minimum convex polygon because it can be compared to other home range studies. Seaman et al. (1999) reported that 87% of the home range studies published in *The Journal of Wildlife Management* from 1980 to 1997 used the minimum convex polygon method, 22% used the harmonic mean estimator, 7% used a kernel method, and 7% used a bivariate normal ellipse. Most authors report a minimum convex polygon because they believe it is more comparable to previous works. However, this is an inadequate justification because comparisons are generally unreliable due to its sensitivity to sample size, inability to calculate multiple centers of activity, and sensitivity to outliers (Seaman et al. 1999). If minimum convex polygon methods are to be used for comparison purposes, sample sizes and a statement of how outliers were treated should be reported.

Table 5.1 highlights the range of estimators described and evaluated in this chapter. Given the strengths and weaknesses among estimators, a variety of analytical techniques can be used in exploratory-type analyses of radiotelemetry data. However, relative to the home range estimators evaluated here, kernel methods (i.e., fixed and adaptive) scored most favorably followed by the cluster analysis technique (Table 5.1). The combination of reasonable sample size requirements, the ability to compute home range boundaries that include multiple centers of activity, computation based on the complete UD, nonparametric methodology, and the lack of sensitivity to outliers contributed to the fact that kernel methods were ranked highest in our evaluation. The cluster analysis method satisfies many of the same criteria as kernel methods (e.g., sample size, autocorrelation, multiple centers of activity) (Table 5.1). However, cluster analysis is a distance-based estimator and does not calculate home range boundaries based on the complete UD. Cluster analyses are a recent addition to the family of home range estimators (Kenward 1987). Although a comprehensive review is warranted, little published literature describing or comparing cluster analysis exists at this time. Kernel-based estimators also have only recently been considered for analyzing radiotelemetry data (Worton 1987, 1989) and have not been extensively reviewed based on recent literature. Therefore, we provide a detailed description and summary of kernel-based estimators.

DESCRIPTION OF KERNEL-BASED ESTIMATORS

Kernel density estimation has a decades-long history as a general statistical technique (Scott 1992). Worton (1987, 1989) first applied kernel methods to estimate the UD and specifically home range size. Since Worton (1987), several studies have evaluated kernel methods (Worton 1989, Naef-Daenzer 1993, Seaman and Powell 1996, Hansteen et al. 1997, Samietz and Berger 1997, Swihart and Slade 1997, De Solla et al. 1999, Ostro et al. 1999, Seaman et al. 1999). Currently, kernel methods set the standard for nonmechanistic estimation of UDs (e.g., estimation based on simple descriptive, statistical models). Key sources of detailed information on kernel estimation include the ecological references cited above, statistical monographs (e.g., Silverman 1986, Scott 1992, Wand and Jones 1995, Bowman and Azzalini 1997), and the primary statistical literature (e.g., Jones et al. 1996a).

The input into a kernel estimator is the measured locations for an individual or other unit of analysis (e.g., group of animals). As with most other home range estimators, the locations are assumed to be independent. The estimated value of the UD at a location point is calculated by:

$$\hat{f}(x) = \frac{1}{nh^2} \sum_{i=1}^{n} K\left[\frac{x - X_i}{h}\right], \tag{5.3}$$

where $\hat{f}(x)$ is the estimated probability density function, or UD, n is the number of locations, h is the smoothing parameter or bandwidth, X contains the x and y coordinates for the n observed locations, x is the point at which the kernel estimate is calculated, and $K(.)$ is the kernel function, a bivariate symmetric function (Worton 1989). Usually the kernel function is a probability density function, such as the bivariate normal (Silverman 1986). The estimated kernel density function can be seen as the sum of n separate kernel functions, each centered at an observed location (Silverman 1986). The general kernel formula can be extended to any number of dimensions, although most practical uses have been limited to univariate or low-dimensional multivariate applications (Silverman 1986).

Typically the estimated kernel density function is evaluated over a fine grid to delineate the UD. The kernel home range estimate is generally reported as the minimum area that includes a fixed percentage of the estimated UD volume. Some software packages estimate the minimum area incorporating the specified percentage of the observed locations (Seaman et al. 1999), which seems to defeat the purpose of estimating the UD at all. In contrast to harmonic mean calculations, the evaluation grid size and spacing has no effect on the estimated UD value at each grid point (Worton 1989).

The bandwidth (or smoothing parameter) value is widely recognized as the critical component in kernel density estimation (Silverman 1986, Worton 1995). The bandwidth controls the width of the individual kernels and therefore determines the amount of smoothing applied to the data (Fig. 5.1). At small bandwidths, the estimated kernel density function partially breaks into its constituent kernels (Fig. 5.1C). Large bandwidths result in all local peaks and valleys being smoothed over into a single surface (Fig. 5.1D). It is critical that researchers recognize the critical influence of the bandwidth value on kernel estimates and justify their choice of smoothing parameter options.

Two major subdivisions of the general kernel technique are the fixed and adaptive kernel methods (Worton 1989). With the fixed kernel method, the same bandwidth is used over the entire evaluation area. As will be discussed, the bandwidth can be chosen by least squares cross validation (LSCV) or other methods that attempt to minimize error between the estimated and true density. Several bandwidth configurations are possible, such as smoothing

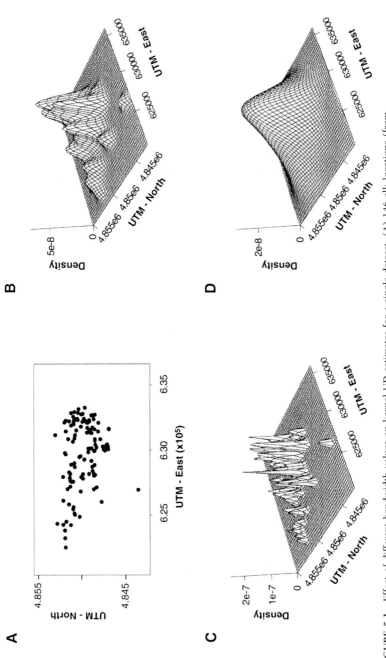

FIGURE 5.1 Effect of different bandwidth values on kernel UD estimates for a single dataset. (A) 146 elk locations (from Millspaugh 1999). Fixed kernel UDs for these locations based on three bandwidth values (h). (B) h = 500. (C) h = 1500. (D) h = 150. (Kernel plotting routines from Bowman and Azzalini 1997).

differently along the x and y axes. With the adaptive kernel method, a local bandwidth is selected for each observation. The local bandwidth is larger in areas with fewer observations than in areas with many observations. As a result, the tails of the distribution are smoothed more than the peaks. In practice, the local bandwidths in the adaptive kernel method are scaled from a global value (Silverman 1986). The global value is selected as in fixed kernel estimation and the local scaling factors are based on fixed kernel pilot estimates of the density at each observed location (Worton 1995, Seaman and Powell 1996). Both fixed and adaptive kernels have been used and evaluated in home range estimation, with the fixed kernel method generally producing more accurate and precise estimates, at least in the outer contours of the home range (Worton 1995, Seaman and Powell 1996, Seaman et al. 1999).

Kernel home range methods offer several advantages over previous estimators. First, kernel estimators are free of distributional assumptions such as those of the bivariate normal method. The kernel method performs well in simulations, even when multiple bivariate normal distributions are combined to produce complex multimodal home ranges (Worton 1995, Seaman et al. 1999). Second, based on simulated data, some forms of kernel methods are superior to other nonparametric techniques for estimating the UD. However, simulations by Seaman and Powell (1996) indicated that superior kernel performance depended on the proper choice of user-defined options (e.g., bandwidth selection) within the general kernel method. Third, kernel density estimation is not a specialized method for home range studies; it is a widely used statistical technique that continues to be actively developed by statisticians. Implementation of kernel methods in analyses of space use and movement data can build directly on more general studies.

Several weaknesses of kernel estimators are evident, specifically the lack of a general variance expression, the assumption of independence, and the high sensitivity to bandwidth choice. It should be noted that these weaknesses, with the exception of bandwidth sensitivity, are not unique to kernel estimators. The flexibility and nonparametric nature of kernel methods are both a major advantage and disadvantage. Kernel estimators assume nothing about the processes affecting space use or movements. This may actually be an advantage, such as when home range estimation is preliminary to hypothesis testing and model selection. With this approach, estimates of home range size or other features of the UD are used as the raw data points for formal analysis and comparison with other studies. This analytical framework will continue to be important despite the increasing dominance of model-driven approaches in ecology (Burnham and Anderson 1998).

Methods for Selecting Smoothing Parameters

The flexibility of selecting smoothing parameters is often viewed as a disadvantage to kernel methods given the wide range of available choices. Applying a range of smoothing parameter values in exploratory analyses is valuable for revealing the structure of a particular data set (Silverman 1986). However, objective methods of choosing smoothing parameters are necessary for making statistical comparisons of multiple home range estimates within and between studies. A large number of bandwidth selection methods have been evaluated with new variations appearing frequently (Wand and Jones 1995). Here, we briefly discuss three bandwidth selection methods. Kernel home range evaluations have focused on two of these methods (Worton 1995, Seaman and Powell 1996, Seaman et al. 1999): the reference or optimum bandwidth h_{ref} (sometimes denoted h_{opt}) and the bandwidth chosen by least squares cross validation, h_{lscv} (h_{cv}). A third method, "solve-the-equation plug-in" bandwidth selection (h_{pi}), has not been evaluated in ecological studies. Variations of this method exist, but here we use h_{pi} to denote the general class.

The reference bandwidth is an estimate of the ideal bandwidth if the true distribution is the bivariate normal (Silverman 1986). Least squares cross-validation (LSCV) and plug-in methods do not assume any particular underlying distribution. LSCV and plug-in methods estimate a bandwidth that minimizes a measure of the discrepancy between the estimated density and true density. LSCV minimizes an objective function based on the mean integrated squared error between the estimated and true density (Silverman 1986). Plug-in methods seek an "ideal" bandwidth that minimizes the asymptotic mean integrated squared error. The equation for this ideal bandwidth contains a function of the unknown density $f(x)$. A pilot bandwidth is chosen so that this function can be estimated; the estimate is then "plugged in" to the equation for the ideal bandwidth, producing h_{pi}. Wand and Jones (1995) and Jones et al. (1996a, b) provide more detailed information about plug-in techniques. LSCV is currently the recommended bandwidth selection method in the ecological literature (Seaman et al. 1999). The reference bandwidth oversmooths for multimodal distributions (Silverman 1986). As a result, it overestimates home range size to a greater degree than h_{lscv} (Worton 1995, Seaman and Powell 1996, Seaman et al. 1999) and is not recommended for general use (Seaman et al. 1999).

Despite its widespread use, LSCV has several drawbacks. Sampling variability is high in that estimation of bandwidth values for multiple realizations of a single density varies greatly (Park and Marron 1990). Simulations indicate that h_{lscv} will most often choose overly small values that undersmooth (Sain et al. 1994). The function for h_{lscv} frequently has multiple local minima, and the recommended practice is to choose the largest local minimizer (Park and

Marron 1990). Plotting the h_{lscv} function over the range of potential bandwidth values is advised, especially if the minimization routine is simply searching for a global minimum. Such output could wisely be incorporated into home range software that offers h_{lscv} as an option.

In some cases, particularly when many of an animal's locations are at or near the same point, the LSCV function descends monotonically to a degenerate bandwidth value of 0 (Silverman 1986) or to whatever lower bound has been specified for the bandwidth search interval (e.g., $0.1 * h_{ref}$; Worton 1995). When this occurs, the LSCV method has essentially failed. Software packages may automatically revert to h_{ref} or use the lower bound of the search interval. Researchers should recognize when this has occurred and be aware that the software's default response may not be desirable. For example, using a plug-in bandwidth selection method or scaling h_{ref} by an appropriate proportion (Worton 1995) may be more justifiable than simply using h_{ref}.

In essence, h_{lscv} produces an unbiased estimate of the ideal bandwidth at the expense of high variability. In contrast, some forms of h_{pi} have much lower sampling variability than h_{lscv} (Jones et al. 1996a). Plug-in methods may have a positive bias in some cases with a tendency towards moderate oversmoothing (Jones et al. 1996a). While the tendency of h_{lscv} to undersmooth is generally seen as a weakness, h_{lscv} can outperform h_{pi} when the true density really does have multiple, relatively closely spaced modes, at least for univariate data (Loader 1999). Our opinion is that the "strength" of h_{lscv} is of minor importance given the typically low sample sizes of many radiotelemetry studies. When h_{lscv} is so low that the estimated UD breaks up into numerous small peaks corresponding to single or very localized clumps of observations, we would have low confidence that h_{lscv} is accurately reflecting the underlying UD, at least for portions of the distribution. However, whether such clumping is "undesirable" depends on the biology of the species and the intent of the investigator (i.e., whether the researcher is more interested in identifying such discrete, localized clumps or in delineating a continuous boundary around the general area of use). Comparisons of h_{lscv} and h_{pi} are needed to examine their performance in estimation of home range size and overlap in relation to sample size.

Fixed vs. Adaptive Kernels

Despite the close relationship between fixed and adaptive kernel methods, these techniques may produce dramatically different estimates (Seaman and Powell 1996, Seaman et al. 1999). Contrary to initial expectations (Worton 1989), the fixed kernel method generally appears to have lower bias and better surface fit than the adaptive kernel for a given bandwidth selection method (Seaman and Powell 1996, Seaman et al. 1999). For example, the percent bias of a 95% contour home range size estimated for complex simulated home ranges based on 50 locations and LSCV bandwidths was 5.5% for the fixed kernel and 36.6% for the adaptive kernel (Seaman and Powell 1996). The unreliability of estimates in the outer contours has significant implications for home range studies. The discrepancy between fixed and adaptive kernel home range size estimates is driven primarily by the higher bias of adaptive kernel methods in estimating the outer contours of the UD (Seaman et al. 1999). The adaptive kernel attaches more uncertainty to the locations at the outer edge of the sample by smoothing the density function to a greater degree. This approach may be appropriate for studies in which estimation errors are greater for locations at the edge of an animal's home range (e.g., if distance from receiver to animal is greater for such locations than for locations near core areas). As reported by Seaman et al. (1999:741), "most studies report results for the 95% home range area, but the peripheral area has the least data to support an estimate, has the least biological significance for the animal, and has the most opportunity to influence numerical results." For this reason, Seaman et al. (1999) recommended that future studies focus on the interior of the animal's home range and noted only minor overall differences in surface fit between fixed and adaptive kernels with LSCV. However, the fixed kernel produced fewer cases of very high error in surface fit.

Worton (1995) commented that the choice of fixed vs. adaptive kernels was unimportant compared to the choice of bandwidth values. However, this comment is based on comparisons made between different scalings of the LSCV bandwidth. For example, for simulated home ranges with two centers of activity, the fixed kernel estimate method with bandwidth equal to 0.8 h_{lscv} performed similarly to the adaptive kernel with a bandwidth equal to 0.6 h_{lscv} (based on 50 locations per simulation). Unfortunately, the ideal scaling of h_{lscv} in these simulations varied with the pattern of the home range (e.g., uniform vs. normal mixtures, unimodal vs. multimodal). To determine the pattern that best matched a real dataset, a researcher would have to visually examine the data and subjectively select a pattern that "looked right." Since objective bandwidth selection is a prerequisite to statistical comparisons, this subjective approach is not desirable. As a result, the better performance of fixed kernels for unscaled h_{lscv} and h_{ref} is arguably more relevant than their similar performance under best-case scenarios with scaled values for these bandwidths.

The choice between fixed vs. adaptive kernel ultimately depends on specific research objectives, sampled distributions, and biology of the animal being studied. For example, if the primary objective is to determine centers of activity, then a fixed kernel approach with a strict minimization of the LSCV score may be appropriate in order to differentiate between discrete areas within the home range. In contrast, where more uncertainty is attributed to locations on the edge of the distribution, the adaptive kernel has an advantage by smoothing the density function to a greater degree. Finally, the previous two considerations are often dependent on the biology of the animal being studied.

Sample Size and Variance Calculations

As with previous nonparametric estimators (White and Garrott 1990), no explicit variance formula is available for kernel home range estimates. This prevents standard *a priori* sample size calculations. However, bootstrap variance estimates can be calculated for home range area (Worton 1995; Hansteen et al. 1997). Seidel (1992) and Worton (1995) used a smoothed bootstrap approach, in which bootstrap samples were simulated from the estimated UD. Hansteen et al. (1997) used the more familiar unsmoothed bootstrap approach in which bootstrap samples were taken from the set of observed locations. The smoothed bootstrap approach is preferable for small sample sizes (Shao and Tu 1995), such as those typical of ecological studies. Discretization or rounding of location coordinates can cause problems for h_{lscv} (Silverman 1986); the smoothed bootstrap approach also has the advantage of producing bootstrap samples with a lower level of discretization (Seidel 1992). However, one approach is not universally better than the other (Shao and Tu 1995).

The smoothed bootstrap approach is straightforward. Given the estimated UD $f(x)$, based on n observations, the bivariate kernel function k, and bandwidth values (h_x, h_y), the procedure for simulating is as follows (note that our notation mixes that of Scott 1992 and Worton 1995):

1) Simulate a bootstrap sample from the estimated UD.
   a) Select an integer j by sampling from the set $\{1, 2, \ldots, n\}$. Then (x_j, y_j) are the coordinates for the jth observed (not simulated) location.
   b) Generate a random deviate $\{(t_x, t_y)\}$ from the kernel k.
   c) Return the simulated observation $(x^*, y^*) = (x_j + h_x t_x, y_j + h_y t_y)$.
   d) Repeat n times, sampling with replacement for step (a).

2) Estimate the parameter of interest (e.g., home range size) for the bootstrap sample. The bandwidth should be estimated independently for each bootstrap sample as opposed to repeatedly using the bandwidth calculated for the "real" dataset. Bandwidth selection, particularly h_{lscv}, is known to have a high sampling variability; this should be incorporated into the overall uncertainty about the parameter of interest.

3) Repeat steps 1 and 2 many times. For example, Worton (1995) formed 1000 bootstrap samples. Calculate the variance and other statistics of interest for the bootstrap distribution of home range size.

Bootstrap variance estimation for home range analysis is currently an underutilized tool. Because point estimates of home range size generally become raw data points for subsequent analysis, the importance of a measure of uncertainty for each point estimate may be unwisely overlooked. Examining some measure of precision along with each point estimate is vital for determining the quality of these estimates. Based on an unsmoothed bootstrap approach, Hansteen et al. (1997) noted drastic negative bias and high coefficients of variation of 0.35–0.71 for 3 real home ranges estimated by the kernel method. However, it is not clear whether the kernel technique was performing poorly for these specific cases, or whether the poor performance was an artifact of the h_{ref} bandwidth and calculation method used (Seaman et al. 1999) or the unsmoothed bootstrap approach. We believe the bootstrapping technique should be implemented as a standard option in all home range software.

Because theoretical variance estimates are not available for home range estimates, bootstrapping can play a valuable role in study design. Bootstrap estimates based on data from a pilot study or previous work on a species in a general area can be used for designing future work. In particular, optimal allocation of resources for a fixed effort could be explored, specifically when there are trade-offs between how many animals to radio-collar and how many locations can be taken from each (Otis and White 1999). Further investigation of sample size/variance relationships for a variety of real datasets could provide general guidance to complement that of Seaman et al. (1999).

Kernel estimation assumes independence between observed locations, a weakness noted by White and Garrott (1990) for several other nonparametric estimators. However, as discussed earlier, the assumption of strict temporal independence has been overemphasized. The emphasis should be shifted to choosing a suitable sampling design over a predefined period of interest (Swihart and Slade 1997, De Solla et al. 1999, Otis and White 1999).

While model-driven analyses of space use and movements are growing increasingly common (see The Future: Modeling the Movement Process), nonparametric estimation of the UD will continue to be important. Continued refinement of the general kernel methods is needed, particularly in bandwidth estimation methods. Both in analysis of real data and comparison of kernel estimators with new techniques, researchers should be aware that different forms of kernel estimators can produce dramatically different results. Researchers should carefully consider which set of options to use and report with justification those choices. Finally, increased use of bootstrap variance estimates is highly recommended, particularly as a step in careful study design that optimizes sampling effort.

ANALYSIS OF SITE FIDELITY

Techniques for the analysis of simple animal movements include a suite of tools designed to assess site fidelity. Often, quantification of an animal's home range ignores the temporal aspect of animal behavior by depicting the spatial arrangement of location points without regard for the sequence in which the locations were recorded. Analysis of site fidelity explores these temporal dynamics of animal behavior and encompasses three general types of movement: (1) migration, (2) dispersal, and (3) localized movement. White and Garrott (1990) defined migration as a regular, round-trip movement of individuals between two or more areas or seasonal ranges, whereas dispersal is one-way movement of individuals from their natal site or an area that has been occupied for a period of time. Localized movements describe the daily movement patterns of an individual, often within a home range.

DISTANCE AND DIRECTION MOVED

Graphic Representation

The most common presentation of simple movement data is the reporting of distance and direction moved. Early techniques were generally restricted to still graphics or plots of location points. As computer graphics improved, sophisticated software was developed to assist with this form of presentation. A limitation to the use of still graphics is the loss of the time dimension (White and Garrott 1990); it does not provide any indication of time spent at a location or direction moved from a location. The use of vector plots or animated graphics overcomes this problem (White and Garrott 1990). With continuing advancements in radiotelemetry equipment, including GPS and satellite technology (see Chapter 4), these types of analyses are becoming more sophisticated. For example, the Raptor Center at the University of Minnesota, St. Paul has initiated a program called "Highway to the Tropics" where rehabilitated raptors are equipped with satellite transmitters and returned to the wild. The Center is able to track individual bird movement via satellite and continuously update maps on the World Wide Web (http://www.raptor.cvm.umn.edu).

Quantification of Distance and Direction Moved

If animals are located intensively enough during long-distance movements, one can measure, describe, and test hypotheses about the distance and directionality of such movements (White and Garrott 1990). Because the measurement of direction (i.e., compass bearing) is circular, normal statistical procedures are not appropriate. Zar (1996) presented a clear overview of the most common statistical procedures used for circular distributions. Kernohan et al. (1994) compared the distance and direction of white-tailed deer (*Odocoileus virginianus*) movements at Sand Lake National Wildlife Refuge, South Dakota from 1992–1994 (Fig. 5.2). Capture location and tag return

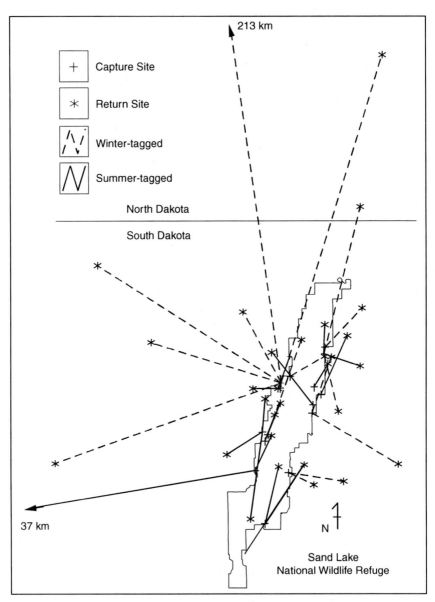

FIGURE 5.2 Direction (°) of movements exhibited by white-tailed deer (*n* = 31) captured at Sand Lake National Wildlife Refuge, South Dakota, 1992–1994. [From Kernohan et al. (1994), with permission.]

information for 298 white-tailed deer were used to compare distance and direction traveled between all-, summer-, and winter-tagged deer by sex and age (Table 5.2). Distance (d_i) between capture location $i(x_i, y_i)$ and tag location (x_{i+1}, y_{i+1}) was calculated as:

$$d_i = \sqrt{(x_i + 1 - x_i)^2 + (y_i + 1 - y_i)^2}, \qquad (5.4)$$

Direction of travel in degrees (a_i) from capture location (x_i, y_i) to tag location (x_{i+1}, y_{i+1}) was calculated as:

TABLE 5.2   Distance and Direction for All-, Summer-, and Winter-Tagged White-Tailed Deer Based on Movements ≥ 2.3 km[a] for Deer Captured at Sand Lake National Wildlife Refuge, South Dakota, 1992–1994

Season/sex/age	n	Distance (km)		Direction (°)	
		Mean	SE	Mean	SE[b]
All deer	31	14.6	6.76	13.0	14.34
Male	13[c]	25.8	15.79	357.6	5.68
Female	17	6.7	1.78	30.6	4.38
Fawn	4[d]	4.4	0.92	—[e]	—
Yearling	13	26.8	15.73	—	—
Adult	13	3.4	1.98	—	—
Summer	16	6.0	2.11	1.2	19.84
Male	6	10.2	5.43	349.9	20.68
Female	9	3.4	0.40	10.2	5.53
Fawn	4	4.4	0.92	—	—
Yearling	5	10.5	6.66	—	—
Adult	6	3.5	0.80	—	—
Winter	15	23.8	13.63	25.2	20.27
Male	7	39.1	28.98	358.5	6.52
Female	8	10.5	3.38	85.3	10.66
Yearling	8	36.9	25.20	—	—
Adult	7	8.9	3.44	—	—

Note: From Kernohan et al. (1994), with permission.
[a]2.3 km represents mean diameter of yearly home ranges of radio-collared deer on Sand Lake National Wildlife Refuge.
[b]Angular error calculated as (angular deviation/$\sqrt{n}$) (Zar 1974).
[c]One observation was omitted from sex comparisons due to lack of classification.
[d]One observation was omitted from age comparisons due to lack of classification.
[e]Directional analyses by age were not conducted.

$$a_i = \begin{cases} \arctan(Y_i/X_i)(180°/\pi) & \text{if } X_i > 0 \\ 180° + \arctan(Y_i/X_i)(180°/\pi) & \text{if } X_i < 0 \\ 90° & \text{if } X_i = 0 \text{ and } Y_i > 0 \\ 270° & \text{if } X_i = 0 \text{ and } Y_i < 0, \end{cases} \quad (5.5)$$

where distance of the X vector was calculated as:

$$X_i = x_{i+1} - x_i \quad (5.6)$$

and distance of the Y vector was calculated as:

$$Y_i = y_{i+1} - y_i. \quad (5.7)$$

The arctan function computed angles calculated in radians that were converted to degrees. This angle (a_i) was then converted to a bearing from true north (b_i) with the equation

$$b_i = 90 - a_i. \quad (5.8)$$

If b_i was negative, 360° was added to the result to make the value positive. Mann–Whitney U test and Kruskal–Wallis analysis of variance were used to compare distance traveled (Kernohan et al. 1994). A Rayleigh's z test (Zar 1996) was used to test whether direction traveled was random or oriented (Kernohan et al. 1994). If direction was oriented, then Watson's U^2 test for nonparametric two-sample testing (Zar 1996) was used to compare direction traveled (Kernohan et al. 1994).

Complete circular statistic references can be found in Mardia (1972) and Batschelet (1981). In addition, Hooge and Eichenlaub (1997) have developed MOVEMENT, an Animal Movement Analysis ArcView Extension that has the capability to perform circular statistic functions necessary for analyzing movement direction. This program also is useful for computing a number of other movement analyses, including site fidelity tests and home range analysis (Appendix A). Pace (Chapter 7) describes techniques for estimating and visualizing movement paths from radio-tracking data.

DISTANCE BETWEEN HOME RANGE CENTROIDS

The distance between home range centroids in consecutive years has been used as a measure of fidelity (Schoen and Kirchhoff 1985, Garrott et al. 1987, Brown 1992). A simple method for computing the home range centroid is to calculate a mean location from all locations contained within the location distribution of interest (Garrott et al. 1987). White and Garrott (1990) suggested use of Hotelling's T^2 statistic for testing locational shifts of the centroids (e.g., mean location). As previously described, a Mann–Whitney U

test can be used on distance data to test for differences (Schoen and Kirchhoff 1985, Garrott et al. 1987). Although this technique is straightforward to compute, it is unable to detect range expansion or contraction when the mean location of the home range or core area remains constant. In addition to range expansion or contraction, a shift in the time spent in various portions of an animal's home range cannot be detected using mean location shift as a measure of site fidelity (White and Garrott 1990). To detect such expansion and temporal shifts, a test for change in the variances and covariances of the x and y coordinates is required (Mielke and Berry 1982). Such tests attempt to compare complete UDs over time and space.

Multiresponse Permutation Procedures

Mielke and Berry (1982) proposed a nonparametric test based on multiresponse permutation procedures (MRPP) to test for changes in an animal's use of space (also see Mielke et al. 1976, Zimmerman et al. 1985, Biondini et al.1988). This method determines whether two or more sets of locations come from a common distribution. Multiresponse permutation procedures are a class of nonparametric tests independent of assumptions regarding underlying distributions or homogeneity of variances and have power for detecting slight differences. The strategy of the MRPP statistic is to compare the observed intragroup average distances with the average distances that would have resulted from all other possible combinations for the data under the null hypothesis (Biondini et al. 1988). Permutation procedures emphasize use of test statistics based on ordinary Euclidean distances (an exponent of 1) as opposed to squared Euclidean distances (an exponent of 2) that are used in most conventional parametric and nonparametric methods (Slauson et al. 1994). Ordinary Euclidean distance-based statistics have greater power (i.e., the probability of rejecting the null hypothesis when it is false) to detect location shifts (i.e., shifts in central tendency) among skewed distributions than do squared Euclidean distance statistics (Zimmerman et al. 1985, Biondini et al. 1988). Slauson et al. (1994) developed BLOSSOM Statistical Software for computing permutation procedures (see Appendix A for software availability).

Home Range Overlap

Home range estimation can be used to measure site fidelity in two ways. The first is to compare overlap of home range estimates from a single animal between two or more time periods (e.g., Kernohan 1994, Phillips et al. 1998). The procedure for this analysis will be discussed in more detail later in this chapter (see Static Interaction). The second method requires comparing a home range estimate from a single animal to a random home range estimate based on random walk paths (Munger 1984, Danielson and Swihart 1987, Palomares 1994). The relationship between the observed and random home ranges can be assessed using the techniques described earlier (e.g., mean location shifts, MRPP). The use of random walk models for developing the underlying distribution from which to compute random home ranges is discussed briefly later in this chapter (see The Future: Modeling the Movement Process–General Movement Models).

ANALYSIS OF ANIMAL INTERACTIONS

Despite the general availability of analytical techniques to assess space use of individual animals, the way two or more animals interact in their environment has received little attention. This is surprising given that space use sharing among animals may be influential in determining individual space use characteristics (Lewis and Murray 1993, Moorcroft et al. 1999).

Animal interaction analyses can be categorized as either static or dynamic (Macdonald et al. 1980, Doncaster 1990). Static analyses measure animal interaction throughout a time interval of interest. An example of a static interaction analysis would involve computing the percent area overlap between two animals' home ranges (e.g., Mizutani and Jewell 1998). In contrast, dynamic interaction analyses compare the relationship among animals at a particular point in time, thus requiring simultaneous or near simultaneous locations for each animal (Minta 1992). An example of a dynamic interaction would be to compare the distance between the simultaneous locations of two or more animals to what would be expected randomly.

Static and dynamic approaches measure different aspects of animal interaction. Dynamic interactions analyses inherently incorporate the temporal nature of the relationship, whereas static analyses measure spatial overlap without considering whether two animals use the same space simultaneously or at different times. It is important to note that an investigator may observe little dynamic interaction between two animals but record great static interaction. However, the converse cannot be true (Doncaster 1990, Minta 1992).

STATIC INTERACTION ANALYSES

Rasmussen (1980) identified three criteria of a desirable static interaction analysis: (1) it should assess the similarity in the location of high-use areas; (2) it should not be affected by sample size; and (3) it should not be affected by the differences in the size of the spatial frame. Seidel (1992) also described the need for a technique that responds to subtle variations in space use patterns.

Rasmussen (1980), Dunn (1979), and Macdonald et al. (1980) developed parametric tests for measuring the dependence in the concurrent movements of two individuals. However, the interaction analyses created by Dunn (1979) and Macdonald et al. (1980) are unrealistic in many situations because they require that each range be distributed about a single center of activity (White and Garrott 1990). Rasmussen (1980) computed an index of similarity in the utilizations of home ranges that tested for correlation in the UDs of each animal's range; positive correlation indicated attraction and a negative correlation indicated repulsion. An estimate of the UD can be obtained using the grid cell method wherein the number of fixes in each cell would quantify animal use patterns (Rasmussen 1980). A Spearman's coefficient (Zar 1996) obtained from the entire grid cells frequented by one or both animals tests the correlation of the two use patterns. One problem with this approach is the designation of grid cell size. Cell size must be large enough that some cells contain several fixes yet not so large that the overall configuration is obscured (Rasmussen 1980). In additional, for all unshared cells, these zero values give a nonlinear relationship between the coefficients and overlap extent.

Perhaps the most widely used and intuitive technique to assess static interaction among two or more individuals (as well as site fidelity as discussed earlier) involves superimposing two-dimensional home range maps (HR_1 and HR_2) (e.g., Mizutani and Jewell 1998). The measure of space use sharing is the percent area overlap between the ranges computed as:

$$HR_{1,2} \frac{A_{1,2}}{A_1} \quad \text{and} \quad HR_{2,1} \frac{A_{1,2}}{A_2}, \tag{5.9}$$

where $HR_{1,2}$ is the proportion of animal 1's home range overlapped by animal 2's home range, $HR_{2,1}$ is the proportion of animal 2's home range overlapped by animal 1's home range, and $A_{1,2}$ is the area of overlap among HR_1 and HR_2. Most commonly, this approach involves calculating minimum convex polygon estimates as the measure of individual animal movements. However, as discussed earlier in this chapter, polygon space use approaches have several undesirable properties, such as correlation of sample size to range size and inability to conform to irregularly shaped ranges. These undesirable properties ultimately limit the utility of this technique for measuring static interactions. However, the largest problem in using polygon techniques, and all non-UD methods in general, is that they do not consider the internal anatomy of the home range; instead, they strictly focus on the outer boundary. We believe that an accurate assessment of static space use sharing requires more knowledge about use patterns within the confines of the outer home range boundary. Seidel (1992) effectively illustrated how space use measures based on polygon overlaps for home ranges that are either irregular or nonuniform may not accurately depict space use sharing.

Data from an elk (*Cervus elaphus*) radiotelemetry study in South Dakota (Millspaugh 1999) further illustrates the potential problem of ignoring use patterns within the confines of the outer home range boundary. Point data were collected from two radio-collared bull elk in South Dakota during fall 1996 using techniques described by Millspaugh (1999) (Fig. 5.3). First inspection of these data illustrates clear spatial separation between these two animals. However, the percent area overlap between these two animals as a function of the computed 100% convex polygons indicated a relatively high measure of overlap. The opposite could also occur. That is, it is entirely possible that the locations from one animal are very similar to those of another, yet many of the peripheral locations are quite separated (Fig. 5.4). This example illustrates the potential problem of ignoring use patterns within the confines of the outer home range boundary when computing space use overlap.

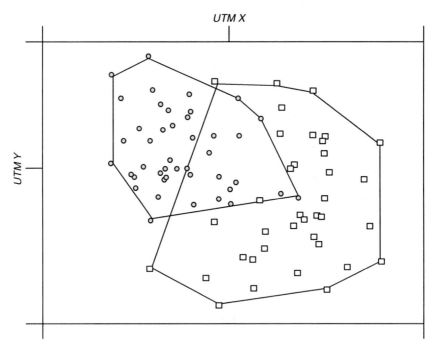

FIGURE 5.3   Point locations of two radio-collared bull elk (filled circles $n = 43$ points and rectangles $n = 37$ points) in Custer State Park, South Dakota during fall 1996.

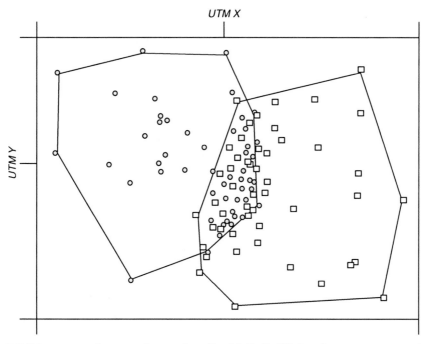

FIGURE 5.4 Point locations of two radio-collared bull elk (filled circles $n = 51$ points and rectangles $n = 49$ points) in Custer State Park, South Dakota.

One might suggest the use of another home range estimator, such as the aforementioned UD approaches, because these estimators consider use patterns by creating the UD and are better able to cope with irregularities in nonuniformly shaped home ranges. However, the comparison is still two-dimensional. This approach does not take full advantage of the information offered by these nonparametric UD techniques, namely, the ability to define use as more than a two-dimensional home range boundary.

In an attempt to overcome some of the problems inherent in the percent area overlap approach, Seidel (1992) proposed that researchers measure the overlap of the UD vs. overlap of the outer boundaries. She proposed the volume of intersection (V.I.) index statistic as a measure of the degree of overlap in shape and location of two or more UDs, measured as:

$$V.\ I.\ Index = \int_{-\infty}^{\infty} \int_{-\infty}^{\infty} [\hat{f}_1(x, y),\ \hat{f}_2(x, y)] dx\, dy, \qquad (5.10)$$

where $\hat{f}_1$ is the UD for animal 1 and $\hat{f}_2$ is the UD for animal 2. The V.I. index ranges from 0 for no overlap to 1 for complete overlap. By incorporating the entire UD in the space use comparison, we explicitly account for the use patterns within the outer home range boundary.

One consideration in computing the V.I. index is the procedure for estimating $\hat{f}_1$ and $\hat{f}_2$. Of the previously mentioned home range estimators, kernel-based estimators are best suited for this analysis because they do not assume the data conform to a prespecified distribution and because they provide true probability density functions (Worton 1995). As discussed earlier in this chapter, the choice of which kernel-based technique to use is a little more difficult (see Description of Kernel-Based Estimates). Worton (1995) reported that neither fixed nor adaptive kernels were best; rather the choice of smoothing parameters is probably most important. Although we recognize the importance of selecting an appropriate smoothing parameter, as previously discussed in this chapter, the work of Seaman et al. (1999) suggested that there may be some important differences in fixed and adaptive kernel estimation of the UD.

Seaman et al. (1999) reported that fixed kernel estimates were least biased in the outer contours, whereas adaptive kernels were least biased in the inner contours. These results have important ramifications for space use sharing estimates. Because space use sharing most commonly manifests in the outer home range boundary (Seidel 1992), fixed kernel estimates are most appropriate for measures of space use sharing using the V.I. index approach (Seidel 1992).

Dynamic Interaction Analyses

Whether the movements of one individual are influenced by those of another is important in movement (Doncaster 1990) and resource selection studies (Millspaugh et al. 1998a). The main advantage of dynamic interaction analyses is that the time series nature of locational data is maintained, which may portray a more accurate representation of species interactions. There are three general groups of dynamic interaction analyses. The first compares the distance of simultaneously taken observations of two individuals (Kenward et al. 1993). For example, within the program RANGES V (Kenward and Hodder 1996), dynamic interaction analyses are used to measure the relationship between n pairs of locations $(x_{1j}, y_{1j}, x_{2j}, y_{2j})$ for each animal to what could be expected randomly. The observed mean distance (D_o) between two animals is calculated as:

$$D_o = \frac{1}{n} \sum_{j=1}^{n} \sqrt{(x_{1j} - x_{2j})^2 + (y_{1j} - y_{2j})^2}. \qquad (5.11)$$

This value is then compared to an expected distance (D_E), calculated as:

$$D_E = \frac{1}{n^2} \sum_{j=1}^{n} \sum_{k=1}^{n} \sqrt{(x_{1j} - x_{2k})^2 + (y_{1j} - y_{2k})^2}, \qquad (5.12)$$

which considers all distances between all recorded observations. A sign test is used to test whether D_o is significantly different from D_E (Zar 1996).

A second dynamic interaction analysis compares the association between two animals based on how often they are observed together. Radio-collared animals are readily and often observed, such as during aerial telemetry flights, where the affiliation among specific animals can be determined. One technique described by Cole (1949), termed the coefficient of association, measures the relationship between two animals as:

$$Coefficient\ of\ association = \frac{2AB}{A + B}, \qquad (5.13)$$

where A is the number of times animal 1 is observed, B is the number of times animal 2 is observed, and AB is the number of times animals 1 and 2 are observed together during a specified time period. Bauman (1998) used this approach to quantify the level of association between radio-collared elk in Wind Cave National Park, South Dakota. In Bauman (1998), Knight (1970), and Varland et al. (1978), a coefficient of association > 0.5 indicated affiliation or fidelity, whereas a coefficient of association < 0.5 indicated little, or no, association or affiliation among two animals.

The third approach compares use of a shared area at the same time. Minta (1992) described a dynamic interaction method to assess spatial and temporal animal interactions summarized as binomial frequencies. As with other dynamic interaction analyses, it is assumed data are independent and observations are collected simultaneously or with a sufficiently short time lag that they are considered simultaneous. Also, it is assumed that there is a use area shared (e.g., home range area used) by two animals. Each time animal locations are sampled, each animal is present or absent from the shared home range area. The "use" frequencies of this shared area are compared to expected frequencies. The null hypothesis tested is that each animal uses its range independent of the other animal, spatially and temporally. In the case when home ranges overlap little, the subset of simultaneous locations that exist in the overlap area may be small. In this case, Minta (1992) recommended pseudo-Bayes' estimators to replace zero cells. Using the resulting spatial coefficients of interaction, the relationship between two individuals can be described as either random, attraction, or avoidance (Minta 1992). In addition, Minta's (1992) coefficient of temporal interaction can be computed to determine whether joint use of the overlap area was random, simultaneous, or solitary. One limitation with this technique is that only assessments of interaction between two animals can be made at a time (Minta 1992), even in the event where more than two animals' space use patterns overlap.

Other dynamic interaction techniques have been described, but they rarely meet the necessary assumptions. Dunn and Gipson (1977) and Dunn (1979) described a dynamic interaction analysis where the movement data of two animals were modeled using the multivariate Ornstein–Uhlenbeck process. However, it is assumed that animal movement distributions are bivariate normal. This assumption is unrealistic in most cases because few animals conform to such distributions (Minta 1992, White and Garrott 1990).

THE FUTURE: MODELING
THE MOVEMENT PROCESS

Technological limitations have been one hindrance to detailed movement-based modeling of radiotelemetry data, at least for many wide-ranging species. Advancing radiotelemetry technology (e.g., the use of GPS hardware/software) allows investigators to automatically collect nearly constant observations (see Chapter 4). Perhaps a larger limitation has been the overriding focus on quantitative techniques for home range estimation rather than rigorous analysis of movements in radiotelemetry studies. Several authors have criticized this emphasis (e.g., White and Garrott 1990, Gautestad and Mysterud 1993, 1995); however, such a focus has not been misplaced. Quantifying the general area used by an animal and the habitats in which it spends most of its time are vital steps toward understanding its ecology. Estimating size, shape, composition, and structure of home ranges will continue to be an appropriate technique for analyzing radiotelemetry data. Limitations arise when home range estimation does not examine meaningful hypothesis about factors underlying an animal's movements and behavior. As with any model, it is paramount to understand its purpose in terms of the biological information desired (Van Winkle 1975). Appropriate estimator(s) should be chosen based on consideration of the criteria presented earlier in this chapter. Understanding the full range of questions to be asked of any dataset must precede actual collection of data. This will ensure that data are complete for the most appropriate analyses, whether home range estimation or otherwise.

A large number of the techniques discussed in this chapter are either nonparametric or based on simple statistical measures (e.g., average distance moved) and descriptions (e.g., dispersal patterns). Generally these methods focus on the analysis of locations and ignore the connection between successive locations (other than the dynamic interaction techniques described earlier in this chapter). The preponderance of simple movement analyses cannot be attributed to a lack of quantitative methods for analyzing movement data. Numerous ecological field studies have used random walks/diffusion approximation models to analyze observed movements (Turchin 1998). Most of these studies have focused on insects or other small animals that could be tracked visually or with simple technologies (e.g., Kareiva and Shigesada 1983, Stapp and Van Horne 1997). In the next decade, radiotelemetry studies will probably focus more on quantifying movements with analysis techniques such as fractal analyses (see Chapter 6) and diffusion modeling. In particular, comparison and testing of mechanistic models will be an increasingly integral part of these field studies (Hilborn and Mangel 1997, Burnham and Anderson 1998, Turchin 1998).

Mechanistic models of the movement process offer the lure of much higher explanatory and predictive power than simple descriptive, statistical models (Moorcroft et al. 1999). Mechanistic models allow researchers to directly examine assumptions about the dominant biological and environmental factors affecting an animal's movements. Unlike nonparametric models, mechanistic models allow direct prediction of how movements will change when environmental factors change (Moorcroft et al. 1999). In addition, there are strong efforts to scale movement-level characteristics up to measures of habitat quality, population-level spatial distributions, and rates of population spread. This scaling is directly incorporated into diffusion approximation models (Turchin 1998).

We categorize movement-oriented quantitative approaches into three non-exclusive levels of increasing complexity. The first approach is descriptive rather than model-oriented. Movement parameters are estimated or described without reference to an underlying model. A second approach uses general movement models as a framework for analyzing field data. A third approach goes further by forming *a priori* mechanistic models based on specific biological attributes of the species under investigation. Model predictions are then compared to observed movement data. These approaches are not mutually exclusive. For example, the random walk/diffusion approximation models frequently used in the second approach may form the engine for the biologically driven models in the last approach (e.g., Moorcroft et al. 1999). Although not a focus of our brief discussion, all of these approaches can address temporal and spatial scales ranging from minute-by-minute travel paths to natal dispersal events.

Descriptive Approaches

Descriptive methods of quantifying animal movements are methodologically similar to many of the techniques discussed in this chapter. Some parameter of an individual's movement pattern is estimated and may become a raw data point in subsequent analyses. A large number of radiotelemetry studies have used straightforward measures such as daily movement distances and dispersal distances, among others. Quantities such as turning angles between successive movements and net squared displacement, which are an integral part of random walk models (Turchin 1998), also can be used as basic summaries of movements without any reference to formal models. Certainly in many situations, investigators may wish to directly test hypotheses about these parameters without any underlying movement model.

One movement parameter in particular, the fractal dimension (d), has received much attention in recent movement studies (Chapter 6, reviewed by Turchin 1996, 1998). The fractal dimension is a measure of the complexity or tortuosity of a movement path, ranging from 1 for a perfectly straight movement path to 2 for a continuous random walk (Brownian motion) that would eventually visit all points in a defined area (Turchin 1996). The utility of d is based on the assumption that it is a scale-invariant feature of a movement path (With 1994), which could allow fairly profound insights into how movement characteristics vary between species, sexes, and other attributes and how movements are related to underlying landscape structure. However, Turchin (1996) questions this assumption and demonstrated that d is often scale dependent. In both correlated random walk models and empirical data sets, d appears to range from 1 at very small scales to 2 at large scales for a given movement path (Turchin 1996). The shape of this scale-dependency curve may vary between species, obscuring the value of interspecific comparisons of d (Turchin 1996). Subsequently, some studies have justified use of d by focusing on a single species and a single spatio-temporal scale (Claussen et al. 1997, Ferguson et al. 1998).

Simple statistical summaries can stimulate insight into how environmental and habitat characteristics affect routine movements. These movement parameters can be used to scale up from movements to population distribution and other higher level processes. For example, Stapp and Van Horne (1997) used powder tracking to trace movement paths by deer mice (*Peromyscus maniculatus*). They tested whether properties of small-scale movements, specifically fractal dimension and a measure of trail directionality, could be used to predict the spatial distribution of individuals. While continuous monitoring of very fine scale movement paths may often not be feasible with radiotelemetry, it does allow investigators to estimate movement parameters at larger spatio-temporal scales. Ferguson et al. (1998) used satellite telemetry to examine seasonal movements by polar bears (*Ursus maritimus*) in the Canadian Arctic. Links between sea ice distribution and polar bear population structure were interpreted based on correlations between the fractal dimension of seasonal polar bear movement paths and the fractal dimension of sea ice. However, caveats of Turchin (1996) seem to apply; unless d is demonstrated to be scale invariant for a particular study, it is difficult to attach a firm biological meaning to correlations between movements and landscape fractal dimensions.

GENERAL MOVEMENT MODELS

Random walk models of various complexities provide the basic framework for modeling animal movements (e.g., Kareiva and Shigesada 1983, McCulloch and Cain 1989, Turchin 1991, Johnson et al. 1992). Diffusion approximations to discrete random walks establish the link between individual movements and population-level spatial distributions (Turchin 1991) or the spatial distribution of locations (Moorcroft et al. 1999). Turchin (1998) provided a thorough guide to such models. Here we discuss several rationales for greater use of these models in radiotelemetry studies.

First, while factors affecting movements may be seen as far more complex than we can hope to model, random walk models provide a useful starting point for examining the movement process. Evaluations of how well random walk models fit the data are based on easily calculated quantities (e.g., average squared move distance, turning angles). For example, Turchin (1998) demonstrated this approach using dispersal data for northern spotted owls (*Strix occidentalis*), concluding that dispersal movements occurred on both a local scale centered on temporary home ranges and on a larger scale. Movements at the larger scale met assumptions of a simple random walk. Similarly, foraging behavior of fallow deer (*Cervus dama*) was accurately modeled by a first-order biased random walk (Focardi et al. 1996). Such generic models are a logical benchmark for examining more complicated models that directly incorporate biological factors (Moorcroft et al. 1999).

A second rationale for using these generalized models is to better parameterize spatially explicit conservation models. For example, Boone and Hunter (1996) used a random walk model to simulate the effects of human land use activities on grizzly bear (*Ursus arctos*) dispersal. Their model incorporated patch permeability estimates for various habitats and ownership categories. Radiotelemetry studies can play a vital role in refining the movement rules in such models (Boone and Hunter 1996).

Finally, diffusion-approximation models directly scale from individual movements to population-level spatial distributions and rates of spread (Turchin 1998). While this scaling has not been the goal of most radiotelemetry studies, it is likely to receive greater attention as radiotelemetry techniques are applied to an ever-increasing diversity of organisms.

BIOLOGICAL MODELS

The most sophisticated modeling approach incorporates the behavioral biology of a species directly into the model structure. Moorcroft et al. (1999) fit a mechanistic home range model to coyote radiotelemetry locations. Their model built on a general diffusion-advection model in which movements have a random, diffusive component and a directional bias inward toward a pack home range center. Behavioral biology was integrated by assuming that directional bias resulted from encounters with foreign scent marks; differential equations describing scent mark accumulation and decay were linked to the pack movement equations. This model fit observed data for several neighboring packs significantly better than the simpler general diffusion-advection model, which did not parameterize the directional bias (Moorcroft et al. 1999).

The explanatory and predictive capabilities offered by detailed biological models come at the expense of computational ease and loss of generality. Even the mathematically sophisticated model of Moorcroft et al. (1999) incorporated only one major behavioral explanation for movements and a single center of activity for each pack. However, for many species our understanding has gone beyond simple descriptions of home range area, and even biologically simplistic mechanistic models can facilitate a deeper understanding of a species' movements.

SUMMARY

Throughout this chapter we have described a wide range of techniques used for analyzing animal space use and movements using radiotelemetry data, beginning with estimation of an animal's home range (including internal areas of more intense use or centers of activity) and progressing to more "advanced" techniques such as modeling movement processes. Throughout this chapter we have presented limitations to using many traditional techniques, ultimately leading to our discussion of modeling animal movements using a combination of advanced descriptive (e.g., fractal dimensions; see also Chapter 6) and visualization (see Chapter 7) approaches, general movement models (e.g., random walk models), and biological models.

Paramount to any radiotelemetry study is the due diligence of considering an appropriate sampling protocol (see Chapter 2). The importance of sample size and autocorrelation of data in estimating home ranges and animal movements cannot be overemphasized. When considering trade-offs between sample size and independence, obtaining an adequate sample size is more important than independence between data points (Andersen and Rongstad

1989, Reynolds and Laundré 1990, McNay et al. 1994, Swihart and Slade 1997, Otis and White 1999). We believe that preliminary data may be used to assign minimum sample sizes (S. Harris et al. 1990), paying attention to the warnings given previously.

Few investigators report sample size used to construct home range estimates (Seaman et al. 1999). In Seaman and colleagues' (1999) review of *The Journal of Wildlife Management* articles published from 1980 to 1997, they found that 21% of the home range manuscripts did not report sample size, 17% reported the total number of data points for the entire study, and 14% reported the minimum number of data points used. Altogether, only 49% of the published home range papers reasonably reported sample size used to construct their estimates. We recommend obtaining a representative sample of 30 or more locations, preferably 50 or more locations, when using home range estimators. These should be seen as minimum sample sizes rather than optima; estimation with fewer than 30 locations can suffer from both imprecision and strong bias.

Although there are critics of the continued use of home range estimators for analyzing radiotelemetry data (White and Garrott 1990, Gautestad and Mysterud 1993, 1995), we believe there is still a place for appropriate use of this technique. For example, analysis of nonuniform use patterns based on intensity of sampled locations described by the complete UD is a valuable technique to analyze space use over time. However, when reporting home range results, papers should include justification for the technique chosen (S. Harris et al. 1990). Our evaluation of twelve home range estimators highlighted several important criteria to consider when selecting and using current methods including sample size and autocorrelation of data, ability to calculate outer boundaries and core areas from complete UDs, and providing means to compare different estimators (Table 5.1). Adaptive and fixed kernel estimators outperformed other estimators evaluated and should be considered in most applications of animal space use and movement analyses where home range estimation is desired. Kernel estimators have much to offer as a general-purpose home range estimator. However, their overall performance can be greatly affected by choice of smoothing parameters and other options. As with any model, it is paramount to understand its purpose in terms of the biological information desired (Van Winkle 1975). Furthermore, an appropriate estimator(s) should be chosen based on consideration of the criteria presented earlier in this chapter (see Summary of Home Range Evaluation).

Any analysis of radiotelemetry data must consider the full range of biological and environmental factors (e.g., terrain effects) that may impact the results. Home range estimation and animal movement analyses must begin to answer difficult questions regarding the internal anatomy of home range boundaries and movement pathways through the judicious use of biological and environmental information. Furthermore, radiotelemetry studies should increasingly go beyond simple, general descriptions (i.e., home range size and shape) and use a model-driven approach to examine the key factors determining why and how an animal uses space. This is not to say that there is not room for analytical techniques that describe home range size and shape, simple animal movements (i.e., migration, dispersal, localized), or animal interactions, whether static or dynamic; rather the challenge becomes one of integrating these approaches into a more rigid study design to compare specific biological hypotheses and models regarding animal space use and movements.

CHAPTER 6

Fractal-Based Spatial Analysis of Radiotelemetry Data

Christian A. Hagen, Norm C. Kenkel, David J. Walker, Richard K. Baydack, and Clait E. Braun

The inherent complexity of ecosystems presents formidable challenges to biologists interested in describing, modeling, and managing animal populations (Milne 1997). Researchers now recognize that a multiscale approach is required to elucidate the spatio-temporal components of ecosystem complexity and to understand animal movement patterns in natural landscapes

(Ritchie 1998, With et al. 1999). Because ecological complexity varies with scale, observations at multiple scales and multiscale approaches to data analysis are required (Johnson 1980, Milne 1997).

Although animal movements are known to be spatially and temporally complex, few studies have examined the multiscale features of movement patterns. Analytical tools traditionally used by wildlife biologists, while useful in summarizing and modeling space use, generally fail to explicitly consider scaling issues (Gautestad and Mysterud 1993). Thus, new approaches are required to characterize the scaling properties of animal distributions. Fractal geometry is particularly suited to this task, as it explicitly takes a multiscale approach (Mandelbrot 1983, Milne 1997).

While fractal geometry has not been widely used to analyze radiotelemetry data, a number of recent studies have used fractals to characterize and model animal movement paths (e.g., With et al. 1999). Movement paths have traditionally been modeled using random walks (Kareiva and Shigesada 1983), with the aim of translating movement data into simple measures of displacement and habitat residency (Turchin 1998). Random walk models assume that movement positions are not spatially autocorrelated. A continuous path must therefore be represented as a simplified, discrete set of uncorrelated moves prior to analysis. An alternative view recognizes that the line segments connecting adjacent steps will continue to reveal a complex, erratic path as observation frequency increases (Gautestad and Mysterud 1993). This is the essence of the fractal approach. Instead of simplifying an inherently complex path as a discrete set of moves, one recognizes the underlying complexity and characterizes its scaling properties using the power law distribution (Viswanathan et al. 1996, Turcotte 1997).

Generally, it is difficult to obtain continuous movement-path data for a variety of reasons. Individuals may be logistically difficult to follow, or the act of following them may alter their behavioral pattern. In addition, movement paths are generally only available for short time intervals and may not include key movements, such as dispersal (Turchin 1998). An alternative approach is to obtain periodic fixes on animal positions using radiotelemetry. The result is a set of points that defines a utilization distribution (Worton 1989; see also Chapter 5). Radiotelemetry data are used in wildlife biology and natural resource management to determine home ranges (White and Garrott 1990), to summarize and model dispersal (Turchin 1998), or to quantify habitat selection (Manly et al. 1993). However, most of the currently available analytical approaches are not spatially explicit, and fewer still consider the multiscale features of animal movement patterns (Gautestad and Mysterud 1993, Viswanathan et al. 1996).

Data from radiotelemetry fixes often reveal that individuals have indistinct home range boundaries and show local variations in their intensity of space use (Worton 1989, Gautestad and Mysterud 1995). Standard techniques, such as home range analysis, cannot fully characterize these complex dispersion patterns (Loehle 1990). As an alternative strategy, Gautestad and Mysterud (1993) suggest a multiscale approach to the analysis of radiotelemetry data. They hypothesize that animal movements result from complex interactions between coarse and fine-grained responses, so that individuals relate to their environment in a multiscale, hierarchical manner. The result is a complex utilization distribution with fractal properties (i.e., scale-free and character-ized by clumps within clumps within clumps; Gautestad and Mysterud 1993).

In this chapter, we propose a method for the multiscale analysis of spatial radiotelemetry data based on fractal geometry and the generalized entropy. In addition, we outline a fractal-based dispersal model known as the Lévy flight. We illustrate the method using radiotelemetry data from a disjunct population of sage grouse (*Centrocercus urophasianus*) in northwest Colorado.

MULTISCALE ANALYSIS
OF RADIOTELEMETRY DATA

Radiotelemetry data are often used to obtain new insights into how a land-scape is used by an individual or population. A set of radiotelemetry fixes is a sample from an underlying spatio-temporal distribution and, thus represents an empirical estimate of the utilization distribution of an individual or popu-lation. Utilization distributions are typically underdispersed (contagious), indicating that some regions are used disproportionately relative to others (Turchin 1998; see also Chapter 5). The degree of contagion is thought to reflect both the scale-invariant spatial complexity of available habitat (Etzen-houser et al. 1998) and the intrinsic behavioral dynamics of animal move-ments (Gautestad and Mysterud 1993).

A number of statistical methods are available to test the null hypothesis of spatial randomness in dispersion data (Upton and Fingleton 1985). The simplest tests consider only nearest neighbors and, thus, assess pattern only at fine spatial scales. More rigorous second-order approaches produce a profile of how spatial pattern changes with scale (Ripley 1977, Kenkel 1993). These methods, while useful in evaluating deviations from spatial randomness, do not explicitly characterize important features of the utilization distribution, such as the degree of contagion. Alternative methods are therefore required to fully explore the scaling features of animal location data.

Fractal geometry (Mandelbrot 1983) provides the tools necessary to char-acterize and model multiscale contagion in radiotelemetry data. Spatial data are said to display fractal properties if the same underlying pattern of con-tagion is resolved on an ever-diminishing scale. The degree of scale-invariant contagion is quantified using a scaling parameter, D, known as the fractal dimension (Schroeder 1991, Kenkel and Walker 1996). A multiscale, fractal-based approach to the analysis of radiotelemetry data is useful in addressing the following questions:

ARE ANIMAL MOVEMENTS STATISTICALLY SELF-SIMILAR?

Self-similarity is defined as invariance against changes in scale. Scale invariance is an attribute of numerous natural phenomena and laws, and is the unifying concept underlying fractal geometry (Schroeder 1991, Kenkel and Walker 1996). As an example, consider a map of the spatial dispersion of radiotelemetry fixes. If a map scale is not indicated, it may be difficult if not impossible to determine whether the map covers 10 m or 10 km, particularly if the same general pattern manifests itself across scales. This intuitive notion of scale invariance provides the rationale for applying the fractal power law distribution to spatio-temporal radiotelemetry data. In practice, statistical self-similarity is demonstrated if the fractal dimension is found to be independent of scale (Turchin 1996, Milne 1997).

Self-similarity represents a fundamental departure from the scale-specific paradigm that pervades much of wildlife biology and theoretical ecology (Gautestad and Mysterud 1993, 1995). The demonstration of fractal scaling in animal movements therefore has far-reaching consequences for the analysis of radiotelemetry data and for modeling movement patterns in natural landscapes (Viswanathan et al. 1996).

IF MOVEMENT PATTERNS ARE SELF-SIMILAR, WHAT IS THE DEGREE OF SPATIAL CONTAGION?

Unlike other statistical distributions, the power law does not include a characteristic length scale and can therefore be applied to scale-invariant phenomena (Turcotte 1997). The fractal dimension D, derived from the power law distribution, is used to quantify the degree of scale invariance. A smaller D value indicates greater contagion (Schroeder 1991).

As a scale-invariant measure of spatial contagion, the fractal dimension provides valuable insight into how an individual or population uses the landscape (With et al. 1999). While home range analysis gives important information on the spatial *extent* of animal movements, it generally provides limited insight into the spatial *dispersion* of those movements. At one extreme, an individual may simply move through the area randomly, eventually visiting all regions of its home range ($D = 2$, which is implicitly assumed in most home range models; see Gautestad and Mysterud 1995). Alternatively, an individual may visit some regions of its home range more frequently than others, and some areas may not be visited at all. This results in a contagious pattern ($D < 2$), suggesting a preference for some areas and underutilization of others. Because individual locations indicate areas where conditions satisfy the requirements of a species (Milne 1997), the degree of spatial contagion may be indicative of underlying physical and biological processes. An individual with a movement pattern of $D = 2$ uses its home range markedly different from one with a movement pattern of $D = 1$.

WHAT FACTORS MIGHT CONTRIBUTE TO THE DEGREE OF CONTAGION OBSERVED?

The fractal dimension can be used as a comparative index of self-similarity to test specific hypotheses related to how individuals and populations perceive landscapes (Crist et al. 1992). For example, it is known that animal movement patterns are often affected by natural and human-induced habitat fragmentation (Storch 1997, Wiens 1997), but to what degree? One could compare the fractal scaling of animal movements in two populations, one occupying a fragmented landscape and the other an unfragmented one. Comparisons could also be made within the same population over time (e.g., comparing movement patterns in harsh vs. mild winters), between sexes or age-classes, or across species (e.g., Etzenhouser et al. 1998, With et al. 1999). Comparative approaches can provide valuable insights into the processes determining spatial contagion on the landscape. Standard statistical methods can be used to compare individual D values from two or more populations (Gautestad and Mysterud 1993). Alternatively, Monte Carlo simulation can be used to estimate confidence limits for measured D values (Loehle and Li 1996).

FRACTAL ANALYSIS OF SPATIAL PATTERN

BOX COUNTING

A number of approaches are available to explore fractal phenomena and estimate the fractal dimension (Frontier 1987, Schroeder 1991, Hastings and Sugihara 1993, Kenkel and Walker 1996, Milne 1997). Here, we consider estimation of the fractal dimension for a point pattern (utilization distribution) derived from radiotelemetry fixes. Box counting is the most commonly used approach for estimating D (Hastings and Sugihara 1993, Turcotte 1997).

Formally, box counting obtains a δ covering of a point pattern. A square box of side length δ is centered on each point, and a count is made of the number of boxes N_δ required to cover the pattern. This procedure is repeated at various values of δ. A simpler approach is to superimpose a grid of non-overlapping boxes over the pattern, and count how many boxes are occupied. The fractal dimension D is given by the limit:

$$D = - \lim_{\delta \to 0} (\log N_\delta / \log \delta), \tag{6.1}$$

where N_δ is the number of boxes of diameter δ containing at least one point. The limit $\delta \to 0$ is not defined for discrete point patterns; instead, one plots $\log N_\delta$ against $\log \delta$. The point pattern is fractal if a straight line is obtained over the range of δ values: the negative of the gradient of this line is the fractal dimension. The grid method is somewhat sensitive to grid location, particularly for small sample sizes ($n < 500$). Grid placement should therefore be randomly replicated to ensure stability of results (Appleby 1996, Milne 1997).

GENERALIZED ENTROPY

The fractal dimension described above considers only presence-absence of points in the boxes: the number of occupied boxes is counted but the number of points in a given occupied box is not considered. By considering the distribution of point counts within boxes, a set of q dimensions is defined that more fully characterizes the fractal pattern. The dimension described above is known as the cluster fractal dimension, which is defined at $q = 0$. More generally, counts of the number of points (n_i) in each of N_δ occupied grid boxes are obtained and expressed as proportions ($p_i = n_i / n$) of the sample size n. These are used to determine the generalized entropy (Renyi 1970) expressed as:

$$I_q(\delta) = 1/(1 - q) \ \log \sum_{i=1}^{N_\delta} p_i^q. \tag{6.2}$$

The generalized entropy defines a family of functions, each of which is referenced by the parameter q. From this, the generalized dimension D_q for the qth fractal moment is given by:

$$D_q = - \lim_{\delta \to 0} [I_q(\delta) / \log (\delta)]. \tag{6.3}$$

A plot of the generalized entropy $I_q(\delta)$ against $\log \delta$ is used to estimate D_q (Hentschel and Procaccia 1983). If a straight line is obtained over the range of δ, D_q is given by the negative of the gradient. Varying q generates a family of generalized dimensions (Table 6.1).

TABLE 6.1   Common Generalized Entropy Functions

q	$I_q(\delta)$	Dimension
0	$\log N_\delta$	Cluster
$\to 1$	$-\sum p_i \log p_i$	Information
2	$-\log \sum p_i^2$	Correlation

Measuring dimension as a function of q reveals the multifractal aspects of a pattern (Stanley and Meakin 1988, Scheuring and Riedi 1994). Consider the effect of varying q. At $q = 0$, each occupied box is weighted equally irrespective of the number of points it contains. For positive values of q, greater weight is given to boxes containing more points, while negative q values result in greater weight given to boxes containing few points. In analyzing statistical fractals, it is generally recommended that $0 \leq q \leq 3$ (Appleby 1996).

DILUTION EFFECT AND MONTE CARLO TEST

For fractal patterns of finite size, the log-log plot (e.g., $\log N_\delta$ vs. $\log \delta$, cluster fractal dimension) will deviate from linear at small δ values since N_δ necessarily approaches the sample size n as δ decreases (Gautestad and Mysterud 1993). In fact, the maximum possible number of occupied boxes is $N_\delta = n$ and is therefore independent of δ. In a slightly different context, Gautestad and Mysterud (1994) refer to this as the dilution effect. Resolving the dilution effect problem requires careful selection of a lower bound for δ that is appropriate to the resolution of the data (i.e., sample size). An objective procedure to determine this lower bound is to generate known fractal patterns (e.g., using the Lévy flight model) of sample size n and derive the δ value at which the log-log plot deviates from linear.

A finite sample will also underestimate the true fractal dimension. For example, a random spatial pattern has a theoretical fractal dimension $D = 2$, but this value is only achieved as n approaches infinity (Gautestad and Mysterud 1993). For an empirical random pattern of finite size, $D < 2$. The smaller the sample size n, the further the deviation from $D = 2$. An observed D value (obtained from an empirical point pattern of size n) must therefore be compared against values obtained from random patterns of the same size to determine whether the observed pattern deviates significantly from random (Hastings and Sugihara 1993:106).

An appropriate null model is that the observed point pattern is statistically random (i.e., all areas of the landscape are used equally). This hypothesis is readily tested using a Monte Carlo procedure (Manly 1997):

1. Generate a random point pattern of n points.
2. Compute D for this randomly generated pattern.
3. Repeat the above steps 100 times to generate a distribution of expected D values under the null hypothesis.

If the observed D value is less than the 100 random D values, the null hypothesis is rejected and one concludes that the observed pattern deviates significantly from random. This is a conservative test; confidence limits based on 100 simulations are considered adequate for tests at the conventional 5% level (see Kenkel 1993).

MODELING FRACTAL PATTERNS: LÉVY FLIGHTS

Animal movements over finite time intervals rarely achieve a strictly random pattern (Gautestad and Mysterud 1993). Animal positions are determined in part by previous locations, since travel distances are constrained by organism mobility. Movement patterns therefore have a memory component. The Lévy flight power law contains such a memory component making it well suited for modeling animal movements. The Lévy flight model can be used to simulate fractal patterns (Mandelbrot 1983), and time series animal movement data (Viswanathan et al. 1996). The model is a special case of the random walk, in which "step" lengths are selected randomly from a power law probability distribution. Conventional random walks, by contrast, assume that step lengths are constant. A fractal point pattern, termed Lévy dust, is obtained by plotting the set of landing points between Lévy flights. Sequentially connecting the points generates a path (Lévy flight random walk) of the same fractal dimension as the corresponding point pattern (Mandelbrot 1983). In this sense, there is a close relationship between the path and point pattern approaches to fractal analysis.

Although mathematically rigorous, Lévy dust cannot be simulated. However, an approximate realization on a two-dimensional torus is obtained from a set of n finite steps of a random walk (Ogata and Katsura 1991). Specifically, each flight is a random vector with polar coordinates $(R; \Theta)$. The azimuth Θ is selected uniformly and independently from the interval $[0, 2\pi]$, while the radius R is chosen independently according to conditional probability:

$$\Pr(R > r | R > r_0) = 1 \text{ if } r \leq r_0, \text{ otherwise } = (r_0/-r)^D, \qquad (6.4)$$

where r_0 is the minimum flight distance, and D is the fractal dimension of the simulated pattern.

This is implemented by generating R values according to the power law distribution:

$$R = r_0(1 - x)^{-1/D}, \tag{6.5}$$

where x is a random uniform value $(0,1)$. Provided that n is very large, the resulting point pattern has fractal dimension D and is self-similar over a range of scales from r_0 to the torus distance. Different degrees of self-similar contagion are readily simulated by varying the parameter D in Eq. (6.5).

EXAMPLE: SAGE GROUSE LOCATION DATA

To illustrate the fractal approach, we examined the utilization distribution of a sage grouse population in northwest Colorado. Sage grouse may travel large distances between winter and summer habitat, but movements of 1–5 km are typical within a season (Beck 1977, Connelly et al. 1988, Hagen 1999). Annual movement patterns are closely linked to the availability and spatial distribution of suitable habitat, as sage grouse generally select large, landscape-scale habitat patches (Patterson 1952, Oyler-McCance 1999). Within these large patches, habitat selection is a function of vegetation composition and quality (Beck 1977, Fischer et al. 1996) and topographic features (Hupp and Braun 1989). This hierarchy in habitat selection makes the species well suited for testing the fractal approach to radiotelemetry data analysis. Here we use the generalized entropy and Lévy flight modeling to test the hypothesis that sage grouse locations exhibit self-similar fractal properties. Specifically, we address the following questions:

1. Is the sage grouse utilization distribution statistically self-similar?
2. If the utilization distribution is self-similar, what is the degree of spatial contagion?
3. What factors contribute to the degree of spatial contagion?

Study Area

The sage grouse population studied occurs within a 1400 km^2 area of the Piceance Basin and Roan Plateau region, northwest Colorado. Suitable sage grouse habitat in this region is highly fragmented, the result of both natural and human-induced processes. The study area is a structural basin dissected by parallel undulating ridges, producing a highly fragmented and scale-invariant habitat landscape. Elevation ranges from 1800 to 2700 m, but sage grouse are generally restricted to middle and upper elevational regions where sagebrush (*Artemisia* spp.) is most abundant. Agricultural practices have altered or degraded sagebrush habitat at the periphery of the study area, and fossil fuel exploration and extraction have resulted in localized habitat degradation. Colonization by pinyon pine (*Pinus edulis*) and Utah juniper (*Juniperus utahensis*), the invasion by nonnative weeds, and domestic cattle grazing have also contributed to habitat fragmentation in the area.

POPULATION SAMPLE

Male and female grouse were night-trapped on or near lek sites during the breeding season, using spotlights and long-handled nets (Giesen et al. 1982). In total, 44 individuals were trapped at six of eight known active leks. Each bird was fitted with a 14-g lithium battery or a 20-g solar-powered radio ($< 3\%$ of body mass). To ensure that the sample was representative of the population, 25 individuals having inadequate relocation information (fewer than 15 radiotelemetry fixes) were excluded from consideration. The remaining 19 individuals had 27 ± 5 relocations and were representative of all six of the sampled leks. We feel that these individuals are representative of the population as a whole, particularly given that sage grouse are capable of traveling considerable distances (Fig. 6.1). Sage grouse locations were documented from April 1997 to December 1998. Attempts were made to relocate radio-marked individuals every week from June to August, and every 2 weeks otherwise.

DATA ANALYSIS

Self-Similarity and Fractal Dimension Estimation

We obtained fractal dimension estimates for the population using the box counting method. A 96×96 grid was placed over the 44×44 km study area and occupancy at box sizes 4, 6, 8, 12, and 16 determined. Smaller box sizes were not used, as Lévy flight fractal simulations at $n = 519$ revealed a strong dilution effect. The cluster ($q = 0$), information ($q = 1$), and correlation ($q = 2$) fractal dimensions were determined from these data.

Random Simulations

We used Monte Carlo simulation to obtain 100 realizations of a random spatial pattern of $n = 519$ points. For each realization, a 96×96 grid and box sizes of 4, 6, 8, 12, and 16 were used to estimate the cluster, information, and correlation dimensions. Upper and lower limits of the D estimates were

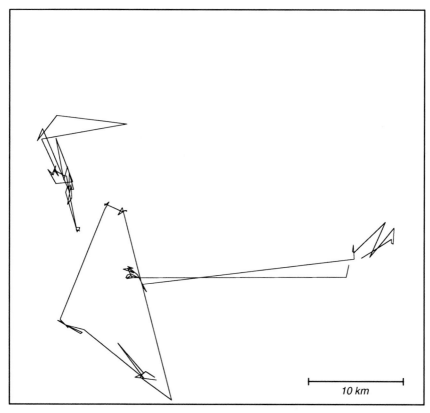

FIGURE 6.1 Movement paths for three individual sage grouse tracked between April 1997 and December 1998 in the Piceance Basin – Roan Plateau region of northwestern Colorado.

determined for the 100 random simulations and compared to the observed D values. In this test, the null hypothesis is that the observed pattern of sage grouse locations is statistically random.

We also statistically compared the observed pattern to a random Lévy dust pattern by simulating Lévy flights at $D = 2$. One hundred realizations of random Lévy dust at $n = 519$ points were obtained and analyzed as above. In this test, the null hypothesis is that sage grouse are randomly foraging on the landscape.

Lévy Flight Model

We tested the appropriateness of the Lévy flight model by comparing the empirical distribution of sage grouse movement distances to that expected from the Lévy flight power law conditional probability, Eq. (6.5). The model was developed at $n = 519$ and $D_0 = 1.06$ (i.e., the empirical fractal dimension of the sage grouse radiotelemetry data).

Results

The observed pattern of sage grouse locations ($n = 519$) displays evidence of strong spatial contagion (Fig. 6.2). Furthermore, the pattern appears scale invariant because the same overall degree of spatial contagion occurs across scales (Fig. 6.3). These empirical observations were confirmed by the fractal analysis. The log-log scatterplots for the q dimensions are linear (Fig. 6.4), and

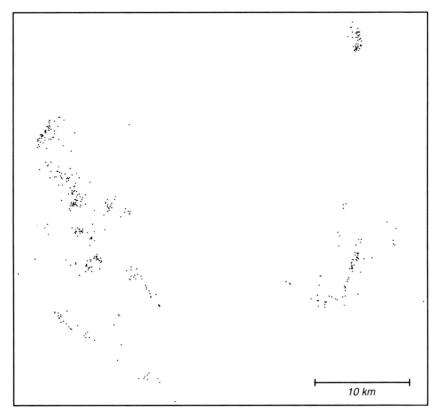

FIGURE 6.2 Spatial pattern of sage grouse relocations ($n = 519$) for all individuals tracked within the study area.

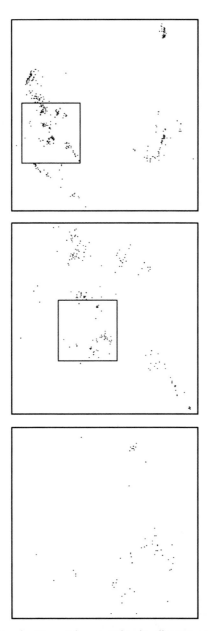

FIGURE 6.3 Sage grouse relocations at three spatial scales, illustrating the self-similar properties of the spatial point pattern. Note how successive magnification (top to bottom) resolves a similar pattern. The middle and lower panels are magnifications of the small box areas in the upper and middle panels, respectively.

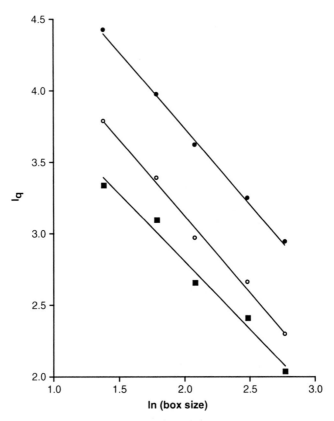

FIGURE 6.4 Power law relationships between I_q and box size (natural log scale) for $q = 0$ (cluster dimension, filled circle), $q = 1$ (information dimension, open circle), and $q = 2$ (filled square, correlation dimension). The fitted lines are principal components.

the Monte Carlo tests indicate significant deviations from both empirical random and random Lévy flight patterns (Table 6.2). For these random patterns, the finite sample size problem ($n = 519$) results in q dimensions that deviate strongly from the theoretical value of $D = 2$. However, the observed q dimensions are much lower still, indicating strong spatial contagion of grouse locations.

The empirical distribution of movement distances adheres closely to the Lévy flight power law model (Fig. 6.5). This confirms that movements are statistically self-similar at $D_0 = 1.06$ (i.e., the frequency distribution of sage grouse dispersal distances is log-log linear). For example, 195 movements were less than 0.5 km, but only 7 were between 5.0 and 5.5 km. Deviations

TABLE 6.2   Observed q-Dimensions (Cluster, Information, and Correlation) for the Sage Grouse Point Pattern ($n = 519$)

Data	$q = 0$	$q = 1$	$q = 2$
	D_q		
Observed	1.06	1.07	0.95
Random[a]	1.59–1.67	1.53–1.63	1.44–1.58
Lévy[a]	1.51–1.63	1.55–1.66	1.51–1.66

[a]Range of 100 simulations, $n = 519$.

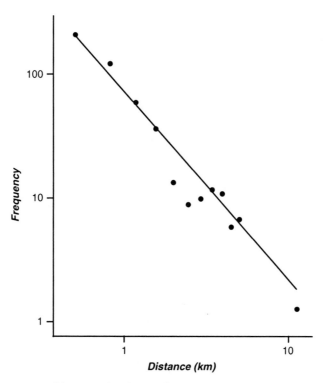

FIGURE 6.5   Empirical frequency distribution of sage grouse movement distances (filled circles) and the theoretical Lévy flight power law model at $D = 1.06$ (line).

from expectation at intermediate distances (3 to 4.5 km, Fig. 6.5) are likely attributable to the small sample size and the comparative rarity of longer distance dispersal events.

For illustrative purposes, we simulated Lévy flight point patterns for fractal dimensions $D_0 = 2.0, 1.5$, and 1.06 at $n = 519$ (Fig. 6.6). The simulation at $D_0 = 1.06$ results in a pattern that is statistically similar to the sage grouse location data (also $D_0 = 1.06$, Fig. 6.2). At $D = 1.5$ the pattern is somewhat less strongly contagious, whereas at $D = 2.0$ points are well dispersed.

Discussion

Self-Similarity

The utilization distribution of the sage grouse population is statistically self-similar, as demonstrated by both the linear log-log plot of spatial location data and adherence of dispersal distances to the Lévy flight model. A statistically self-similar pattern is consistent with hierarchical habitat selection, an idea that has allowed researchers to synthesize resource use studies into a single approach (Johnson 1980). Just as Johnson's (1980) selection order has unified resource selection studies, self-similarity has the potential to unify the study of movements through time and across individuals. Self-similarity may be an adaptive strategy for optimal foraging in habitats where resource availability is also scale invariant (Viswanathan et al. 1996) and may be of adaptive significance in avoiding predators (Bascompte and Vila 1997). While Johnson (1980) provides a quantitative approach to the problem of scaling and habitat selection, his approach is aspatial. Our results indicate that a full understanding of the hierarchical nature of habitat selection requires examination of the scaling properties of location data. The fractal power law distribution is appropriate here because, scale-invariant patterns cannot be properly characterized using a Poisson random walk model (Viswanathan et al. 1996).

Degree of Contagion

The sage grouse movement data are highly contagious across scales ($D_0 = 1.06$). Unlike standard home range estimators, the fractal dimension explicitly characterizes the scaling properties of a species' utilization distribution (Gautestad and Mysterud 1993). Utilization distributions are often characterized by voids (i.e., unvisited areas within the home range), but these are not fully characterized in standard two-dimensional home range analysis (see Chapter 5). Fractal analysis, by contrast, explicitly quantifies the spatial features of voids in the species home range. In this sense, home range and

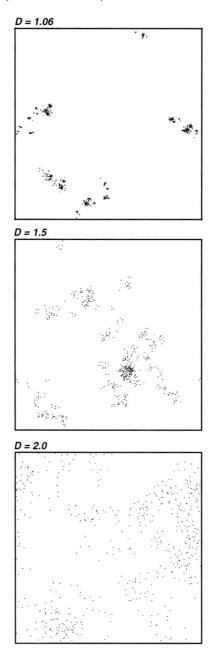

FIGURE 6.6 Example Lévy flight point patterns at $n = 519$, for three cluster fractal dimensions ($D = 1.06$, $D = 1.5$, and $D = 2.0$).

fractal analysis are complementary approaches to characterizing utilization distributions.

Factors Affecting Spatial Contagion

The observed contagion of sage grouse locations is best understood in terms of the spatial configuration of available habitat, landscape physiography, and distribution of leks. Sage grouse generally select relatively large patches of sagebrush at the landscape level (Patterson 1952, Oyler-McCance 1999), while movements within patches are determined by a suite of finer scale variables (Hupp and Braun 1989). While apparently suitable sage grouse habitat (sagebrush/deciduous shrub, sagebrush, and grassland communities) comprises about two-thirds of the Piceance Basin study area, only a very small proportion of this is consistently used by the population. An observational study in the North Park region of Colorado found that sage grouse use only 7% of the available winter habitat (Beck 1977). Landscape-scale habitat features appear to be important in explaining these findings. For example, the central region of our study area consists of long but very narrow (< 0.5 km) patches of sagebrush surrounded by agriculture and pinyon-juniper woodlands (Fig. 6.7). Such a spatial configuration is apparently unsuitable for sage grouse, as few birds used these patches and no birds were recorded traversing this area.

Self-similarity in the sage grouse utilization distribution indicates that movements within large sagebrush patches are also highly contagious. Preferred sage grouse habitat (sagebrush and grassland) is patchy across scales, and habitat use analyses (using the selectivity index; Manly et al. 1993) revealed strong habitat selection by sage grouse at all spatial scales (Hagen 1999). This result suggests that sage grouse movements are strongly affected by scale-invariant features of the habitat (Viswanathan et al. 1996, Etzenhouser et al. 1998, Ferguson et al. 1998). Physiography may also play a role, as sage grouse are most often found on the upper reaches of sagebrush-covered ridges of the study area (Hagen 1999). In addition, seasonal congregation at lek sites may contribute to coarse-grained spatial contagion (Bradbury et al. 1989). What emerges is a picture of a population subjected to numerous scale-invariant factors that affect movement patterns across all spatial scales.

FUTURE DIRECTIONS

Many ecologists have embraced fractal geometry as the rationale for spatial extrapolation and interpolation of natural phenomena (Hastings and Sugihara 1993, Milne 1997). While a strong case can be made for using the power law

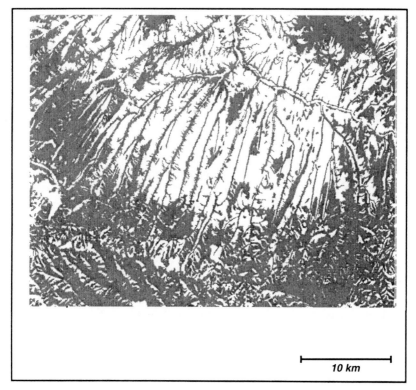

FIGURE 6.7   Spatial distribution of sagebrush habitat (gray) in the main portion of the study area.

distribution to quantify animal movements, does this application have a more fundamental basis? Certainly both the Gaussian and power law distributions have wide applicability in describing and modeling nature (Turcotte 1997). If spatio-temporal events are statistically independent, the central limit theorem provides the basis for application of the Gaussian distribution; examples include classic home range analyses and random walk models. Unfortunately, animal movement data must be simplified to achieve statistical independence (Swihart and Slade 1985a, Turchin 1998), suggesting that this basic assumption is untenable. By contrast, scale invariance of spatio-temporal events provides the rationale for applying the power law distribution (Gautestad and Mysterud 1993). In describing and modeling animal movements, we believe that scale invariance is a more reasonable assumption than statistical independence (Viswanathan et al. 1996).

Fractal geometry provides new and valuable insights into animal movement data that cannot be obtained from traditional home range analyses, first-order spatial statistics, and habitat selection estimation. The fractal dimension quantifies and characterizes the habitat available to a species at different scales, in both space and time (Milne 1997). In this respect, fractal analysis can be viewed as complementary to more traditional approaches. The Lévy flight model produces statistically self-similar clusters of used habitat and voids of unoccupied habitat (Viswanathan et al. 1996), a pattern that wildlife biologists may find intuitive and appealing. Scale-invariance of utilization distributions thus has far-reaching consequences for radiotelemetry data analysis and the modeling of animal movements.

A limitation of the fractal approach is that a large sample size is required to overcome the dilution effect. Point pattern analysis requires a minimum of about 500 radiotelemetry fixes (although thousands are needed to completely overcome the dilution problem). Similarly, movement path analysis requires finely detailed temporal data to properly characterize a path's fine-scale features. Insufficient sampling exacerbates the dilution effect and may lead to the possibly erroneous conclusion that a fractal structure is not statistically self-similar (cf. Panico and Sterling 1995, Turchin 1996). Recent advances in global positioning system tracking and microchip technology have made it possible to collect large amounts of spatially accurate and temporally continuous location data from marked individuals (Cohn 1999). Large data-sets will allow us to explore the self-similar scaling features of animal movements with more confidence.

Many ecological studies have used fractal geometry as a descriptive tool, but most have stopped short of prediction (Kenkel and Walker 1996, Milne 1997). A major challenge for the near future is to determine the processes underlying observed animal movements. Recent studies have hypothesized that movements are matched to the self-similar complexity of habitat and resource availability in space and time (e.g., Viswanathan et al. 1996, Etzenhouser et al. 1998), but the specific mechanism by which this occurs is unclear. A second challenge is to incorporate self-similarity and hierarchical scaling into wildlife biology models and management tools. There has been much theoretical progress in this area (e.g., Ritchie 1998, With et al. 1999, references therein), but such models require field-testing and validation. While it is difficult to predict with certainty the future of fractal analysis in wildlife biology, we are confident that self-similarity and the power law will have important roles in describing and modeling animal movements.

SUMMARY

Animal location data are commonly obtained by tracking radio-marked individuals. Recent technological advances in radiotelemetry offer the potential to collect large amounts of spatially accurate location data, allowing researchers to formulate new and fundamental questions regarding animal movement patterns and processes. We outline a fractal-based approach for analyzing spatial location data that uses Lévy flight modeling and generalized entropy. In this approach, each animal location is viewed as a sampled coordinate from an underlying spatio-temporal distribution. The set of radiotelemetry fixes constitutes a constellation of points known as a utilization distribution. Fractal analysis quantifies and characterizes the degree of spatial contagion of a utilization distribution and determines whether the distribution is statistically self-similar. As such, it offers valuable insight into how a population uses the landscape. The method is illustrated using radiotelemetry data ($n = 519$ locations) from a disjunct population of sage grouse in northwest Colorado.

ACKNOWLEDGMENTS

This research was supported by Colorado Federal Aid in Wildlife Restoration Project W-167-R through the Colorado Division of Wildlife, and by a Natural Sciences and Engineering Research Council of Canada research grant to N.C. Kenkel. We thank M. Commons, A. Lyon, S. Oyler-McCance, S. Peter, K. Petterson, K. Potter, E. Rock, and A. Welch for assistance in the field, and R. Pace and an anonymous reviewer for comments on a earlier draft of the manuscript.

CHAPTER 7

Estimating and Visualizing Movement Paths from Radio-Tracking Data

RICHARD M. PACE III

Insight into hierarchical scaling in ecology has stimulated interest in animal movement data to measure response to spatial scale (Turchin 1998). Animal movement patterns are thought to reflect both resource use and distribution patterns. Animals may forage in a relatively small geographic area (habitat patch); move among various resource patches (forage, cover, water, breeding sites) within a day, season, or year; and move seasonally or permanently to other groups of patches (dispersal). Testing hypotheses in this topic area requires that movement data be acquired and compared under different experimental conditions (e.g., two levels of environmental heterogeneity). Radio-tracking seems to be a reasonable means to acquire movement data at many scales and may be the only practical means to gather movement path data on some vertebrates. Unfortunately, radio-tracking data if gathered via triangulation or satellite tracking can be imprecise and inaccurate relative to the scale at which one wishes to study movement phenomena (White and Garrott 1990).

Sampling designs to acquire movement data via radio-tracking must consider both frequency and precision of location estimates. Sampling and analysis considerations for studying broad time and space movement patterns, such as annual home range and seasonal fidelity within a home range using radio-telemetry data, have been given substantial attention (see Chapter 5). However, daily or briefer movement patterns have received less attention. The purpose of this chapter is to describe some of the sampling considerations and analytical tools helpful in acquiring and evaluating narrowly scaled movement data.

SOURCES OF VARIATION

Location data useful in describing movement patterns during brief time intervals are necessarily spatially correlated: they represent sets of points (states) along a continuous course of travel, hereafter called a *movement path*. Movement characteristics of interest include velocity, degree of directionality (random vs. one-direction), and amount of turning or tortuosity (Turchin 1998). The degree to which movement characteristics vary among individuals or sampling events for the same individual should be considered process variation, and it is the analysis of that variation that may better our understanding of animal behavior. Confounded with process variation is sampling variation: that amount of uncertainty about the true path due to location frequency and precision of location estimates.

Use of radio-tracking to periodically locate an animal along its movement path renders a continuous behavior into a set of discrete points. The frequency with which point locations are recorded (sampling frequency) relative to the speed of the animal directly influences one's ability to reconstruct the actual path from these data (Fig. 7.1). Turchin (1998) discussed the problem of selecting a sampling frequency, mostly with reference to error-free location data. Basically, one must sample often enough to capture the salient features of a movement path while guarding against over sampling to a level that obscures the movement patterns of interest (e.g., between patch movements). Over-sampling can produce too much detail on small-scale movements of lesser interest. Small-scale movements can be thought of as process noise, and their effect on analysis is lessened by reduced sampling frequency. Clearly, noise is scale dependent. To an investigator studying between-patch movements, movements within a patch may be considered noise, whereas another investigator may have strong interests in the manner with which organisms move about in a patch.

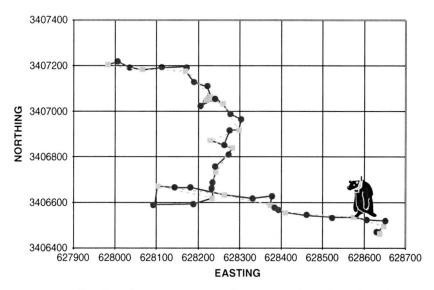

FIGURE 7.1 Effect of sampling rates on a perceived movement path. Circles mark true locations recorded every 1.25 min along a continuous movement path and show some effects due to discretization. Squares mark the true point locations recorded every 5 min along the path. Hence these two sampling rates produce two differing realizations of the same phenomenon.

The imprecision inherent in radio-tracking via triangulation or by satellite also contributes noise to movement path data. Even a radio-location acquired via homing followed by plotting on a map is subject to error due to map precision, plotting precision, and transcription (see Chapter 3). Uncertainty in the estimates of location produces point data that rarely lie on the true movement path. Therefore, it is inefficient to gather location estimates so frequently that the average distance between successive locations is likely less than our uncertainty about individual estimated locations. Often, the scale of interest for movements (tens to hundreds of meters) may be very close to the precision of typical location estimates. Process and sampling variation noise will be indistinguishable and combine to make measurements more variable. Our perceptions of an animal's travels along a movement path gathered via error-prone radio-tracking data will be biased high for distance and angular deviation (Table 7.1). Proximity to habitat features can also be distorted by location errors along the path, a characteristic shared with resource use studies that rely on radio-tracking data. Hence, researchers are cautioned that their choice of telemetry system elements as it pertains to system precision and accuracy may greatly affect the value of their data for understanding movement path-related behaviors.

TABLE 7.1 Comparison of Travel Path Characteristics Measured Against Point Locations Sampled at Two Frequencies (1/1.5 min and 1/5 min) and Two Levels of Location Error (Raw VMMLE[a] Location Estimates every 5 min and Three-Period Moving Window)

Method	Mean distance (m)[b]	Maximum distance (m)	Angular deviation	Angular concentration n
True locations				
Every 1.25 min	121	—	119.2	0.115
Every 5 min	103	287	103.6	0.195
Radio-tracking				
Raw VMMLE	149	527	126.4	0.878
Moving window VMMLE	93	278	118.9	0.116

[a]VMMLE, von Mises–based maximum likelihood estimates.
[b]Mean distance for path sampled every 1.25 min was multiplied by 4 to normalize it relative to others.

IMPROVING ACCURACY AND PRECISION

It is precisely because radio-tracked locations along a movement path are closely spaced in time and not independent that leads to methods of improving their accuracy and precision. Information found in adjacent location data or bearings used to estimate those locations can be incorporated in new, smoothed estimates. Anderson-Sprecher (1994) built robust estimates of wildlife location by using state-space time-series modeling that incorporated dependence among successive observations. His approach resulted in improved (closer to the true location) individual estimates along a path. However, at least two difficulties arise from using the Anderson-Sprecher (1994) estimation procedures. First, his iterated, extended Kalmann filter–smoothing estimators require a moderately complex set of computations not readily accessible to wildlife ecologists. Second, the state-space model incorporated in these estimators assumes a random walk by the animal, and a random walk means no periodically directional movement. I have frequently observed animals tracked during half-day sessions showing periodic directional behavior. Likewise, Turchin (1998) recognized that simple random walks were incomplete descriptors of most movement paths because they lacked a dominant compass direction.

Pace (2000a) proposed a simple moving window approach to smooth the noise about individual location estimates by pooling bearing data collected close in time. That approach also decreased individual location error and resulted in an improved path depiction (Figs. 7.2 and 7.3). In the following, I demonstrate the use of the moving window approach in two field trials and

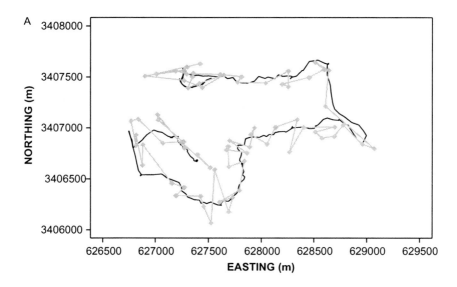

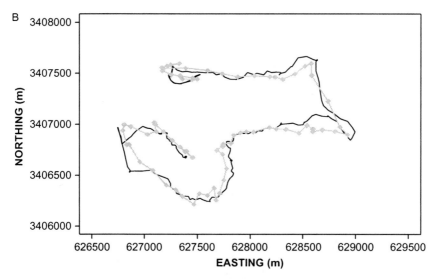

FIGURE 7.2   Predicted movement paths (diamonds) based on time-independent VMMLEs (A) or approximately 10-min moving window VMMLEs (B) overlaid onto true paths of a meandering biologist. Radio-tracking data were simulated based on two towers with bearing SD = 3°.

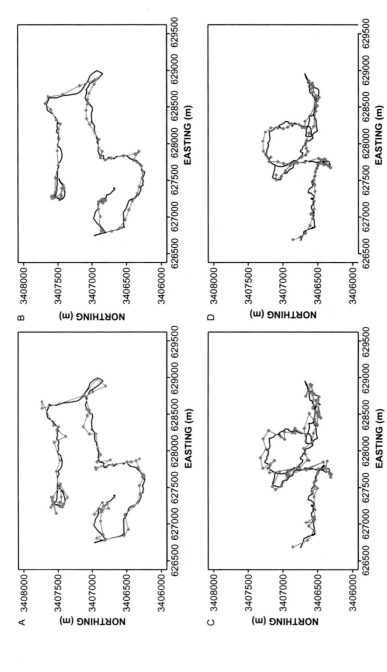

FIGURE 7.3 Predicted movement paths (diamonds) based on time-independent VMMLEs (A and C) or 10-min moving window VMMLEs (B and D) overlaid onto true paths of a meandering biologist. Radio-tracking data were simulated based on three towers with bearing SD = 3°.

compare performance against time-independent location estimates. Further, I use subsampling procedures to characterize the accuracy and precision of both individual location estimates and movement paths and place both in the context of a habitat map.

DEMONSTRATION

RADIO-TRACKING

A radio-tracking system established to collect movement path data might consist of two or more receiving stations (towers) of known location from which azimuths to the target animal are measured periodically. For example, in southern Louisiana my students and I have used two or three people with truck- or boat-mounted antenna stations (mobile stations) simultaneously measuring bearings every 5 min to describe the movement paths of coyotes (*Canis latrans*) and black bears (*Ursus americanus*). Using three people is advantageous because the time series of location data will not be interrupted (although location estimate precision is temporarily diminished) if one tracker moves his system to improve his reception or overall antennae configuration. The result of a tracking session is a multivariate time series of sets of azimuth-station pairs that must be converted to (X, Y) coordinates by use of an estimator such as von Mises-based maximum likelihood estimator (VMMLE) or Andrews estimator (Lenth 1981). GPS radio-collars can be programmed to collect location data frequently enough to describe movement paths. In this case, data already exist as a time series of error-prone (X, Y) coordinates. Factors limiting the number and quality of point locations within a tracking session differ greatly between GPS and traditional VHF systems (Table 7.2), but both types of data may benefit from time series smoothing approaches described below for tracking via triangulation.

Standard procedures for estimating movement paths from radio-tracking data would use VMMLE or Andrews (Lenth 1981) estimates to combine each set of simultaneously received azimuths to produce an estimated location for that time. I refer to these as time-independent estimates because the estimates are independent (ideally) of estimates calculated for adjacent times. With two or three bearings per location estimate, these estimated locations are generally of low precision and quite often skewed (White and Garrott 1990, Pace 2000b). One alternative is to use an iterated extended Kalmann filter–smoother to estimate movement path in a state-space analysis (Anderson-Sprecher and Ledolter 1991). For GPS-based systems, the time-independent estimates constitute the raw data from the collar.

TABLE 7.2 Comparison of Factors Limiting the Quality and Quantity of Point Locations Collected from Multitower Triangulation and GPS[a] Collar-Based Tracking Systems

Factor	Multitower triangulation	GPS collar-based
Quantity	Available continuous man-hours	On-board data storage
	Time required for 1 observer to record 1 bearing	Data transference capabilities (recovery download vs. remote download)
Quality	Tower directional precision	GPS collar design (number of satellites used, estimation algorithms, etc.)
	Number of towers and configuration (including maximum reception distance) Bearings per location	Standard vs. differentially corrected locations
	Experience of personnel	

[a]GPS, global positioning system.

MOVING WINDOWS

A simple modification of the above time-independent estimates is to use all bearings measured during a reasonably short interval of time (a window) in the VMMLE or Andrews procedures to produce each estimated location. The subsequent location along the movement path is estimated by moving the window one time step. That estimate should be considered a smoothed estimated location for the average time of the combined bearings. For tracking coyotes and bears, we used 10 min as the interval over which we combined bearings. Thus, if three observers each recorded one bearing to the target animal at 5-min intervals during a 6-h tracking session, say, 12:00, 12:05, ... , 18:00, then the total of nine bearings recorded at 12:15, 12:20, and 12:25 would be combined to estimate that animal's location at 12:20. Similarly, bearings recorded at 12:20, 12:25, and 12:30 would be pooled to estimate location at 12:25, and so on.

Moving windows could also be used to smooth GPS collar-based tracking data. Here (X, Y) pairs are averaged within suitably small time blocks to filter some of the noise found in these data. Note that GPS-based locations that are not differentially corrected contain a nonrandom error in location that will not be filtered by this procedure.

There are weaknesses of the moving window due to the repeated use of the same data for different location estimates. Clearly, using data (e.g., azimuths) recorded before and after the time of interest to estimate location of a moving animal injects an element of uncertainty into that estimate. This problem occurs whenever a location is estimated via triangulation using anything other than simultaneously recorded azimuths. With a moving window, the presence of unbalanced numbers of azimuths at each time within the window further confuses the time specificity of a particular estimate. Another issue is the lack of independence among *estimates of location*, which is a different issue than lack of independence among locations. If one uses a moving window that captures three time points, then data acquired at any one time will influence three location estimates. Hence, if one wanted to develop inferences relative to individual locations along the path, it might be desirable to use only locations estimated from independent data sets (in the three-time-point example one would only use every third location estimate). Also, any undetected, spurious bearing will potentially influence multiple location estimates. However, in the triangulation setting, the increased sample size affords the tracker more realistic opportunities to detect outliers directly or dilute their influence by using robust estimators.

MOVEMENT PATHS AND MOVING WINDOWS

To examine the performance of moving window estimation in describing movement paths, I used four simulated tracking sessions and one actual tracking session of two field-trial movement paths (Pace 2000a). Movement paths resulted from my attempt to mimic the movements of black bears as observed in my other studies (unpublished data). Over a 6-h period, I meandered through mostly forested bear habitats while wearing a radio-collar and traveling at approximately the same speed and total distance as a bear. My locations along each path were recorded and stored every 1.25 min using a military-grade GPS receiver ($\pm 3\,\text{m}$). The radio-tracking system consisted of three operators using two truck-mounted and one stationary null-peak antennae arrays. Operators were instructed to record bearings to a primary target (me) and a stationary beacon every 5 min. Operators used two-way radios to coordinate activities and mobile operators were free to move to improve their reception to the moving beacon (me) or overall tracking configuration. Repeated bearings on the stationary beacon were used to calibrate antennae and estimate bearing precision.

For both observed and simulated tracking sessions, I calculated VMMLE estimates using both single-time ($n = 2$ or 3 towers) and 10-min moving window ($n = 6$ or 9, 2, or 3 towers times 3 bearings each) bearing sets. I calculated distances between estimated locations and their concomitant true location in time along the movement path and plotted empirical distribution functions for both time-independent and moving window–based estimates.

RESULTS

Using a moving window through time to pool sets of azimuths was clearly superior to time-independent location estimation (Figs. 7.2–7.4). Median distances between estimated locations and their same-time-equivalent true locations were 79 m, 64 m, 48 m, and 42 m for window-based estimates compared to 108 m, 114 m, 58 m, and 55 m for time-independent estimates of movement paths for the path 1 with two towers, path 2 with two towers, path 1 with three towers, and path 2 with three towers, respectively. Pooling bearings through time acted to smooth perceived paths by removing some of the abundant "white noise" that often occurs at the same or larger scale than behavioral phenomena such as movement paths (Figs. 7.2 and 7.3). Smoothing had a profound effect on perceived movement path statistics calculated from radio-tracking location estimates. For example, the time-independent estimate for movement path 1 when tracking was accomplished with two towers had angular concentration of 0.088, angular deviation 126, linearity of 0.065, and mean distance between locations of 150 m. These compared unfavorably to the window-based estimator that had angular concentration of 0.11, angular deviation of 120, linearity of 0.096, and mean distance between locations of 93 m, which were quite close to the actual path statistics.

DISCUSSION

Sample sizes as well as instrumentation influence the accuracy and precision of location estimates. In the case of GPS-based tracking collars, the only control the biologist has is to select the best equipment. Sample size is directly under the control of the biologist. Typical protocol for radio-tracking via triangulation require sample sizes of two to four azimuths, even though large gains in precision have been demonstrated for increased sample sizes (White and Garrott 1986). Thus, much of the smoothing effect of a moving window estimator is in the reduction of sampling error variance.

Beyond showing how one might use a moving window to smooth data along a travel path, I hope that this demonstration makes telemetry users aware of the influence that sampling error can have on perceptions of biological process variability. Results from many of the newer analytical approaches for analyzing movement paths (Turchin 1998, Nams 1996) will provide little information about an animal's behavior unless the sampling error in the perceived path (white noise) is nearly an order of magnitude below the behavioral process variance. The effect of a high sampling error variance to process variance ratio would be to make a path observed during a short-duration tracking session, such as the one demonstrated, statistically

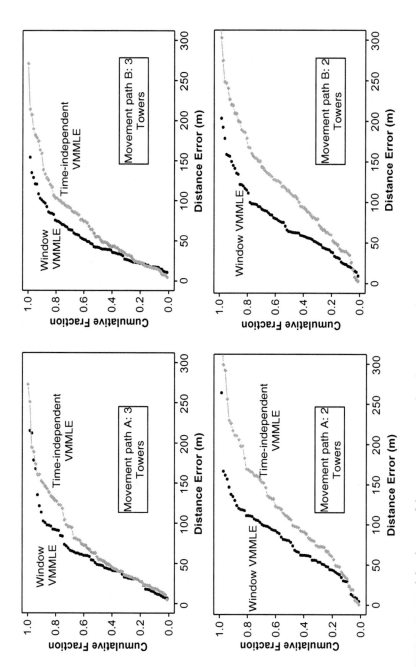

FIGURE 7.4 Empirical distributions of distance errors incurred in the estimation of two movement paths from simulated radio-tracking based on two or three towers with bearing SD = 3°.

indistinguishable from a correlated random walk regardless of whether or not directional movements occurred.

Too much smoothing can result from too long of a moving time window. The point at which this occurs will be a function of the speed of the subject animal and the precision of the tracking equipment in the particular tracking context. Users should note that for radio-tracking via triangulation, only very small gains in estimate precision occur for each added azimuth above 5 or 6 (White and Garrott 1990). As the time width of the moving window increases, estimated locations will be less representative of any single instant in time and will begin to smooth the movement process as well as the sampling process.

VISUALIZING PATHS

Desktop GIS systems are readily available and give investigators powerful tools for visualizing movement paths in the context of available habitat data, landscape features, and movement scale. The (X, Y) coordinates (points) that constitute the path are brought into the GIS system as a data layer. That layer can be either the point data or a polyline that connects the points in their proper time sequence. These data can be displayed overlain on a habitat map or viewed with other location data. They can also be analyzed using tools native to the GIS or additional tools created specifically for telemetry data such as the ARCVIEW extension ANIMAL MOVEMENT (Hooge and Eichen-laub 1997).

REALIZED PRECISION

A complicating factor in the use of radio-tracking to describe animal movement paths is the white noise added to movement data by telemetry error. Although using smoothing approaches can reduce sampling process error, it cannot remove it. Some attempt should be made to examine the scale of location error. One method is to simulate the radio-tracking scenarios used to collect locations along the path including the measurement error or uncertainty about each estimated location. An investigator can use this realized precision to judge the effect of potential errors in location on interpretations of measured paths.

The basic idea for simulating telemetry error comes from a modification of the subsampling approach for habitat selection under triangulation error (Samuel and Kenow 1992). That is, one takes the attributes of their telemetry system (number of towers, their locations and angular precision) and sub-samples around each estimated location making up the movement path set.

When radio-tracking via triangulation, the investigator should use the measured variance of observed bearing measurements to simulate the tracking process directly and not assume that VMMLE location estimates are bivariate normal (Pace 2000b). Using only the towers involved in producing the estimated location, calculate the azimuths from the towers to the estimated location. Then, create at least 100 new sets of azimuths (the same number and tower arrangement used to produce the original estimate) using the appropriate measurement-error model (Pace 1988). For example, below I assumed that the bearings from all towers were independent and normally distributed about a true bearing with SD $= 3°$, which seemed to fit our tracking data well. One may want to add some likelihood of outliers, vary the precision among towers, or incorporate any other factors that seem to mimic the actual error-generating process for a particular tracking scenario. Plotting subsamples provides a scaled view of realized precision for that tracking effort.

ERROR CONTEXT

There are several ways to visualize telemetry error in the context of scale and habitat features. The simplest approach is to plot all of the subsamples along with the estimated path (Fig. 7.5). This view quickly gives the investigator a feel for location precision. One should remember that the spread of subsamples would be centered on the *estimated* location, not the true location of the animal, which is unknown. Another visualization useful for viewing path error comes from randomly sampling the subsampled points and displaying them as though they were realized movement paths (Fig. 7.6). This display gives a better sense of the influence of realized telemetry precision on the whole path.

With fairly precise telemetry equipment, reasonable bearing sample sizes, and frequent locations, multipoint and multipath views tend to overemphasize extremes in error and underrepresent highly concentrated replicates. This is because subsampled points are very close together, converging to a single point as map scale increases. One way around this limitation is to estimate the two-dimensional distribution of subsamples about each estimate. Readily available tools for this, familiar to those who regularly analyze radio-tracking data, are various kernel estimates (Worton 1989). Computer programs for calculating kernel estimates are generally available, and some have been incorporated into extensions for GIS software like ARCVIEW. Contour plots of kernel density estimates are easily produced for display in a GIS as either contours (polygon data) or as density grids (Fig. 7.7). The culmination of this exercise is to pool all the kernels together in an *a posteriori* path probability

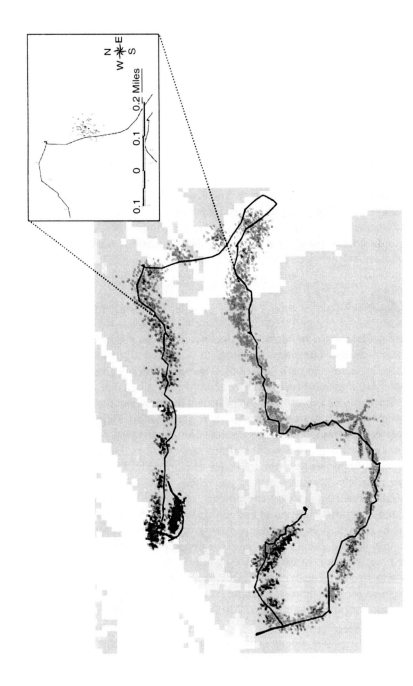

FIGURE 7.5 Graphical depiction of the scale of telemetry precision observed in a tracking session designed to emulate movement paths of bears in Louisiana. Subsampled points were plotted along with the true path and overlaid onto a habitat map used to assess habitat use.

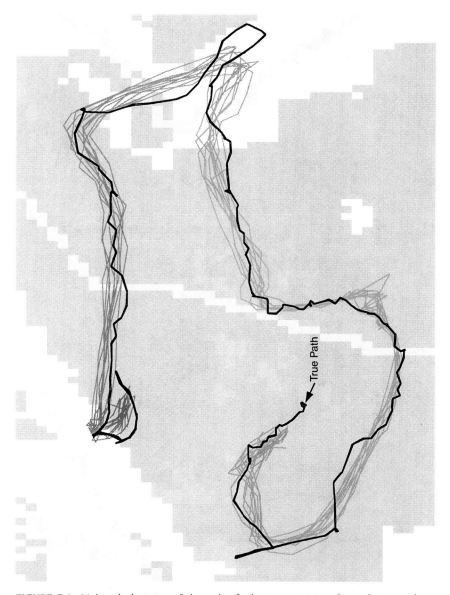

FIGURE 7.6 Multipath depiction of the scale of telemetry precision observed in a tracking session designed to emulate movement paths of bears in Louisiana. Paths were constructed from subsampled points about each estimated location and plotted along with the true path and overlaid onto a habitat map used to assess habitat use. Only 10 replicates were plotted here, but many more are needed to portray the extremes in variability.

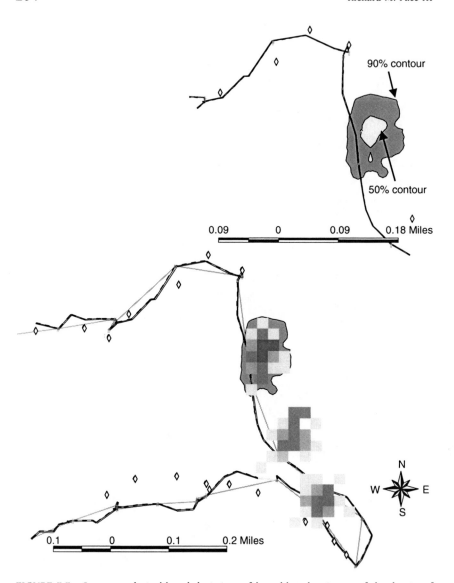

FIGURE 7.7 Contour and pixel-based depictions of kernel-based estimates of the density of subsampled points about each location and plotted along with the true path. The darker shaded squares represent proportionately higher concentrations of subsampled points. The lighter line is the true time-sampled path and diamonds are the window-based estimated locations.

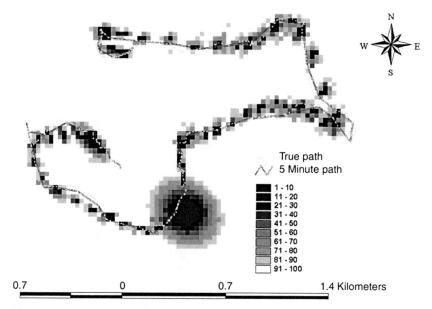

FIGURE 7.8 The set of pixel-based kernels from combined subsampled points about each location along a perceived movement path. The darker shaded squares represent proportionately higher concentrations of subsampled points. Also plotted are the true movements at 48 point locations per hour (True path) and 12 point locations per hour (5 Minute path).

plot (Fig. 7.8). Here all the density plots are overlain and the maximum grid probability value is retained.

Error context should not be interpreted as biological process. That is, the density plot described above is not the same as a utilization distribution. It is an attempt to incorporate our uncertainty about the actual movement path into any assessment of the biological process that we are trying to study. For example, our view of habitat use along the path will incorporate the uncertainty of location estimates by including the likelihood each point location estimate was actually in an adjacent habitat type. Such an approach is advantageous for assessing use of narrow linear features that would only rarely contain an error-prone point estimate.

FUTURE DIRECTIONS

Satellite-based (Argos-like and GPS) and ground-based automated radio-tracking systems will continue to improve both in estimated location precision and data acquisition capabilities (rates, storage volume and transmission reliability). This will greatly increase the opportunity and desire to gather more travel path data. Increased accuracy and precision will allow researchers to examine fine-scaled movement processes that are presently obscured by sampling process variation inherent in radio-tracking. Increased location acquisition rates will allow biologists to design movement path tracking sessions so that a portion of high acquisition rate capability that would probably oversample the movement process (Turchin 1998) could be used to further improve precision of individual location estimates.

CONCLUSIONS

Recording and analyzing wildlife movement paths is an important aspect of behavioral studies and may be helpful in conservation planning. Radio-tracking is the only practical method to gather movement path data on many vertebrates. With some exceptions, individual location estimates have low accuracy and precision and may poorly represent a movement path. Imprecision in movement paths should be considered when calculating movement path characteristics and interpreting movement path connections to habitat mosaics. By combining readily available methodologies and tools such as moving windows to pool data, VMLLEs, subsampling procedures, kernel estimates, and a GIS, investigators can improve accuracy and precision and gain insight of movement paths relative to other geographic features. For example, moving windows in time can be used to pool bearings collected during brief intervals and these pooled sets of bearings can be used to estimate locations along the travel path. The resulting increased sample size improves location accuracy and precision with the overall effect of more closely tracking the true paths. Subsampling procedures can be used to simulate data that reflect telemetry system precision. Multipoint, multipath, and density-based representations of simulated points can be displayed as map layers overlaid on a habitat map to give precision context.

ACKNOWLEDGMENTS

This work was sponsored by the Louisiana Cooperative Fish and Wildlife Research Unit, Louisiana Department of Wildlife and Fisheries, Louisiana State University Agricultural Center, USGS Biological Resources Division and Wildlife Management Institute. I am grateful for field assistance given by my students, D. Hightower, M. Giordano, and J. Pike.

PART V

Resource Selection

CHAPTER 8

Statistical Issues in Resource Selection Studies with Radio-Marked Animals

Wallace P. Erickson, Trent L. McDonald, Kenneth G. Gerow, Shay Howlin, and John W. Kern

Radio Tracking and Animal Populations
Copyright © 2001 by Academic Press. All rights of reproduction in any form reserved.

Radiotelemetry studies of animals are often designed to provide insights
into resource selection so that managers can obtain, protect, and restore
resources used by animals. For example, studies have been conducted to
determine the importance of old growth forests for northern spotted owls
(*Strix occidentalis caurina*) in the Northwest United States (Forsman et al.
1984), to study seasonal habitat selection by pine marten (*Martes americana*)
in a Maine forest where timber harvest has not been conducted for more than
35 years (Chapin et al. 1997), and to study the relationships between land
cover types and moose (*Alces alces*) habitat use on the Innoko National
Wildlife Refuge in west central Alaska (Erickson et al. 1998).

A common approach to the study of resource selection using radiotelemetry
data involves a comparison of resource use to resource availability. Resource
selection occurs when resources are used disproportionately to availability
(Johnson 1980). The purpose of this chapter is to provide a review of the
study designs, statistical issues, and analytical techniques for studying
resource selection and to provide practical guidance for biologists, resource
managers, and others conducting studies of resource selection via radiotelemetry. We focus on statistical issues of scale, techniques for defining
resource use and availability, pooling observations, independence of relocations, and variable and model selection and how these factors affect
inference in resource selection studies. We review resource selection analysis
procedures and provide examples of compositional (Aebischer et al. 1993),
log-linear (Heisey 1985), and logistic regression (Manly et al. 1993) analyses.
We believe the latter two methods are valuable due to their flexibility in
handling continuous and discrete covariates and for their utility in predicting
and mapping probabilities of use through the resource selection function
(Manly et al. 1993).

Other general reviews of methods and issues in resource selection include
those of Alldredge and Ratti (1986, 1992) who compared four statistical
methods of analysis using simulation techniques and data from 10 radio-
marked partridge (*Perdix perdix*), White and Garrott (1990) who provided a
chapter summarizing resource selection techniques using radiotelemetry data,
Thomas and Taylor (1990) who categorized experimental designs in resource
selection studies, and Manly et al. (1993) who focused on the use of logistic
regression and log-linear modeling to quantify resource selection. A volume of
Journal of Agricultural, Biological and Environmental Statistics devoted to

resource selection studies included a paper by Alldredge et al. (1998) that reviewed and compared designs and analyses for resource selection studies based on assumptions, type of data collected, and distributional requirements.

COMMON ASSUMPTIONS IN RESOURCE SELECTION WITH RADIO-MARKED ANIMALS

Alldredge et al. (1998), Manly et al. (1993) and others have outlined common assumptions in resource selection studies involving radio-marked animals. These are repeated below, and most are discussed in more detail later in this chapter.

1. Radio-marked animals are a random sample of the population. An attempt should be made to radio-mark individuals throughout the study area that can be treated as independent experimental units (see Chapter 2).

2. We assume telemetry locations are independent in time (see Chapter 5).

3. One radio-marked animal's resource use is independent of all other radio-marked animals. A common violation of this assumption is when animals are gregarious or territorial.

4. Resource availability does not change during the course of the study.

5. Used resources are classified correctly. Telemetry error may result in misclassification of true resource use. Also, we assume that locations are unbiased with respect to resource types. In radiotelemetry studies, forest cover and other factors may influence the incidence of failed locations, thus potentially biasing estimates of resource type use.

INFERENCE FROM RESOURCE SELECTION STUDIES

In the most basic sense, resource selection analyses are used to identify resources that are used disproportionately to availability. Inferences toward identifying optimum habitat, carrying capacity, understanding survival requirements, and other questions are not directly revealed by resource selection analyses. Resources identified as selected in one study area are not necessarily preferred. If less preferred resources are the only ones available, then these may be selected. Maps that predict high probability of use (e.g., Clark et al. 1993, Erickson et al. 1998) for certain areas do not necessarily define optimal habitat. Competition with other species, degradation of preferred resources, and other factors can influence resource selection and limit inference toward resource preference.

In most cases, the goal of a resource selection study is to make statistical inferences to a population of animals for which the radio-marked sample is assumed to represent. This is achieved by considering the radio-marked animal as the experimental unit to avoid pseudoreplication (Hurlbert 1984), thus reducing dependency problems when individual relocations are treated as the experimental unit (Aebischer et al. 1993, Otis and White 1999). When animals are used as the experimental unit, they are treated as a random effect in the analysis and inferences are then made to the larger population of animals. Some statistical procedures are specifically set up to consider the animal as the experimental unit (e.g., compositional analysis, Aebischer et al. 1993; Friedman's test, Friedman 1937; Johnson's PREFER method, Johnson 1980) while making inferences to the entire population (Table 8.1). For other analytical procedures, such as the χ^2 analysis (Neu et al. 1974, Byers et al. 1984), the relocation is considered the experimental unit. If data are pooled across animals, then animals are considered a fixed effect, and inference is limited to the set of radio-marked animals. If the animal is a fixed effect, there remain important independence assumptions to consider (see Independence Assumptions, this chapter).

Pooling observations from several radio-marked animals is often the only option when sample size is limited; but pooling has important consequences for inference. Biases in estimates of selection patterns and associated variances can exist when data are pooled across time, across animals, or across other factors that may affect selection (e.g., pooling across two separate study areas) (Schooley 1994). Arthur et al. (1996) conducted analyses that considered polar bears (Ursus maritimus) as a fixed effect, where data were pooled across individuals. Inferences were therefore limited to the set of bears under study. In this example, pooling was appropriate due to the small sample of radio-marked animals. When sample size is sufficient, taking into account both the number of animals and relocations per animal, and the desired objective is to infer average selection patterns for the entire population, then analyses that consider the animal as the unit of replication should be considered.

Most resource selection studies are observational, which directly affects inference. In observational studies, levels of the covariates (i.e., resources) of interest are not randomly assigned to the experimental units as is done in manipulative experiments (Montgomery 1984). Consequently, in observational studies apparent selection for a resource may be due to its correlation or association with unmeasured variables. With replication across varying conditions (e.g., different study area, variety of time periods), observational studies can be used to build a "preponderance of evidence" argument for selection of a particular resource. Variability in resource selection between study areas or animals alone may provide useful insight regarding the effects of

TABLE 8.1 Summary of Characteristics Associated with Common Statistical Analysis Methods of Resource Selection (Y = Yes, N = No)

Characteristics	Neu et al. method	Johnson's method	Friedman's test	Compositional analysis	Logistic regression	Log-linear modeling	Discrete choice
Applicable to design I studies, individuals not marked, and study area (i.e., population level) availability	Y	N	N	N	Y	Y	Y
Applicable to design II studies, individuals marked and study area availability	N[a]	Y	Y	Y	Y	Y	Y
Applicable to design III studies, individuals marked, availability defined for each animal	N[a]	Y	Y	Y	Y	Y	Y
Relies on assumption of temporal independence of relocations	Y	N	N	N	Y/N[b]	Y/N[b]	N
Assumes independence among animals	Y	Y	Y	Y	Y	Y	Y
Allows for categorical covariates (e.g., gender, subgroups)	N	N	N	Y	Y	Y	Y
Allows for continuous covariates (e.g., distance to roads)	N	N	N	N	Y	N	Y
Assumes the sample of animals is representative of the population, inferences are to average selection of the larger population from which the sample was obtained	N	Y	Y	Y	Y[c]	Y[c]	Y

[a] Neu et al. method could be applied by pooling data, but this is not recommended.
[b] Independence important if data are pooled across animals, not important when animal used as unit of replication.
[c] Inference is to the larger population if animal is the unit of replication (i.e., data are not pooled).

resource availability on selection (McClean et al. 1998). Experimental manipulation of resource availability would greatly enhance inference and provide more confirmatory results regarding preference (for definitions of use, selection, and preference, see Chapter 12).

Using geographic information system (GIS) technology, researchers have used the results of resource selection studies to predict the probability of use by animals (e.g., Clark et al. 1993, Erickson et al. 1998) and to estimate population abundance across landscapes (Boyce and McDonald 1999). Boyce and McDonald (1999) convert relative probabilities of use given resource characteristics of the pixel to density when the total number of animals in the population is known or estimated (see Example 4, below). This approach takes resource selection studies an extra step by exploring demographic parameters.

We recommend that researchers clearly state the statistical inference that can be made from their study. This involves reporting the time frame of the study (e.g., restricted to daytime selection only), scale of study, and the group of animals to which inferences are drawn (i.e., inferences to entire population or only the radio-marked sample). Inference in resource selection studies is often tied to the design of the study. Below we describe four study designs examining resource selection using radio-marked animals.

STUDY DESIGNS

Although the principles of good experimental design are available in the statistical and biological literature (Montgomery 1984, Manly 1992, Hurlbert 1984), these principles are often not considered in resource selection studies. Expenses associated with radiotelemetry equipment, logistics to gather information (e.g., aircraft), and data processing (e.g., GIS) may affect the study design by limiting the number of animals monitored or the number of locations per radio-marked animal. Furthermore, with the exception of Alldredge and Ratti (1986, 1992), guidance on sample size requirements for the number of animals and number of relocations from field studies has been missing until recently (see Chapters 2 and 11).

Manly et al. (1993) provided a unified approach to the design of resource selection studies. They expanded on the design categorizations provided by Thomas and Taylor (1990) who divided resource selection studies into three categories (designs I, II, and III) depending on whether resource use and availability data are collected at the level of the individual or the population. Below, we summarize the three study designs identified by Thomas and Taylor (1990) and add a fourth study design that considers availability for each individual point of use (Arthur et al. 1996; see also Chapter 9).

Design I

In this design, animals are not uniquely identified, and availability is defined at the population level (e.g., the study area). An example of this design would be the Neu et al. (1974) study of moose (*Alces alces*) resource selection in relation to edge and burn habitats. Moose and moose tracks were identified from the air across the entire study area and the habitat at each observation was recorded. Individual moose were not identified, and a single moose may have been counted more than once. The Neu et al. (1974) technique uses a standard χ^2 analysis to compare the expected use of habitats to the observed counts. Logistic regression (Manly et al. 1993) and log-linear modeling (Heisey 1985) could also be used in this design (see Example 2) as can discrete choice modeling (Cooper and Millspaugh 1999; see also Chapter 9) (Table 8.1).

This design and the analytical approach used by Neu et al. (1974) are not recommended in radiotelemetry studies (Otis and White 1999) due to violation of independence assumptions and pooling of observations across animals which may "average" out individual selection patterns.

Design II

In this design, resource use is identified for each individual, but resource availability is defined at the population level. This design is common in radiotelemetry studies where relocations of animals are used to describe resource use, and a sample or census (e.g., via GIS) is obtained to describe resource availability. For example, Roy and Dorrance (1985) compared habitat availability within coyote (*Canis latrans*) home ranges with habitat availability in the entire study area. In this design, inferences are made to the average selection pattern for the population of animals if each animal is considered the unit of replication (animal is considered a random effect). Analysis approaches for this design include Friedman's test (Friedman 1937), Johnson's method (1980), compositional analysis (Aebischer et al. 1993), discrete choice modeling (Cooper and Millspaugh 1999; see also Chapter 9), logistic regression (Manly et al. 1993), log-linear modeling (Heisey 1985), and multiple regression (see Chapter 12) (Table 8.1).

Inferences would be made only to the animals in the sample if relocations were pooled among the radio-marked animals (i.e., animal is considered a fixed effect). When pooling, temporal independence among observations would be required for this fixed effects analysis to be valid (see Chapter 5). Analytical procedures for this design when data are pooled across animals include the χ^2 analysis (Neu et al. 1974), logistic regression (Manly et al.

1993), and log-linear modeling (Heisey 1985). Due to temporal independence of relocation assumptions, we typically do not recommend this approach unless sample size (number of animals or number of locations per animal) is limited. Guidance on adequate sample size for the number of animals and the number of locations has been given by Alldredge and Ratti (1986, 1992) and Leban et al. (Chapter 11), but the statistical power for expected effect sizes (i.e., strength of selection) will vary by study. Additional analyses comparing how each individual selects resources in comparison with the entire sample of animals can also be made.

DESIGN III

Both resource use and resource availability are identified for each radio-marked animal. In a study of gray squirrel (*Sciurus carolinensis*) selection of discrete land cover types, Aebischer et al. (1993) identified habitat availability within minimum convex polygon (MCP) home ranges as determined for each radio-marked squirrel. Use by each individual was defined as the number of relocations in each of the land cover types. Analytical approaches for this design include Friedman's test (Friedman 1937), Johnson's method (1980), compositional analysis (Aebischer et al. 1993), discrete choice modeling (Cooper and Millspaugh 1999; see also Chapter 9), logistic regression (Manly et al. 1993), log-linear modeling (Heisey 1985), and multiple regression (see Chapter 12).

DESIGN IV

In design IV, resource use is defined separately for each individual and availability is uniquely defined for each point of use. In design III, availability is defined by one measure of availability (e.g., MCP as used by Aebischer et al. 1993). However, new analytical techniques, such as discrete choice modeling (Cooper and Millspaugh 1999; see also Chapter 9), allow researchers to quantify resource availability for each point of use. Arthur et al. (1996) defined availability separately for individual polar bear locations and used a maximum likelihood estimator to estimate the selection function. Availability for individual relocations was estimated by a buffer with a radius equal to the expected distance a polar bear was likely to travel between relocations. They pooled data to estimate an overall selection function due to small sample sizes, limiting their inference to the sample of animals while relying on the independence of relocations assumption. Cooper and Millspaugh (1999) used a similar approach for defining availability. Such definitions of resource availability are flexible, both spatially and temporally, for each relocation point. The above discussion on resource availability leads us to consider the question: what is an appropriate definition of resource availability?

SCALE AND RESOURCE AVAILABILITY

Due to the effects of scale on selection analyses (Porter and Church 1987), one of the most important decisions and often-debated issues in resource selection studies is how to define resource availability (see Chapter 9). Specifying resource availability forces researchers to explicitly define the spatial scale at which selection is studied (Levin 1992). Inference of selection patterns is limited to the scale designated by the researcher. Johnson (1980) provided a formal framework for scaling in resource selection studies from the geographic range of the species (first-order selection), to selection of home range within their geographic range (second-order selection), to selection of particular areas within their home range (third-order selection), to selection of particular food items (fourth-order selection). This framework explicitly incorporates large- and small-scale resource features that may influence resource selection. Johnson's scales are biologically based, reduce some of the arbitrariness in defining availability, and should be considered when conducting resource selection studies. However, Johnson's scales do not completely remove subjectivity; a researcher must still carefully consider how best to compute a home range estimate for example, which is not an easy task (see Table 5.1).

When evaluating what scales to consider in the analysis, researchers should consider that some species might select at a landscape scale only (Hansen et al. 1993), whereas other species (e.g., generalists) may select for more local vegetation features (Pearson 1993). In most cases, selection by a species may occur at more than one scale (Levin 1992, Pedlar et al. 1997). Advancements and availability of GIS make investigations of more than one scale more straightforward than in the past. Within a GIS framework, Marzluff et al. (1997b) investigated three scales of resource selection including home range selection within the study area (Johnson's second order), within home range selection (Johnson's third order), and forage site selection (similar to Johnson's fourth order) for golden eagles (*Aquila chrysaetos*) in the Snake River Birds of Prey National Conservation Area, Idaho. Forage site selection was investigated by comparing high-use core areas to characteristics of sites where eagles attempted to capture prey.

Researchers should also consider the resolution of their data when determining what scales to incorporate. In large/coarse-scale selection studies, it is common to use the boundaries of the study area as a definition of availability (e.g., Erickson et al. 1998). In studies involving LANDSAT TM and other remote sensing data, the boundaries of the "scenes" may be used to define the boundaries of available resources because the adjacent scenes may not be available, but this is extremely arbitrary. Use of study area boundaries to define availability has been shown to produce spurious results in simulations when habitats are aggregated (Porter and Church 1987). If study area boundaries are used to define availability, we recommend varying the boundaries to evaluate the sensitivity of the analysis to boundary definitions. More appropriately, if a researcher is interested in a study area definition of availability, the study area boundaries could be based on the distribution of radio-marked animals under study (e.g., McClean et al. 1998).

Problems also exist when home range estimates are used to define resource availability. The boundary of the home range estimate can vary depending on the technique and number of data points used to model the home range (Seaman et al. 1999; see also Chapters 2 and 5). Animals with few relocations (<50) typically have smaller estimated home ranges than animals with more relocations (>50) if minimum convex polygons are used (see Chapter 5). In contrast, if kernel density estimators are used, animals with fewer relocations (<30) may have larger estimated home ranges than animals with more relocations (>30) (Seaman et al. 1999; see also Chapter 5) because kernel-based ranges may be overestimated with less than 30 data points (Seaman et al. 1999).

When using a $(1 - \alpha)\%$ home range estimate (e.g., 95%) to define availability, one undesirable property is that home range estimates do not include all relocations. Some of the relocations describing use are outside the boundary of the $(1 - \alpha)\%$ confidence bound contour. As a result, some of the relocation points would not be included to characterize use. The implications of excluding excursive points should be evaluated further.

Other methods for defining availability have been used when availability is expected to change during the period of study, or when animals do not have an opportunity to travel across their entire home range between relocations. In a study of female polar bear habitat selection of ice cover types, Arthur et al. (1996) defined availability based on a buffer centered at the bear's previous location with radius equal to the maximum distance a bear was likely to travel between relocations. As described above, this approach explicitly defines availability for each point of use based on movements of the individual but require more complicated analytical methods than many of the standard analytical approaches (e.g., χ^2 analyses or logistic regression).

Results of any resource selection study need to be interpreted in terms of what was considered available. One approach to evaluate robustness of the results is to model selection at different levels of availability (Johnson 1980). Models and relationships that do not change when availability is varied would be most reliable. With GIS data, varying resource availability is straightforward, but there must be some biological relevance to resources considered available. McClean et al. (1998) evaluated the effect of different buffer sizes around locations for defining habitat availability of radio-marked Merriam's wild turkey (*Gallopavo merriami*) in the Black Hills, South Dakota. Their results demonstrated that selection may depend on buffer size. We recommend that researchers consider the objectives of their study and consider assessing selection at more than one level of availability. All levels, whether points around a buffer or a study area level of availability, must have a biological basis specific to the animal under study.

Although generally not as controversial, researchers also must consider how to define "use" in resource selection studies. Below we describe methods to define resource use.

RESOURCE USE

Habitat use in radiotelemetry studies is typically defined by the characteristics of the habitat covariates at relocations (i.e., coordinates, points). Often use is determined by importing relocations within a GIS coverage. In some applications (e.g., Arthur et al. 1996, Aebischer et al. 1993), the discrete characteristics of the point or pixel (land cover type, aspect) are used as categorical variables. Other continuous variables, such as proximity to features, also may be derived from relocations. In other applications, variables derived on a plot centered at the relocation may be used (Marzluff et al. 1997b). In plot-based sampling, the characteristics of the plot centered at the relocation are computed. For example, Erickson et al. (1998) used percentages of land cover classes within 200-m radius buffers of moose observations to define habitat use. Use may be defined at the point level for fine-scale analyses (e.g., forage site selection), or on an area basis at other scales (e.g., home range selection within a study area) (Marzluff et al. 1997b) in the same studies.

For raster coverages such as slope and land cover classes, the corresponding values of the pixel or raster are used to describe use for that relocation. For example, a record describing use by an animal may contain elevation, slope, and a land cover type. In some cases, an area (e.g., a polygon) may be used to describe use. In home range selection, the proportional occurrence of resources within the estimated home range boundaries may be used to characterize use (Aebischer et al. 1993, Kernohan et al. 1998). For vector and raster coverages such as roads, water bodies, and land cover types, the distance of the relocation to these features may be derived for each location within a GIS. Quantitative variables that describe landscape patterns have been developed (Turner and Gardner 1991, Li and Reynolds 1994) and should be considered as "used" when these patterns are expected to be important for an animal.

The appropriate size and shape of the area to consider for defining use depends on the scale of study, patch sizes, variable types, biology of the animal under study, and relocation error. If home range selection is the study object- ive, then the obvious choice of area definition is a home range estimator to define use and possibly availability. In variables that depend on the area surveyed, such as diversity, samples of average "home range" may be used to define availability. When buffers around relocations are considered, then we recommend using buffer sizes greater than average relocation error. We also recommend that researchers use different buffer sizes to test sensitivity. As technology improves the accuracy of location data (see Chapter 4), effects of relocation error will be decreased. Defining use on an area basis may be preferable because relocation error may affect location estimates; also, selec- tion may occur on a mosaic of habitats (Rettie and McLoughlin 1999).

TELEMETRY ERROR

Relocation points representing use are not measured without error. The effects of relocation error on the analysis of resource selection have been extensively discussed (Nams 1989, White and Garrott 1990, Samuel and Kenow 1992, Rettie and McLoughlin 1999; see also Chapters 2, 3, and 10). In modeling resource selection relationships, relocation error has the general effect of decreasing the power of the test because this measurement error adds additional variation into the analysis. Habitat characteristics, such as patch size, can affect the degree of power reduction because relocations in smaller patches would be expected to be misclassified as to their associated habitat type more often (Nams 1989). Relocation error can also produce biased estimates of selection (Samuel and Kenow 1992, Rempel and Rodgers 1997) because telemetry error may depend on the habitat and terrain.

Relocation error in telemetry locations is usually less for global positioning system (GPS) collars than for other telemetry location methods (Moen et al. 1996, 1997, Rempel et al. 1995) but still may bias resource selection analyses. For GPS collars, the frequency of failed location attempts increased and the precision decreased in hardwood canopies in comparison with other habitats (Moen et al. 1997), but see Chapter 4. If the frequency of failed location attempts could be estimated, then a researcher could adjust for this bias (Johnson et al. 1998). Bias adjustments may be possible in simple studies involving only a few discrete habitats but would require large numbers of data to map these frequency probabilities across a study area (Johnson et al. 1998).

VARIABLE AND MODEL DETERMINATION

In multivariate techniques such as logistic regression (Manly et al. 1993) and discrete choice modeling (Cooper and Millspaugh 1999; see also Chapter 9), a researcher must consider appropriate model fitting procedures and which covariates to include in the analysis. Model building is an art (McCullagh and Nelder 1989). The goal of building models should be to develop parsimonious, applicable models. We should not report a single model at the expense of excluding other reasonable fitting models. Instead, we recommend that more than one model be reported if other models fit the data well. Model adequacy should be checked through appropriate goodness-of-fit tests. In light of model selection limitations, professional judgment is important in establishing predictors of selection. The variables to consider should be limited to a reasonable set of possibilities, with decisions based on the knowledge of the animal under study. This can be done through literature review and information from experts in the field. Certain exploratory data analysis methods, such as univariate calculations and graphical displays of individual variables (selection ratios for categorical variables, t-tests comparing use and availability for continuous/ordinal variables), can help reduce the number of candidate variables (Hosmer and Lemeshow 1989).

Resource selection studies can be categorized as exploratory or confirmatory, with most studies being primarily exploratory. When the researcher defines a small set of *a priori* testable models, with a reasonably small set of variables, then analyses and results may be considered more confirmatory (Burnham and Anderson 1998). With a controlled experiment (e.g., feeding trials) where the treatments of interest can be randomly assigned to experimental units, this confirmatory inference is strengthened. When the number of possible covariates is large and collinearity among these covariates exists, then the analyses and results should be considered exploratory. The primary reason for this distinction is that the statistical tools for developing a "best" model only provide guidance for developing parsimonious models that fit the data. These methods have been shown to include nonsense variables unrelated to the response as components of the best model (Burnham and Anderson 1998). The opportunity for spurious results increases with additional covariates. Therefore, we recommend (1) that the candidate set of variables and models considered be defined *a priori*, (2) that the variables considered have a biological basis for inclusion, and (3) that data reduction techniques (Hosmer and Lemeshow 1989) be used prior to data analysis.

Often the set of covariates may still be large. It is common for researchers to derive many potential predictors of resource selection from GIS coverages. In many cases, more than 10 variables may be included as covariates; when only considering main effects, this yields a possible total of $1023 = 2^n - 1$ models.

In nearly all cases with this many possible models, this approach should be considered exploratory (Burnham and Anderson 1998) and the results should be followed by more confirmatory designed studies, based on *a priori* hypotheses.

A large number of model selection methods exist for deriving a "best" model(s), including stepwise procedures (Hosmer and Lemeshow 1989), best or all possible subsets methods (Furnival 1971), methods using Akaike's information criteria (AIC) (Burnham and Anderson 1998) and the Bayesian information criteria (BIC) (Box and Tiao 1992). A very promising approach to understanding the relative importance of competing models is model averaging (Burnham and Anderson 1998), which uses a weight for competing models based on a weighted average of the criteria used (e.g., AIC). This can be extended to determine the relative importance of a particular variable in the model set and to determine the presence of more than one reasonable model. The model averaging approach needs to be adapted to resource selection studies with radio-marked animals when average selection relationships are developed using the animal as the experimental unit.

Another approach for reducing the dimensionality of the data is to first perform principal component analysis and use the scores on the first few principal components as variables in subsequent analysis and modeling efforts (Pielou 1984). Principal component axes are independent, an important statistical consideration when the raw variables are highly correlated. Rotenberry and Wiens (1998) used this approach to define forage patch selection of shrubsteppe species by comparing the scores of random patches and selected sites. We believe that this approach is useful but interpretation of the principal components can often be complicated and subjective because the resulting principal components are now functions of several variables (Johnson 1998).

INDEPENDENCE ISSUES

INDEPENDENCE AMONG ANIMALS

As outlined above, the analytical procedures used in resource use/availability studies assume that animal locations are independent both temporally and spatially. Independence among animals is assumed in all procedures when inferences are made to the average selection of the population (e.g., compositional analysis and Friedman's test; Table 8.1). This assumption is met if the sample of radio-marked animals is truly a random sample of the population of animals under study. This is almost never the case, but it is typically assumed that the radio-marked animals adequately represent the population under study. If animals in the sample are not independent (e.g., mother/offspring), meaning that their use patterns are related, then results may be biased (Alldredge and Ratti 1986, 1992). This dependence is typically not tested and is usually ignored or assumed negligible. The effect of ignoring dependence among animals, if it exists, is to increase type I error rates relative to the nominal levels (i.e., standard errors of coefficients in models are biased low). A bias in point estimates could be expected as well, especially if the number of relocations varies greatly among animals.

If relocations for pairs of animals are sampled at similar times, dependence among pairs of animals can be evaluated by estimating the temporal cross-correlation (Diggle 1990) between the bivariate time series that describe each animal's locations or by estimating Moran's I (Moran 1950) in three spatio-temporal dimensions (x coordinate, y coordinate, and time). If high correlations exist between pairs of individuals, then one conservative approach would be to eliminate one of the pair of individuals exhibiting the spatial correlations (Millspaugh et al. 1998a). Alternative approaches, such as modeling this spatial autocorrelation and adjusting estimates and variances of selection, have been considered (Augustin et al. 1996, Hoetting et al. 2000). However, we are not aware of the use of these alternative approaches when individual animals are considered the unit of replication and inferences are to average selection for a larger population.

INDEPENDENCE OF RELOCATIONS OF THE SAME ANIMAL

As with animal movement analyses (see Chapter 5), many of the standard procedures used to test for resource selection (e.g., χ^2 tests, log-linear, logistic regression) assume temporal independence of relocations when the relocation is used as the experimental unit (Table 8.1). This assumption is violated if relocation data are collected too frequently (Swihart and Slade 1985a, 1986, 1997). For example, the dataset on ring-necked pheasant (*Phasianus colchicus*) selection from Aebischer et al. (1993) contained relocations gathered three times per day over a 10-day period. The individual relocations, especially on a given day, would most likely be correlated given the short duration of time between locations. When relocations are treated as the experimental unit (i.e., Neu et al. technique), the dependencies typically result in underestimating the true variance, which inflates the type I error rate. When individual animals are treated as the experimental unit as was done by Aebischer et al. (1993), the dependencies between relocations are not an issue because we are interested in the trajectory of space use by an animal. Only when relocation data are pooled across individuals or when inferences are made regarding an individual's selection pattern are the dependencies in relocations important. Even when relocations for each animal can be considered statistically independent, if analyses are conducted to make inference to a larger population of animals, using the points as the experimental unit and pooling data constitutes pseudoreplication (Hurlbert 1984, Otis and White 1999).

Approaches to reduce dependencies in relocations include increasing the time interval between successive observations or to subsample the data at sufficient time intervals to meet the independence assumption. Tests of independence of successive relocations of an animal have also been derived (Swihart and Slade 1985a, 1997) and could be applied to evaluate independence (see Chapter 5). Using logistic regression and relocations as the experimental unit, adjustments to the coefficients and standard errors can be made using autologistic models (Augustin et al. 1996, Hoetting et al. 2000).

ANALYZING RESOURCE USE RELATIVE TO AVAILABILITY

Now that we have discussed issues of use, availability, scale, and assumptions, we turn to specific analytical techniques commonly used in resource selection studies. Several techniques such as Friedman's test (Friedman 1937, Conover 1999), Johnson's PREFER method (Johnson 1980), and the chi-square analyses (Neu et al. 1974, Byers et al. 1984) have been extensively reviewed. However, these techniques are becoming less prevalent due to the increased use of other techniques such as compositional analysis (Aebischer et al. 1993), discrete choice modeling (Cooper and Millspaugh, Chapter 9) and generalized linear modeling approaches (e.g., log-linear modeling and logistic regression; Manly et al. 1993). This shift in emphasis is likely due to advancements in GIS and because former methods only consider discrete resource types (e.g., land cover type), while some of the latter methods consider multiple continuous and discrete covariates (e.g., logistic regression and discrete choice modeling). These multivariate techniques explicitly consider resource selection a multivariate process. Also, these recently applied techniques have become more readily available in the mainstream statistical computer packages (Appendix A). Many of the most common techniques to analyze resource selection data have been previously reviewed and summarized. Alldredge and Ratti (1986, 1992) compared the Neu et al. (1974) technique, Johnson's method (Johnson 1980), the Friedman (1937) test and the Quade (1979) method. These reviews list important assumptions and briefly describe the analytical procedures for each. White and Garrott

(1990) provide an in-depth description of the Neu et al. (1974), review the Marcum and Loftsgaarden (1980) technique, Johnson's (1980) approach, log-linear models (Heisey 1985), and the Friedman test (Friedman 1937).

Rather than repeat the cogent remarks of Alldredge and Ratti (1986, 1992), White and Garrott (1990), and others who have described, reviewed, and compared many of these techniques, our emphasis will be on more recently applied analytical techniques that have not been described in other reviews. In particular, our brief review will focus on compositional analysis (Aebischer et al. 1993), and to a lesser degree on mapping techniques such as the Mahalanobis distance (Clark et al. 1993), and polytomous logistic regression (North and Reynolds 1996). Cooper and Millspaugh review discrete choice modeling in Chapter 9.

COMPOSITIONAL ANALYSIS

Compositional analysis is an extension of multivariate analysis of variance (Aebischer et al. 1993). The technique uses categorical covariates, requires multivariate normality, and uses the animal as the experimental unit. Instead of relying on individual points to define use, resource use is defined as the proportion of resources within the estimated home range boundary (or other such trajectory of space use). Using the animal as the experimental unit circumvents problems related to sampling level (Kenward 1992), the unit-sum constraint (i.e., avoidance of one resource type invariably leads to selection of another) (Aebischer et al. 1993), and differential use by groups of animals (Aebischer et al. 1993). Despite these advantages, there are several assumptions underlying compositional-based analyses (Aitchison 1986). These include spatial independence among radio-marked animals (described above), compositions from different animals must be equally accurate, and multivariate normality. To account for unequal compositions, the log-ratios derived in the analysis should be weighted (Aebischer et al. 1993:1321). Failure to meet multivariate normality assumptions will influence the significance values, but randomization procedures can overcome this problem (Manly 1997). Another disadvantage includes the need to add an arbitrary constant to data if use for a particular resource type is 0 (Pendleton et al. 1998).

Compositional analysis can be conducted easily in standard statistical packages (Appendix A) using multivariate analysis of variance (MANOVA). With use being defined as the proportional occurrence within the home range, availability is also defined as proportional occurrence, but at a larger scale. Aebischer et al. (1993) recommend a two-stage analysis, corresponding to Johnson's (1980) second and third orders. Tests for overall selection are indicated by Wilk's lambda statistic (or other MANOVA overall test). If selection is indicated, contrasts comparing individual resource types can be conducted using t-tests or randomization tests.

We now describe and provide an example of compositional analysis; we adopt notation from Aebischer et al. (1993). Given D resource types, an individual's proportional resource use is described by $x_1, x_2, \ldots, x_n$ where x_i is the proportion of the individual's trajectory or home range in resource i. Note that the proportions sum to 1 (unit-sum constraint); a set of components summing to 1 is a composition (Aitchison 1986). It can be shown that for any component x_j of a composition, the log-ratio transformation $y_i = \ln(x_i/x_j)$ renders y_i linearly independent. Given this fact, the differences $d_i = \ln(x_{ui}/x_{uj}) - \ln(x_{ai}/x_{aj})$ are calculated for the ith animal (Table 8.2). For the test of overall selection, Wilk's lambda statistic (Λ) is calculated using MANOVA techniques. One method for generating an approximate P value is to compare $-N^* \ln (\Lambda)$ to a χ^2 distribution with $D - 1$ degrees of freedom (D is number of resource types). SAS provides an exact test and P value using the F distribution (SAS Institute 1999). If there were no selection, one would expect d to be distributed multivariate normal with mean 0. If the overall test indicates selection, then t-tests or randomization tests for comparing resource types can be conducted based on the means and variances of the d_is across animals. Note that compositional analysis is very similar to Johnson's ranking method (1980) which relies on the form $d_i^* = \text{rank}(x_{ui}) - \text{rank}(x_{ai}) - \text{rank}(x_{uj}) + \text{rank}(x_{aj})$. Using logarithms of proportions uses more information than the use of ranks. Note also that the equation for d_i can be written as $d_i = \ln(x_{ui}/x_{ai}) - \ln(x_{uj}/x_{aj}) = \ln(w_i) - \ln(w_j)$, the difference in logs of the selection ratios.

TABLE 8.2 Layout of Matrix Used to Establish Habitat Rankings in Compositional Analysis[a]

Habitat	Habitat (denominator)				Pos. values
(no.)	1	2	...	D	(total)
1		$\ln (x_{U1}/x_{U2}) - \ln(x_{A1}/x_{A2})$	...	$\ln (x_{U1}/x_{UD}) - \ln(x_{A1}/x_{AD})$	r_1
2	$\ln (x_{U2}/x_{U1}) - \ln(x_{A2}/x_{A1})$		...	$\ln (x_{U2}/x_{UD}) - \ln(x_{A2}/x_{AD})$	r_2
.	.	.	.	.	.
.	.	.	.	.	.
.	.	.	.	.	.
D	$\ln (x_{UD}/x_{U1}) - \ln(x_{D2}/x_{A1})$	$\ln(x_{UD}/x_{U2}) - \ln(x_{AD}/x_{A2})$	...	.	r_n

[a]The number of positive values in each row ranks the habitats in increasing order of relative use. Data from Aebischer et al. (1993).

TABLE 8.3   Hypothetical % Use and % Availability for Five Radio-Marked Individuals Analyzed in Example 1

	% Use				% Home range			
Animal	Habitat A	Habitat B	Habitat C	Habitat D	Habitat A	Habitat B	Habitat C	Habitat D
1	73	14	1	12	20	2	1	77
2	57	24	1	18	21	14	1	64
3	48	14	30	8	8	6	59	27
4	15	50	34	1	10	53	32	5
5	41	14	44	1	12	9	77	2

EXAMPLE 1: COMPOSITIONAL ANALYSIS OF A HYPOTHETICAL DATASET

We present an example of compositional analysis based on hypothetical data from five radio-marked animals and four resource types. Assume this is a design III study, with availability defined within home range estimates for each of the five individuals, and use determined by the proportion of relocations of each individual. Table 8.3 summarizes the hypothetical use and availability estimates for each individual.

Table 8.4 summarizes the values of $d_i = \ln(x_{ui}/x_{ai}) - \ln(x_{uj}/x_{aj})$ using resource D for the reference category. MANOVA is used to test for overall selection using SAS's PROC GLM (SAS Institute 1999). Wilk's lambda statistic is 0.017 ($P < 0.0257$), indicating significant selection. The ranking of the resources would be A, B, C, and D based on the average of the d's across the five animals, with the reference category (resource D) receiving a value of 0. t tests would be conducted to indicate which resources are significantly

TABLE 8.4   Differences in Log Ratios Calculated from Data in Table 8.3 Comparing Habitat Use within Home Ranges to Availability Defined by the Home Range

	Difference in log ratios (d)		
Animal	$A/D(d_1)$	$B/D(d_2)$	$C/D(d_3)$
1	3.154	3.805	1.859
2	2.267	1.808	1.269
3	3.008	2.064	0.540
4	2.015	1.551	1.670
5	1.922	1.135	0.134

different than others. To compare resources A, B, and C to resource D, a one-sample t-test (or randomization test) is conducted comparing the mean of the d_i's to the value 0. To compare resources A, B, and C, one-sample t-tests (paired) are used to compare the average of the differences $(d_i - d_j)$ for the ith and jth resource.

For example, the comparison of resource A and D is conducted by a t-test comparing the mean for d_1 to 0. The mean for d_1 is 2.473 with P value < 0.001, indicating significantly higher selection of resource A over D. The comparison of resources A and B is conducted comparing the mean of the differences between d_1 and d_2 across the five animals. This mean is 0.401 with a P value $= 0.139$, indicating no significant difference in selection between resources A and B. Table 8.5 summarizes the results of these comparisons.

MAHALANOBIS DISTANCE

Advancements and widespread availability of GIS platforms have greatly enhanced our ability to map and predict animal use patterns (Clark et al. 1993, Andries et al. 1994, Mladenoff et al. 1995, Knick and Rotenberry 1998; see also Chapter 12). Predicting species occurrence or use has the potential to be important to land managers by planning reserves or minimizing impacts from human activities.

The Mahalanobis distance statistic is a multivariate technique useful with GIS data for mapping resource-animal relationships (Clark et al. 1993). The technique assumes that (1) animals distribute themselves optimally throughout the study area and (2) resource selection can be adequately approximated by a multivariate mean and variance. The mean vector of used resources is

TABLE 8.5   Means, Standard Deviations (SD), and t-Test Results for Making Pairwise Comparisons of Habitat Types (One-Sample and Paired t-Tests with d.f. $= 4$)

Comparison	Relation to differences d_i	Mean	SD	P value	Interpretation
A vs. D	d_1	2.473	0.511	0.0001	Habitat A selected over habitat D
B vs. D	d_2	2.072	0.919	0.003	Habitat B selected over habitat D
C vs. D	d_3	1.094	0.660	0.010	Habitat C selected over habitat D
A vs. B	d_1-d_2	0.401	0.558	0.139	No difference
A vs. C	d_1-d_3	1.379	0.718	0.005	Habitat A selected over habitat C
B vs. C	d_2-d_3	0.978	0.725	0.021	Habitat B selected over habitat C

compared to random points throughout the study area and is represented by the Mahalanobis distance. Maps of probable use are created by computing the Mahalanobis distance in a GIS throughout the study area. The technique has been applied to female black bear (*Ursus americanus*) (Clark et al. 1993) and black-tailed jackrabbit (*Lepus californicus*) habitat use (Knick and Dyer 1997, Knick and Rotenberry 1998).

The technique has some important advantages and disadvantages. In contrast to logistic regression (Manly et al. 1993) and other methods for mapping resource selection using a GIS, only used resources need to be identified. This method and polytomous logistic regression (described below) are unique among selection analyses because the basis for the model uses only mean characteristics and variability of the use patterns. There is no need to compute resource availability, which helps circumvent an important and elusive aspect of some resource selection techniques. For this reason, the approach may be useful compared to logistic regression and other mapping techniques if defining availability is problematic. In addition, the Mahalanobis distance is not prone to problems of covariance among variables that might be problematic for multiple regression techniques (Knick and Rotenberry 1998). Despite these advantages, Knick and Rotenberry (1998) discussed limitations of the Mahalanobis distance technique and recommended its use when the various landscapes are well sampled (when sampling of use is distributed across the study area), and the landscape does not change during the period of study. Furthermore, although the technique does not require defining availability for estimating the mean vector, the mapping procedure requires defining a study area boundary, which may influence the results.

POLYTOMOUS LOGISTIC REGRESSION

When data consist of the number of relocations in a block of area and continuous covariates are measured on the blocks, then extensions to the methods described previously are necessary. Multiple regression (McCullagh and Nelder 1989), Poisson regression (Hastie and Pregibon 1992), or polytomous logistic regression (North and Reynolds 1996) can be applied in this situation. North and Reynolds (1996) recommend the use of polytomous logistic regression over binary models (presence/absence) because presence/absence information provides little information about the value of the site.

As with the Mahalanobis distance statistic, polytomous logistic regression avoids assumptions about no-use sites and only incorporates measures of use. Therefore, the technique does not rely on the assumption that used sites are suitable habitats and "unused" sites are unsuitable habitat (North and Reynolds 1996). Uncertainties in identifying species presence or absence (Johnson 1980) make such assumptions questionable, although reasonable in many radiotelemetry studies. Instead of comparing use to availability, comparisons of categorical use intensities are compared within the polytomous logistic regression framework. North and Reynolds (1996) blocked the study area or home range into equal sizes and categorized the use within the blocks as low, medium and high. Then using polytomous logistic regression, a generalized linear model (McCullagh and Nelder 1989), they related the ordinal dependent variable (amount of use) to the habitat-related covariates. Ordinal polytomous regression allows use of both continuous and categorical variables, as does logistic regression (Manly et al. 1993) and discrete choice modeling (Cooper and Millspaugh 1999; see also Chapter 9), and does not require a constant covariance structure (North and Reynolds 1996). Others have recommended using the utilization distribution to quantify the amount of use in blocks/grid and then using multiple regression to develop selection relationships (see Chapter 12). One consideration when blocks or grids are used is that the researcher must specify grid size and the spatial extent of the grids, which may influence results. We recommend researchers using this technique evaluate the robustness of the procedure by using different size grids and vary the spatial extent of grids if this extent is defined arbitrarily (e.g., study area boundary).

Below we provide examples of resource selection analyses using log-linear modeling and logistic regression primarily due to a lack of applied literature on these methods for resource selection in radiotelemetry studies. They are now more readily available in standard statistical packages and may prove useful because of the interpretation of the odd's ratios and associated confidence intervals. Here we provide an analysis of a moose data set using Neu et al. (1974), assuming the observations came from a single radio-marked moose in Minnesota. The next section considers an example that estimates an average selection function across multiple individuals.

EXAMPLE 2: MOOSE HABITAT USE IN MINNESOTA

The original Neu et al. (1974) study design would be categorized as design I because use is not identified for individual animals and availability is measured at the study area level. In our hypothetical reanalysis we assume that data are collected on one individual animal (a single radio-collared moose) and availability is measured within the moose's home range (design III). This analysis makes inference to the one individual and assumes independence of relocations.

Poisson Regression (Log-Linear Modeling)

We first analyzed selection using a log-linear model (Heisey 1985). Poisson regression can be used here because the covariates are categorical. We will see in the next section that logistic regression can also be used in this case. Poisson regression (McCullagh and Nelder 1989) relates the logarithm of mean count to a linear combination of other covariates. If Y_i is the ith value of a count variable, $\mu_i = E[Y_i]$, and $X_{i1} \cdots X_{ip}$ are known habitat variables. Poisson regression assumes: (1) Y_i is a Poisson random variable with mean parameter μ_i and (2) $\log(\mu_i) = \beta_0 + \beta_1 X_{i1} + \cdots + \beta_p X_{ip}$. If Y_i were assumed to be a normal random variable with mean μ_i and $\log(\mu_i) = \beta_0 + \beta_1 X_{i1} + \cdots + \beta_{pXip}$, we would be conducting regular (Normal theory) multiple regression in log scale. Nearly all concepts from regular regression theory have direct analogs in Poisson regression theory (McCullagh and Nelder 1989). The analogue of residual sum-of-squares in Poisson regression is called the deviance and is a function of the data and predicted values (like residual sum-of-squares). In practical terms, deviance measures how closely the linear model fits the data. If one linear model with explanatory variables $X_{i2} \ldots X_{ip}$ has deviance D_1 and another model with $X_{i1} \cdots X_{ip}$ has deviance D_2, the difference in deviance, $D_2 - D_1$, asymptotically follows a χ^2 distribution with one degree of freedom. Using the χ^2 distribution, the difference in deviance provides an approximate test of the significance of the missing term, X_1. A better test (i.e., better asymptotic properties), and one that accounts for more variation than expected under the Poisson assumption, is obtained by dividing the difference in deviance by an estimate of mean residual deviance (i.e., D_2/residual degrees of freedom or the Pearson χ^2/residual degrees of freedom) and compare the resulting statistic to an F distribution (McCullagh and Nelder 1989, Venables and Ripley 1994). For more details on Poisson regression and generalized linear models, refer to McCullagh and Nelder (1989).

To analyze the moose data provided by Neu et al. (1974) in Table 8.6 using Poisson regression, we set Y_i equal to the count of relocations of the moose on the ith habitat type (e.g., $Y_1 = 25$) and model the log of the expected value of Y_1 as a function of the explanatory variables appearing in Table 8.6, X_1, X_2, X_3, and X_4. To conduct the Poisson regression, it is necessary to assume that Y_i values are independent. Independence of Y_i is reasonable provided relocations were far enough apart in time.

If there were no selection, we would expect the number of relocations observed in each type of habitat to be proportional to the amount of habitat. Assuming B_i is the amount of habitat i, we incorporate this expectation into the analysis by fitting an intercept and offset term of the form $\log(\beta_i)$ in the Poisson regression model. The no-selection model is:

TABLE 8.6 The Response, Y_i, Offset, B_i, and Explanatory Variables Used in the Poisson Regression Analysis of Moose Data

Habitat	Sample count Y_i	Habitat proportion B_i	Burn interior X_{i1}	Burn edge X_{i2}	Outside edge X_{i3}	Edge X_{i4}
In burn, interior	25	0.340	1	0	0	0
In burn, edge	22	0.101	0	1	0	1
Out burn, edge	30	0.104	0	0	1	1
Out burn, interior	40	0.455	0	0	0	0
Total	117	1.000				

$$\mu_i = B_i e^{\beta_0} \Rightarrow \log(\mu_i) = \log(B_i) + \beta_0. \tag{8.1}$$

McCullagh and Nelder (1989) refer to $\log(B_i)$ as an offset whereas Manly et al. (1993) refer to $\log(B_i)$ as a base rate. When the no-selection model is fit to the data in Table 8.6, $\hat{\beta}_0 = 4.762$ with deviance 35.40 on 3 d.f. The predicted mean counts, $\hat{\mu}_i$, from the no-selection model are 39.78, 11.81, 12.17, and 53.24 for burn interior, burn edge, outside burn edge, and outside burn away from edge, respectively. These predicted counts are what we expect under no selection because they equal the proportion of acres in each habitat times the total number of relocations. For example,

$$39.8 = 0.340 \times 117 = B_1 \sum_{i=1}^{4} Y_i. \tag{8.2}$$

Full selection of habitat types is allowed by adding the indicator variables X_{i1}, X_{i2}, and X_{i3} to the no-selection model. The estimated full-selection model is

$$\mu_i = B_i e^{\beta_0 + \beta_1 X_{i1} + \beta_2 X_{i2} + \beta_3 X_{i3}} \Rightarrow \log(\mu_i) = \log(B_i) + 5.383 - 0.179 X_{i1}$$
$$+ 0.907 X_{i2} + 1.188 X_{i3}. \tag{8.3}$$

The standard errors given by the Poisson regression routine are

$$se(\hat{\beta}_0) = 0.1581, se(\hat{\beta}_1) = 0.2550, se(\hat{\beta}_2) = 0.2654, \text{and}$$
$$se(\hat{\beta}_3) = 0.2415. \tag{8.4}$$

The full-selection model contains four parameters and consequently fits the four observed counts perfectly. The predicted counts from the full-selection model are the observed counts: 25, 22, 30, and 40. When this happens, we call the model saturated because its deviance is 0 with no residual degrees of freedom. The difference in deviance between the no-selection model and the full-selection model is 35.40 with 3 d.f. ($P < 0.0001$, χ^2 test with 3 d.f.). Strong evidence of greater use than availability is present.

Assuming $\hat{\beta}_0^*$ is the estimated intercept from the no-selection model, the ratio of observed count to expected count for habitat type i is

$$\exp(\hat{\beta}_0 + \hat{\beta}_i)/\exp(\hat{\beta}_0^*) = \exp(\hat{\beta}_0 - \hat{\beta}_0^* + \hat{\beta}_i). \qquad (8.5)$$

For example, the ratio of observed count to expected count for the inside burn edge type is $\exp(4.476 - 4.762 + 0.907) = 1.86$. Manly et al. (1993) call these selection ratios.

If no selection occurs, the selection ratios are 1.0. The selection ratios for the moose data are 0.63, 1.86, 2.46, and 0.75 for burn interior, burn edge, outside burn edge, and outside burn away from edge, respectively. Manly et al. (1993) arrive at the estimated resource selection function (RSF) by scaling the selection ratios so that they add up to 1. For the moose data, the estimated RSF is 0.11, 0.33, 0.43, and 0.13 for burn interior, burn edge, outside burn edge, and outside burn away from edge, respectively. The point estimates indicate selection for the edge of the burn, both inside and outside.

The ratio of two selection ratios, or odd's ratio, represents the relative desirability among habitats. For example, given equal access, inside the burn near edge is $1.86/0.75 = \exp(\hat{\beta}_2) = \exp(0.907) = 2.48$ times more likely to be selected by this moose than outside the burn far away from the edge. Approximate confidence intervals on odd's ratios are easy to compute if one of the habitat types is coded with all 0's in the analysis (referred to as "dummy coding" in Hosmer and Lemeshow 1989). Levels coded with all 0's are sometimes called reference levels, and in this analysis the reference level is the category outside burn far from the edge. Reference levels are a statistical artifact that makes comparison of selection between levels of a variable convenient. Note that the equation for selection of the reference level is $\exp(\beta_0)$ because the reference level has been coded with all 0's. Assuming the other levels of the variable have been coded with 1's and that β_i is the coefficient for the dummy variable associated with level i, the odd's ratio for selecting level i relative to the reference level is $\exp(\beta_0 + \beta_i)/\exp(\beta_0) = \exp(\beta_i)$. An approximate confidence interval for the odd's ratio of habitat type i relative to the reference level is conveniently calculated as $\exp(\hat{\beta}_i) \pm 2se(\hat{\beta}_i)$. Odd's ratios and confidence intervals for comparing one nonreference level with another nonreference level are possible but are slightly more difficult because they involve more coefficients and covariance terms.

In the moose example, an approximate confidence interval for the odd's ratio of inside the burn near the edge compared with outside the burn far from the edge is 1.46 and 4.21 (without rounding errors). Thus, we are 95%

confident that inside the burn near the edge is between 1.46 and 4.21 times more desirable than outside the burn far from the edge. As an example involving nonreference levels, consider that near the edge outside the burn was $\exp(\hat{\beta}_3 - \hat{\beta}_2) = 1.32$ (95% confidence interval $= (0.76, 2.32)$) times more likely to be selected than near edge inside the burn, where the confidence interval was estimated as:

$$\exp[(\hat{\beta}_3 - \hat{\beta}_2) \pm 2\sqrt{\mathrm{var}(\hat{\beta}_3) + \mathrm{var}\hat{\beta}_2) - 2\mathrm{cov}(\hat{\beta}_3, \hat{\beta}_2)]}. \qquad (8.6)$$

The covariance term, $\mathrm{cov}(\hat{\beta}_3, \hat{\beta}_2)$, can be estimated from the variance covariance matrix reported by the Poisson regression routine and is equal to 0.025 for this example.

Strong evidence suggests selection for edge over other types. To test selection for edge specifically, we add X_{i4} to the no-selection model. The estimated edge selection model is $\log \mu_i = \log(B_i) + 4.4038 + 1.1322 X_{i4}$ with deviance 1.51 on 2 degrees of freedom. The reduction in deviance χ^2 is $35.40 - 1.51 = 33.89$ on 1 d.f. ($P < 0.0001$). Therefore, nearly all the selection in the full-selection model is for the edge of the burn. The predicted counts from this edge selection model are 27.80, 25.62, 26.38, and 37.20 for inside burn interior, inside burn edge, outside burn edge, and outside burn far away, respectively. Edge, both inside and outside the burn, was preferred approximately 3:1 over nonedge areas, i.e., $3.10 = \exp(1.1322) = (25.62/11.81)/(27.80/39.78) = (26.32/12.17)/(37.20/53.24)$.

Logistic Regression

In this section we analyze the same data using logistic regression. We present this analysis to illustrate the equivalence of the logistic regression approach to the Poisson regression approach when all variables are categorical. To facilitate the logistic approach, we assume that the habitat types on the Little Sioux burn study area are computerized in a GIS and that a random or systematic sample of 200,000 points was placed across the area. In reality, the sampling of Little Sioux burn would not be necessary because the habitat proportions are known; however, we wish to illustrate the justification for logistic regression and its relationship to the Poisson analysis in this case, and to illustrate the approach when continuous covariates are considered. To that end, we assume that 68,025 of 200,000 random points fell in the burn interior, 20,222 in burn near edge habitat, 20,830 in outside burn near edge habitat, and 91,040 in outside burn far from edge habitat. In practice, we would not obtain these numbers exactly. These numbers represent the expected long-run average of the number of points in each habitat class assuming the sampling was repeated a large number of times. Consider the data in Table 8.7. The "Sample count" column is the number of relocations of the moose in each habitat type. The "used + available" column is the number of relocations in each habitat type plus the expected number of random points in each type. To conduct the logistic regression analysis, we assume the numbers in the used column are the number of Bernoulli "successes" and the numbers in the "used + available" column are the number of Bernoulli "trials."

The no-selection model fits an intercept in the logistic model. The no-selection model is:

$$\log\left(\frac{\tau_i}{1 - \tau_i}\right) = \beta_0, \tag{8.7}$$

where τ_i is the probability of success on each of the Bernoulli "trials" in habitat type i. Using data from Table 8.6, the estimated parameter from the no-selection model is $\hat{\beta}_0 = -7.4439$ and yields predicted counts of 39.77, 11.82, 12.18, and 53.23. Each predicted count equals $\hat{p}_i n_i$ where $\hat{p}_i = 1/[1 + \exp(-7.4439)]$ and n_i is the used + available count. The binomial deviance from the no-selection model is 35.372 on 3 d.f., as compared to 35.398 from the Poisson no-selection model. The difference in these deviances is negligible. If the size of the available sample were infinite, the deviance from the logistic model would equal the deviance of the Poisson model.

The full-selection model is fit by adding X_{i1}, X_{i2}, and X_{i3} to the no-selection model. The estimated full-selection model is

$$\log\left(\frac{\tau}{1 - \tau}\right) = -7.730 - 0.179X_{i1} + 0.907X_{i2} + 1.188X_{i3}. \tag{8.8}$$

Again, the full-selection model is saturated (i.e., deviance $= 0$) and the likelihood ratio χ^2 is 35.372 on 3 d.f. ($P < 0.0001$).

TABLE 8.7   The Number of "Successes" (Used), Number of "Trials" (Used + Available), and Explanatory Variables Used in the Logistic Regression-Analysis of Moose Data[a]

Habitat	Sample count Y_1	Used + available	Burn interior X_{i1}	Burn edge X_{i2}	Outside edge X_{i3}	Outside interior X_{i4}
In burn, interior	25	68,025	1	0	0	0
In burn, edge	22	20,222	0	1	0	0
Out burn, edge	30	20,830	0	0	1	0
Out burn, interior	40	91,040	0	0	0	1
Total	117	200,117				

[a]A random sample of 200,000 available points is assumed.

The estimated RSF is computed from the logistic regression model by omitting the intercept and scaling the numerator of the logistic function (Manly et al. 1993:126). Omitting the intercept, the numerator of the logistic function is $\exp(-0.1787) = 0.836$ for burn interior, $\exp(0.9073) = 2.477$ for burn edge, $\exp(1.1882) = 3.881$ for outside burn edge, and $\exp(0) = 1$ for outside burn far away from edge. Dividing each of these numbers by their collective sum results in the same estimated RSF obtained from Poisson regression, $0.11 \, (= 0.836/7.595)$, 0.33, 0.43, and 0.13, respectively, for each habitat.

Standard errors of the logistic regression coefficients can be read directly from the logistic regression output. For these data,

$$se(\hat{\beta}_o) = 0.1581, se(\hat{\beta}_1) = 0.2549, se(\hat{\beta}_2) = 0.2655, \text{and } se(\hat{\beta}_3) = 0.2416. \quad (8.9)$$

These standard errors differ by at most 0.0001 from the standard errors obtained using Poisson regression. Selection ratios, odd's ratios, and confidence intervals are computed the same way as in the Poisson regression analysis. If the intercept is omitted when computing selection ratios, the standard error of the reference level is the standard error of the omitted intercept. In these data, the reference level is outside burn far from edge and the selection ratio is $\exp(0) = 1$ with approximate 95% confidence interval equal to

$$\exp(0) \pm 2se(\hat{\beta}_0) = (0.73, 1.37). \quad (8.10)$$

EXAMPLE 3: LOGISTIC REGRESSION AND INFERENCES TO A POPULATION

Both the log-linear and logistic regression model approaches for the moose example above assume independence of relocations and, because of our formulation design, make inference to a single animal. To make statistical inference to a population, coefficients and odds ratios can be averaged across the set of sample animals as is recommended for selection ratios (Manly et al. 1993) and in profile analysis (Morrison 1976). We recommend this approach, which effectively uses the animal as the experimental unit. Variability would then be estimated across the n coefficients where n is the number of animals. Define βij as the logistic regression coefficient (see Example 1) for the ith animal and jth variable and $\beta \cdot j$ as the average coefficient for the jth variable:

$$\beta_{\cdot j} = \sum_{i=1}^{n} \beta_{ij}. \quad (8.11)$$

The standard error for this estimate is:

$$
\mathrm{se}(\beta_{.j}) = \sqrt{\left(\frac{\sum_{i=1}^{n}(\beta_{ij} - \beta_{j})^2}{n-1}\right) \bigg/ n}. \tag{8.12}
$$

This formulation treats the animal as a random effect as discussed in the section on inference.

We illustrate the approach of using logistic regression and drawing inferences to the population using data collected on brown bears (*Ursus arctos horribilis*) in Alaska. The data consist of relocations of 25 female brown bears with cubs collected during the summer season over a 4-year period. Only bears with at least 20 relocations were included. Minimum convex polygon estimates were used to define availability for each bear (design III). This example is used for the sole purpose of illustrating the approach of developing logistic regression models for individual animals and then averaging these coefficients across bears for purposes of making inferences to average selection.

GIS was used to derive several continuous covariates identified as potential predictors of within home range selection. Given the large number of possible predictors of selection (initially 25), we consider this analysis exploratory. Paired *t*-tests were initially conducted comparing the mean for each covariate calculated for the home ranges (availability) and for the relocations (use). The sample size for the *t* test was the number of bears (25) with data consisting of the mean of the used relocations and random points within each home range. Any variable significantly different ($P < 0.10$) between use and availability was retained. Further reductions in the candidate set were made by eliminating variables highly correlated with other variables. If correlations were 0.70 or greater for a given pair of variables, only one was retained (selected by biologist). Variables retained are listed in Table 8.8.

A backward elimination stepwise procedure was used to select a final model, although other selection procedures (e.g., forward stepwise) could have been used. The standard error of the average was used to determine if each coefficient was significantly different from zero ($n = 25$, d.f. $= 24$, $\alpha = 0.10$) using a *t*-test. We removed the variable with the highest P value from the model and refit the new model as above, until each variable left in the model was significantly different from zero. Table 8.9 contains the final model selected by the stepwise procedure.

These models are for illustration of the methods. Currently we have not investigated the effect of different home range estimators on our estimates of selection patterns, the effect of our arbitrary sample size cutoff ($n = 20$ relocations), and the arbitrary selection of 500 random points to describe the covariates for each bear. If relocation data are instead pooled across

TABLE 8.8 Description of Variables Available for Modeling Brown-Bear Resource Selection after Considering Results of Paired t-Tests and Correlation Analyses[a]

Variable	Description
ATRAIL_C	Distance to recreation trail
COVER_C	Distance to forest or shrub cover
DEV_C	Distance to human development on the Kenai Peninsula
DEV_MN	Mean density of human developments on the Kenai Peninsula
ELEV_MN	Mean density of elevation within 100 m
HROAD_C	Distance to high-use road
HTRAIL_C	Distance to high-use recreation trail
KESTM_C	Distance to salmon spawning stream on the Kenai Peninsula
KEXT_C	Distance to resource extraction site on the Kenai Peninsula
LROAD_KM	Density of low-use roads
LSITE_C	Distance to low-use recreation site
SSTMH_C	Distance to high-potential salmon spawning stream
SSTMH_KM	Density of high-potential salmon spawning streams
SSTML_KM	Density of low-potential salmon spawning streams

[a]Distances are in hundreds of meters and densities are in counts or meters per square kilometer.

TABLE 8.9 Coefficients for the Final Logistic Regression Model of Home Range Selection for Bears in the Summer with Cubs[a]

Parameter	n	Estimate	SE	t	P	Interpretation
COVER_C	25	−0.5742	0.1777	−3.2310	0.0036	Probability of use decreases as distance COVER_C increases
ELEV_MN	25	−0.0045	0.0015	−3.0492	0.0055	Probability of use decreases as density ELEV_MN increases
KESTM_C	25	−0.0364	0.0138	−2.6313	0.0146	Probability of use decreases as distance KESTM_C increases
LROAD_KM	25	−0.0216	0.0104	−2.0702	0.0494	Probability of use decreases as density LROAD_KM increases

[a]Variable description are given in Table 8.8.

animals and effective sample size is the number of animals multiplied by the number of relocations, then the animal is considered a fixed effect, and inferences would be to the set of animals in the sample (e.g., Smith et al. 1982, Arthur et al. 1996). Often this may be the only way to provide a reasonable analysis for a small dataset, but independence of relocations is assumed and statistical inferences are to the fixed set of animals under study and not a population of animals.

EXAMPLE 4: RELATING RESOURCE SELECTION TO THE DENSITY OF ANIMALS

We now illustrate the method proposed by Boyce and McDonald (1999) for calculating population density by type of habitat using resource selection functions and data on population size for the area. Consider the case of discrete habitat classes. Define $w(x_i)$ as the relative probability of use for the ith habitat estimated using logistic regression or by selection ratios (Manly et al. 1993) and $A(x_i)$ as the area of the ith habitat type. The relative use $U(x_i)$ is calculated by $U(x_i) = w(x_i)A(x_i)/\sum_j w(x_j)A(x_j)$, summed over the number of habitats, $j = 1, 2, \ldots, m$. The expected number of animals in the ith habitat is $N_i = N^*U(x_i)$ and the density of animals in the ith habitat is $D_i = N^*U(x_i)/A(x_i)$. For example, if $N = 1000$ animals exist in a population, the relative probabilities of selection for four types of habitat for an individual are $w(x_1) = 0.1$, $w(x_2) = 0.1$, $w(x_3) = 0.3$, and $w(x_4) = 0.5$. If the area of the four types are $A(x_1) = 100$ square miles, $A(x_2) = 200$ square miles, $A(x_3) = 300$ square miles, and $A(x_4) = 400$ square miles, then relative use for the first habitat is $U(x_1) = 0.1 * 100/[0.1 * 100 + 0.1 * 200 + 0.3(300) + 0.5 * (400)] = 0.0315$. Relative use values for the other habitats are $U(x_2) = 0.0625$, $U(x_3) = 0.281$, and $U(x_4) = 0.625$. Applying the formulas for the expected numbers in the ith habitat yields $N_1 = 31.25$, $N_2 = 62.50$, $N_3 = 281.25$, and $N_4 = 625$ with corresponding densities $D_1 = 0.313$, $D_2 = 0.313$, $D_3 = 0.938$, and $D_4 = 1.563$. These density estimates by habitat would be applicable for the time the survey was conducted. The primary utility of this approach is that it provides a way of determining and presenting resource selection results in terms of the density of habitats as opposed to simply relative probabilities or ranks of use.

FUTURE DIRECTIONS

A greater emphasis on presenting results in terms of confidence intervals in ecological studies (Johnson 1999) is expected to continue and will be observed in resource selection studies. Confidence intervals on odd's ratios and maps depicting predicted use will include estimates of variability and will be more useful to managers than significance tests.

Most resource selection analyses used to make inference to a population of animals assume a common selection function for the individuals. Inference is then based on an "average" model of selection. If selection is conditional upon availability, then this assumption may not be met. For example, two animals' selection of resources may be very different if the amount of forage is greatly different in the two home ranges. Mysterud and Ims (1998) propose a method for studying an animal's selection of resources as a function of availability. Selection of resources is modeled separately for each animal and the selection relationships are allowed to vary, as availability varies from animal to animal. This approach has merit when looking at functional response to habitats across a population of animals or across study areas. The authors described the methods using only a dichotomous predictor variable, so methods must be expanded to more complicated models. These investigations into how selection varies among individuals, study areas, and other factors will enhance how transportable specific resource selection models are to other populations of animals and other study areas. Studies must be conducted with more replication over both time (e.g., seasonally, annually) and space (e.g., landscapes, other populations of animals) while using experimental manipulation of habitats to assess selection and to strengthen statistical inferences beyond those made from observational studies.

The rapid increase in GIS technology, including smaller scale imagery and faster computing times, should have a positive effect on resource modeling and mapping capabilities. More precise and higher resolution data that lead to vegetation species maps instead of broad land cover categories should also lead to more useful models of resource selection if selection is affected by small-scale habitat features. Statistical modeling techniques and the mainstream statistical computing packages need to be adapted to better handle the large volumes of data that can be generated from GIS and GPS technology. Furthermore, the effect of employing large numbers of random or systematic points sampled using the GIS to describe availability in logistic regression must be studied much more extensively. This issue may be most apparent when points are considered the experimental unit and when model selection methods are needed to reduce the number of variables observed from a GIS.

The effect of spatial autocorrelation on inferences for resource selection has not been well studied. Future research should focus on the effects of spatial correlation and the effectiveness of autologistic (Augustin et al. 1996) and pseudo-likelihood methods (Wolfinger and O'Connell 1993) for compensating for those effects.

SUMMARY

We reviewed the statistical issues and analyses in resource selection studies involving radio-marked animals. Table 8.1 contains a summary of these issues for the most common analytical procedures. We discussed statistical inference in resource selection studies and recommend that variation among individual animals be considered when sample sizes for the number of animals and the number of relocations are sufficient (see Chapter 11). Using the animal as the experimental unit can often circumvent the problem of dependence among relocations of the same animal. However, often sample sizes are limited in resource selection studies, so that pooling of observations may be required; in this case, the independence assumption should be considered.

We recommend (1) using a biological basis for defining resource availability (e.g., expected travel distance between subsequent relocations, average home range), (2) considering various definitions of availability within a given scale (e.g., different home range estimators) as a sensitivity analysis, (3) analyzing selection at multiple scales (e.g., home range selection, within home range selection), and (4) using the animal as the experimental unit.

We identified problems due to telemetry relocation error, such as reduced power of statistical tests and biases in estimated selection patterns, and recommend that researchers consider biologically meaningful buffers around relocations to help overcome error. Improvements in telemetry technology, such as differential correction and GPS collars, reduce the effects of telemetry error; however, failed signal detection can still bias results.

Well thought out *a priori* models should be developed to alleviate problems in model selection. Designed experiments when landscapes and habitats can be manipulated should be considered. When large numbers of predictor variables are considered and studies are observational, analyses should be considered exploratory. Univariate comparisons of variables, such as use vs. availability comparisons, confidence intervals and correlations among predictor variables, graphical presentations, professional knowledge and expertise, and model selection criteria, should be used to guide biologists in developing reasonable and useful exploratory models.

We presented examples of a log-linear and logistic regression analysis approach to resource selection and recommend these techniques because of the expanded availability of the methods in standard statistical packages, as well as the flexibility of these techniques in considering continuous and discrete covariates and their predictive and mapping abilities, especially with GIS. When availability is difficult to define, mapping techniques and polytomous logistic regression may be appropriate alternatives. Also, if data are fine scale and there are opportunities to define availability for each relocation point (e.g., Arthur et al. 1996, Cooper and Millspaugh 1999; see also Chapter 9), approaches such as discrete choice modeling may be appropriate.

A simple example illustrating Boyce and McDonald's (1999) approach to estimating density by type of habitat using logistic regression and information on population size was presented. This method and example illustrates the relationship between the resource selection relative probabilities and animal density by habitats.

ACKNOWLEDGMENTS

We thank the Interagency Brown Bear Study Team (IBBST), Sean Farley, chairperson, for the use of the brown-bear data. We are grateful to Josh Millspaugh, Steven Knick, and John Marzluff for comments on the manuscript. We also thank John Payne and David Yokel (BLM), Robert Skinner, and Ed Merritt (USFWS), and Fritz Reid, Robb Macleod, and Dick Kempka (Ducks Unlimited), for funding and support that greatly contributed to the development of this manuscript.

CHAPTER 9

Accounting for Variation in Resource Availability and Animal Behavior in Resource Selection Studies

ANDREW B. COOPER AND JOSHUA J. MILLSPAUGH

In its barest essence, resource selection studies focus on how animals make choices, with the goal of understanding why those choices were made. Volumes of literature have been written on the factors influencing where and when animals choose to forage, rest, or reproduce (e.g., Smith and Fretwell 1974, Persson and Greenberg 1990, Leclerc 1991, Lubin et al. 1993, Welham and Ydenberg 1993, Abrams 1994, Owen-Smith 1994). During the past two decades, many statistical techniques have been developed for analyzing resource selection data (Manly et al. 1993). Such techniques include χ^2 goodness-of-fit tests (Neu et al. 1974, Byers et al. 1984), log-linear models (Heisey 1985), logistic regression (Thomasma et al. 1991), discriminant function analysis (Dunn and Braun 1986), rank-order tests (Johnson 1980), and compositional analysis (Aebischer et al. 1993). The choice of which technique to use is a function of the data collected, objectives of the research, and assumptions inherent in both the data and the analyses (Alldredge and Ratti 1986, 1992).

Two important assumptions contained in the aforementioned techniques are that the availability of resources is constant over time and that each individual has equal access to those resources deemed available to that individual. Violations of these assumptions have been discussed in the area of habitat selection (Johnson 1980, Porter and Church 1987). Habitat selection studies assume either that all habitats in the study site are available to all individuals, or that some spatially or temporally defined subsets of the area are available to different individuals. These subsets could be defined by the individual animal's home range or by subdividing the observations and definitions of availability by season (Johnson 1980, Thomas and Taylor 1990). For example, compositional analysis (Aebischer et al. 1993) allows for individually defined definitions of availability of categorical resources, but this definition of availability is held constant over a period of time, usually the entire study period, to allow for multiple resightings of the individual. In each case, animals are assumed to have an ideal free distribution (Fretwell and Lucas 1969) in that all habitat patches deemed available are equally accessible to an individual at all times. This assumption applies to both the spatial and temporal aspects of the study. Any change in resource availability over time violates this assumption because not all resources are available to the individuals at all times in the same proportion. Along similar lines, this assumption could be violated by a factor such as competitive exclusion, which may prevent some individuals from using some areas defined as available.

Along with the complications of defining resource availability, there is also an often-unexplored aspect of resource selection relating to animal behavior. In the case of habitat selection, animals may choose different habitat patches for different behaviors (see Chapter 12). When researchers aggregate the data from a range of behaviors, then what is actually estimated is the joint selection of *both* an activity *and* a resource. Resources related to common behaviors will appear frequently in the dataset, resources associated with rare behaviors will appear infrequently in the dataset, yet all the selected resources will be compared to a single definition of availability. This results in resources chosen for rare behaviors being discounted in the analysis simply because these rare activities appear less often in the dataset.

Consider the hypothetical case of an animal that always uses a single specific resource for each activity, but that resource differs for each activity. As such, the resource chosen most often is simply the resource used in the most frequent activity, and the resource chosen least often will be the one used in the least frequent activity. If the researcher ignores behavior and if the availability of the least chosen resource is at all common, then many of the standard techniques will predict avoidance of that resource because it was chosen less frequently than expected by its availability. This is despite the fact that the least chosen resource is *always* chosen for a specific activity. Such an analysis can lead to grave mistakes in management if that least frequent behavior, and thus the least chosen resource, is of critical importance to the individual. This problem does not disappear when one moves away from a categorical definition of resources to resources defined by a suite of continuous and categorical variables. In fact, it becomes even more troublesome. The significance of any attribute would be less a function of its true importance than a function of both the frequency of the behaviors and the range of values this attribute has for chosen and nonchosen resources. Though all resource selection analysis techniques, including discrete choice models, can be applied without behavioral data, only by separating the different behaviors can the researcher begin to understand the true motivations behind the animal's decision making process. As we demonstrate in our case study, differences in the frequencies of the various behaviors need not be extreme for analyses to run awry.

Discrete choice models are one solution to the problems described above. Developed more than 25 years ago by econometricians, discrete choice models permit the researcher to define resource availability separately for each animal observation and allow for a hierarchical structure, so that the selection of resources across multiple behaviors may be explored in a single analytical framework (Cooper and Durant unpublished data, Ben-Akiva and Lerman 1985, Cooper and Millspaugh 1999). This chapter illustrates how to account for variation in resource availability and animal behavior, and describes both the theory and application of discrete choice models to studies of wildlife resource selection. As a case study, the multinomial logit form of the discrete choice model, which is a more complex version of logistic regression, is applied to summer, diurnal, microsite resource selection by elk (*Cervus elaphus*) in Custer State Park (CSP), South Dakota.

METHODS

THE THEORY UNDERLYING DISCRETE CHOICE MODELS

Discrete choice models are derived from economic utility theory, where "utility" is synonymous with satisfaction. These models assume that an individual gains satisfaction from selecting a given resource. The true nature of this satisfaction goes undefined in the model but may include such factors as net energy intake or protection from predators. Satisfaction is assumed to be a function of the attributes of the resource. In the case of resource selection, such attributes may include canopy closure, forage quality or quantity, and density of hiding cover. When presented with a set of resources, the model assumes that an individual will choose the resource that maximizes this satisfaction. This satisfaction-maximizing assumption provides the theoretical framework for the resource selection process in discrete choice models.

In most statistical packages, the equation defining satisfaction takes a linear form, though the variables describing the attributes could be transformed (e.g., log transformation) or occur as an nth-order polynomial (Ben-Akiva and Lerman 1985). The utility provided by resource i to individual j (U_{ij}) thus takes the form:

$$U_{ij} = \mathbf{B}'\mathbf{X}_{ij} + e_{ij} = b_1 x_{1j} + b_2 x_{2j} + b_3 x_{3j} + \cdots + b_m x_{mj} + e_{ij}, \qquad (9.1)$$

where $\mathbf{X}_{ij}$ is a vector of length m of the continuous and/or categorically defined attributes of resource i as perceived by individual j, $\mathbf{B}$ is a vector of length m of estimable parameters that determine each attribute's contribution to utility, and e_{ij} is an error term. It is important to note that when resources can be uniquely defined (e.g., a predator chooses to stalk either deer, elk, moose, or rabbit), then the vector, $\mathbf{B}$, may actually be different for each resource and therefore take the form of a matrix with each row, B_i, corresponding to each uniquely defined resource. In this case, a resource-specific constant should be included in the utility equation. In the case of habitat selection, however, resources are rarely defined uniquely; usually researchers define the patch solely by its associated attributes such as canopy closure, vegetation type, slope, aspect, etc. In this case, a single vector, $\mathbf{B}$, will be used to calculate the utility of all patches.

Interaction effects between two or more attributes of the resource such as the density of hiding cover and distance from roads may be incorporated into the utility function. Along similar lines, one can incorporate interactions between attributes of the individual or the setting surrounding the choice and attributes of the resource. One such interaction between an individual's attributes could be the individual's sex and each resource's proximity to human developments (e.g., roads). When we refer to the setting surrounding the choice, we are not talking about the attributes of the resources available but rather variables that do not differ across resources but vary among observations such as the time of day or time of year. Thus, one can test to see if the effect of canopy closure changes at different times of day or in different types of weather. However, for reasons described below, attributes of the individual and of the choice setting cannot occur in the equation solely as main effects. Interaction effects are common in generalized linear models (e.g., linear regression, logistic regression) and allow the researcher to account for the fact that the effect of one variable may depend on the value of another variable (McCullagh and Nelder 1989). As with logistic (Thomasma et al. 1991, Manly et al. 1993) and log-linear resource selection models (Heisey 1985, Manly et al. 1993), interaction effects allow researchers to better approximate and understand how these variables affect the selection process, though estimating such interactions requires increased sample sizes.

Because the researcher has limited information about an individual animal's utility and characteristics, there are unobserved components of the utility function. Two animals of the same sex, age, and size may choose different resources because of different past experiences or variability in tastes between individuals. To account for these unobservables, an error term is included in the utility function, making utility itself a random variable (Manski 1977). Please note that work is currently under way to develop "canned" software packages that can directly incorporate some of these issues. With utility now a random variable, one can explore utility, and therefore choices, in a probabilistic sense.

The theoretical framework for the selection process in discrete choice models is that an individual will choose the resource that provides the maximum utility or satisfaction. Therefore, the probability of choosing resource A equals the probability that the utility derived from resource A is greater than the utility derived from each of the other available resources. This requires the researcher to specify what resources are available to the individual temporally and spatially. In mathematical notation, the probability of individual j choosing resource A $[P_j(A)]$ out of the set of i resources can be written as

$$P_j(A|i) = \Pr(U_{Aj} > U_{ij} \forall i) = \Pr(\mathbf{B}'\mathbf{X}_{Aj} + e_{Aj} > \mathbf{B}'\mathbf{X}_{ij} + e_{ij} \forall i). \qquad (9.2)$$

Rearrangement of terms produces the equation:

$$P_j(A|i) = Pr(\mathbf{B}'\mathbf{X}_{Aj} - \mathbf{B}'\mathbf{X}_{ij} > e_{ij} - e_{Aj} \forall i). \qquad (9.3)$$

Because the attributes of an individual, such as age and sex, will not change over the choice process, the main effects of such attributes will drop out of Eq.

(9.3), but the interaction effects will remain. Therefore, one could explore the effect of the interaction of an individual's age and canopy closure (e.g., Millspaugh 1999), but one could not determine only the effect of age on the selection process.

If utility was directly measurable and was a continuous random variable, thus not requiring the probability statements in Eqs. (9.2) and (9.3), one could assume that each of the error terms in Eq. (9.1) is independently and identically distributed as a normal random variable and perform standard linear regression (Neter et al. 1990). Since utility is not directly measurable and we can only observe the outcome of the probability statements of Eqs. (9.2) and (9.3), standard linear regression cannot be applied. However, we can still assume that the error terms are independently and identically distributed as normal random variables and use the probit model. Probit models are rarely used when more than two choices are available because of severe computational problems (Ben-Akiva and Lerman 1985). With N total choices available to the individual, solving Eq. (9.3) as a probit model would require solving an $N - 1$ order integral.

If the difference in the error terms for each of the choices $(e_{ij} - e_{Aj} \forall i)$ is distributed as a logistic random variable, then the multinomial logit form of the discrete choice model arises (McFadden 1981). This assumption is equivalent to each of the error terms (e_{ij}) being distributed as an independent Type I (Gumbel) extreme value random variable. This distribution closely approximates the normal error distribution assumption with the advantage of producing easily estimable solutions in a maximum likelihood framework (Ben-Akiva and Lerman 1985). Also, Ben-Akiva and Lerman (1985) demonstrate that when only two choices are available to the individual and choices are the same across individuals, the multinomial logit form reduces to the logistic resource selection equation described by Thomasma et al. (1991) and Manly et al. (1993).

Given the Type I extreme value distribution assumption for Eq. (9.3), the probability of individual j choosing resource A rather than any other of the i resources available takes the form:

$$P_j(A|i) = \frac{\exp(B'X_{Aj})}{\sum_{\forall i} \exp(B'X_{ij})}. \tag{9.4}$$

This is identical to the first-order statistic for a series of random variables (the U_{ij}'s) that are distributed as a Type I extreme value. The parameters in Eq. (9.4) (the B's) are estimated by a maximum likelihood equation that includes the resources chosen and not chosen across all j individuals (see Data Requirements and Parameter Estimation below). Based on this equation, the probability of choosing resource A is a function of the attributes of resource A and the attributes of all other resources available to that individual at that particular time and place, termed the *choice set*. Though the choice set provides a definition of availability that is as arbitrary as in all other resource selection techniques, it gives researchers complete flexibility in defining resource availability over time and space for each individual observation. In controlled experiments, the researcher can explicitly define the choice set presented to each individual to reduce extraneous variability.

When looking at selection spatially, researchers could assume that the animal has *a priori* knowledge of all resource attributes. In this case, the choice set would contain all resources the individual could reach in a given time period. This would be defined by a circle centered on the individual's initial location with a radius equal to the distance an individual could travel in that time between observations (i.e., Arthur et al. 1996). With advances in geographic information systems (GIS) and computer simulation programs, the choice set could take on more complex shapes. For example, different types of terrain may permit different travel speeds, and different travel speeds will result in noncircular boundaries defining availability. Potential geographic barriers, such as large rivers, ravines, and cliffs, may also be taken into consideration.

Alternatively, the researcher could assume the individual has no *a priori* knowledge. This choice set would contain only those patches that the individual could have sampled based on its initial location, where it was finally located, and the given time period. A simple version would be an ellipse bounding all possible paths between the initial and final locations, given the time duration between observations. If one were performing relatively continuous tracking of individuals, those patches within visual distance of the individual could define the choice set. Many GIS packages now contain "visibility" routines that calculate areas that are visible from a specific location as a function of digital elevation and, in some cases, vegetation.

The boundaries of each individual's choice set at each particular time must be determined based on the researcher's knowledge of the system and the biology of the animal. Different individuals, or the same individual over time, may have completely different choice sets, overlapping choice sets, or even identical choice sets depending on the individuals' distribution over the landscape over time. Only those resources the individual could have actually chosen at that particular time should be included in the choice set. If one is making daily observations of the individual, that individual's annual home range may not be an appropriate definition of the choice set because not all resources within its annual home range are truly available on a given day due simply to travel time necessary to reach other areas within the home range. On the other hand, researchers should not rule out resources simply because they fall outside of the individual's estimated home range. An individual's home range is estimated based on what resources were used. There are reasons why a particular individual did not choose those resources that fall outside its estimated home range, and it may be of interest to determine what those reasons are. If home ranges arise in nature in part from competition between individuals, then variables that account for the presence of competitors should be included in that resource's attributes. One way to do this, assuming that there is at least some overlap of home ranges, would be to include an indicator variable for each resource that equals 1 if that resource lies within another animal's home range, and thus may be defended, and 0 otherwise. If home ranges arise in part due to animals having a tendency to travel only a limited distance from a given, fixed area (i.e., from a den site or nest) even though the individual could, in theory, travel much farther, then each resource should have as one of its attributes the distance from that resource to the fixed area and not be limited to only those resources found within the estimated home range. Researchers will encounter problems if animals are resighted only on an infrequent basis because the size of the choice set will increase nonlinearly with respect to time between observations. In theory, the choice set could increase well beyond the boundaries of the study area and thus beyond the researcher's ability to collect the necessary data.

The model as described thus far assumes that the resource selection process is associated with a specific behavior. In order to account for multiple behaviors, the resource selection question must be nested within a behavioral model (Fig. 9.1). It is important to note that the nested structure of the model assumes not a sequential decision process (choose a behavior then choose a resource), but rather the interrelatedness of the joint decision of behavior and resource, both of which are conditional on what resources are available. The nesting is accomplished through the use of conditional probability statements. The probability of choosing resource A is now conditioned not only on the choice set of i resources available but on the behavior w, where w is a subset of W possible behaviors:

$$P_j(A|i) = \sum_{\forall w \in W} [\Pr(A|i, Y_j = w)^* \Pr(Y_j = w|i)]. \qquad (9.5)$$

As will be discussed in greater detail later, the entire model will consist of $W + 1$ discrete choice models: one discrete choice model for the probability of choosing a resource given each of the w behaviors, and one discrete choice model for the probability of choosing a behavior.

To estimate the probability of choosing resource A given the set of i resources, we must now define utility equations for the resources and the behaviors, and each resource will have a different utility equation for each behavior (Fig. 9.1). To denote this, we now index the coefficients for the utility by the activity, B_w, but continue to suppress the previously mentioned

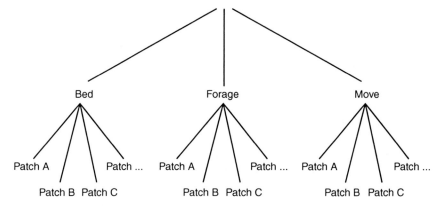

FIGURE 9.1   Resource selection nested within a behavioral model. In this case, an individual chooses a particular habitat patch in which to either bed, forage, or move.

resource-specific index i, which may or may not exist depending on whether the resources are uniquely defined. Each resource utility equation will have the same form as in Eq. (9.1). For each behavior w out of W possible behaviors, however, the utility equation will consist of a constant, attributes of the behavior, and an independent variable that is called the inclusive value. The inclusive value for behavior w for individual j, $\hat{V}_{jw}$, equals the expected maximum utility derived by individual j from choosing any of the available resources for behavior w (McFadden 1978), and equals:

$$\hat{V}_{jw} = \ln\left(\sum_{\forall i} \exp(\mathbf{B'}_w \mathbf{X}_{ij}) \right). \qquad (9.6)$$

In other words, the inclusive value is the maximum expected value for the lower set of branches in Fig. 9.1. Thus, if there are four uniquely defined resources and three behaviors, there will be 12 resource-related utility equations and three behavior-related utility equations. But, as is the case with most resource selection research, if the resource is not uniquely defined there would be only four equations: one resource-related utility equation under each of the three behaviors and one behavior-related utility equation that includes the expected maximum value of the resource-related utility equation given the resources available in the choice set (i.e., the inclusive value).

Since behaviors are uniquely defined and there is a behavior-specific constant in each of the behavior utility equations, we can explore the effects of the attributes of the individual and of the setting surrounding the choice. This is accomplished in a similar fashion to how we have these variables interact with the attributes of the resource. However, because we have a behavior-specific constant, we can have the individual's attributes and the setting's attributes interact with these constants and thus allow for such variables as age, sex, and time of day to directly affect behavior. In essence, we are allowing each behavior to have a different coefficient for a given individual-level or settings attribute. If the resources were uniquely defined (e.g., choosing to stalk elk, deer, or moose as opposed to selecting a habitat patch that is defined solely by its attributes), then individual-level attributes, such as sex and age, and the attributes surrounding the choice setting, such as time of day, could be directly incorporated into the resources' utility equations as well.

Once parameterized, the discrete choice model can be used to estimate the importance of the attributes in the selection process, along with the probability of choosing a given resource. With the utility function defined in Eq. 9.1, regardless of the error distribution assumption, utility is assumed to be compensatory with respect to its attributes. By increasing one attribute of a given resource while decreasing another attribute of that same resource in the proper proportion, it is possible for the utility of that resource to remain unchanged. Because of this property, we can calculate the marginal rates of substitution (MRS) between independent variables in the utility function (Cooper and Millspaugh 1999). The MRS allows researchers to quantify how individuals will trade changes in one attribute for changes in another. To calculate how changes in attributes will affect selection probabilities for resources, direct and indirect elasticities can be computed. For more information on MRS and elasticity, see Cooper and Millspaugh (1999).

Data Requirements, Parameter Estimation, and Hypothesis Testing

The data necessary to estimate the multinomial logit form of the discrete choice model is similar to that required for the standard logistic regression model. In logistic regression, the dependent variable typically takes on the value of 0 or 1. Whether we define our dependent variable to equal 1 for a used resource and 0 for an unused resource or vice versa is immaterial; there is neither scale nor order to the dependent variable. The multinomial logit version of the discrete choice model is just a more complex version of logistic regression. Instead of the dependent variable having only two values, it can have multiple values. But again, these values have neither scale nor order. If there are five resources in the choice set, then the dependent variable will take on the values of 1–5, with each value corresponding to a particular resource. Which resource is given which value is immaterial.

To define the choice set, the researcher must first have the location of the individual at some point prior to the choice event along with the location of that individual when the resource was chosen. With these two points, the researcher can now define the boundaries of the choice set based on the assumptions described earlier (e.g., *a priori* knowledge vs. no *a priori* knowledge). Once the choice set is defined, the researcher must collect data on the attributes associated with the resource that was chosen and the resources that were not chosen, along with data on what the animal was doing. These attributes can be categorical variables or continuous variables. Note that behavioral data are not required to run a discrete choice model; however, in order to gain insight into the true motivations behind resource selection, researchers should collect behavioral data. McFadden (1978) demonstrated that consistent estimates for the parameters of the utility function and the choice probability function can be obtained by randomly selecting without replacement a subset of size 4 of all alternatives within each choice set. This approach is analogous to the data collection technique used for logistic regression analysis (Manly et al. 1993) where the researcher collects data from random locations within the bounds of what has been defined as available. Only in the case of discrete choice models, this would be done within each choice set, which is our definition of availability.

One of the primary trade-offs that the researcher should consider before beginning the experiment is that of breadth of data vs. depth of data (see Chapter 2). Because time is a limited resource, the more variables on which the researcher collects data, the fewer choice events she or he will have time to observe. Standard resource selection models suffer from the same problem; however, with the multinomial logit form of the discrete choice model, the optimum sample size with regard to parameter estimate efficiency depends on the values of the unknown parameters (Ben-Akiva and Lerman 1985). Daganzo (1980) suggests that researchers perform a preliminary study to obtain rough estimates of the parameters and then use these to calculate sample size. Researchers should therefore limit the range of attributes they wish to explore based on relationships previously identified by others, relationships that seem plausible based on the biology of the animal, and, of course, relationships of interest to the researcher. It should be noted, however, that the time spent collecting attribute data can be greatly decreased with the application of GIS maps, so long as they are accurate with respect to the attributes of the locations at the time those locations were chosen.

A similar trade-off exists between the number of behaviors classified and the number of observations per behavior. Researchers should define the behavioral categories based on their knowledge of the biology of the animal and the question that is being asked in the study. In most cases, a simple active vs. inactive definition will not be appropriate because the different active behaviors may depend on different resources (e.g., feeding vs. moving/migrating). Data collection should be spread out randomly over time so that observations of each behavior will appear in the data in proportion to their actual occurrence. If one has narrowly defined behaviors and some are rare, then a large sample size will be required to obtain parameter estimates for resource selection functions for that behavior. If behavior is closely tied to time of day, then the researcher could stratify the data collection based on time of day. However, if one does this, estimation procedures different from those discussed here are required because the behaviors will not be sampled in proportion to their natural occurrence; they will be biased. Interested readers are referred to Ben-Akiva and Lerman (1985) for further details on estimating stratified data.

A final trade-off that should be considered is the number of individuals tagged vs. the number of observations per individual (see Chapter 2). In all of the readily available "canned" software packages, it is assumed that all observations are independent. Thus, we assume that two observations of the same individual are independent of one another, which is not the case in a practical sense. Work is currently under way to develop software that incorporates mixed-effects into discrete choice models, thus allowing one to directly account for repeated measures of the same individual. There are two ways in which this assumption can be violated. The first is that the choice made in the previous observation affects the choice made in the current observation. To avoid violating this assumption, researchers could follow the methods described in Swihart and Slade (1985a, b, 1986; see also Chapter 5). The other way this assumption can be violated is by variation in individual tastes such that the choice pattern of one individual is systematically different from the choice pattern of another individual, with these differences not accounted for by the measurable attributes of the individual (e.g., age, sex). However, this is a relatively minor issue because what we are interested in are the overall average patterns. Randomly selecting individuals to receive tags and tagging numerous individuals will help minimize the effects of violating this assumption. Because of this, researchers should ensure that the trapping technique is not biased for or against one behavior or another. For example, if an attractant is used in the trap that favors aggressive individuals, then the behavior of the tagged population may not be representative of the whole population. As a rule of thumb, when we are interested in the average patterns, more individuals with fewer observations per individual is best.

To understand the estimation routine for a nested discrete choice model, we return to Eq. (9.5). Both the $\Pr(A|i, Y_j = w)$ and the $\Pr(Y_j = w|i)$ portions of

Eq. (9.5) can be set up as discrete choice models. In the first case an animal is choosing a resource from a fixed set of resources; and in the second case an animal is choosing a behavior from a fixed set of behaviors. The $Pr(A|i, Y_j = w)$ portion will be a series of discrete choice models, one for each of the w behaviors out of the possible set of W behaviors. The first step in analyzing the data for a discrete choice model is to segregate the data by behavior. For each behavior, the researcher will have attribute data on the resource chosen and the resources not chosen, or some subset thereof, for each individual each time the behavior was observed. Utility equations are parameterized for choosing a resource given each behavior by applying iterative maximum likelihood techniques to the probability statement depicted in Eq. (9.4) for each of the behavior-based datasets. McFadden (1974) demonstrated that maximum likelihood estimation would produce consistent, asymptotically normal, and asymptotically efficient estimates of the parameters in Eq. (9.4) under a very wide range of assumptions. Many statistical programs can do this, though some are better than others, and readers are referred to Mannering (1998) and Larson (Appendix A) for a review of these.

For those who cringe at the words "iterative maximum likelihood techniques," remember that Eq. (9.4) is just a more complex version of logistic regression. When the choice set is the same for all individuals at all times and consists of only two options, Eq. (9.4) is in fact logistic regression. As a result of this relationship, the basic approach to model fitting is *identical* to that for multivariate regression (Neter et al. 1990) or generalized linear models (McCullagh and Nelder 1989) where variables and interaction effects can be systematically added to or deleted from the equation based on a variety of statistical tests. As with the log-linear and logistic regression models, sample size will determine the total degrees of freedom for the model and, thus, the number of attributes and attribute interactions that can be explored. Because of this, the use of categorical rather than continuous variables for the attributes will greatly reduce the number of attributes that can be included in the model. In discrete choice models, each choice event constitutes a single sample, and the total sample size equals the number of choices observed. A single sample will include the attributes of the chosen resource and of the nonchosen alternatives from the individual's choice set.

The two main tests used to determine the significance of variables in fitting the multinomial logit model are the asymptotic t-test and the likelihood ratio test (LRT) (Ben-Akiva and Lerman 1985). Any statistical package that can estimate multinomial logit equations will provide the necessary output for these tests, namely, the t-statistic and the log-likelihood of the model. The asymptotic t-test is used primarily to determine whether a particular parameter value is significantly different from zero in a statistical sense. In general, if the coefficient is not significantly different from zero, then that variable can be removed from the equation. Sokal and Rohlf (1995) recommend an α-level of 0.05 for main effects and 0.10 for interaction effects due to the increased variability associated with interaction effects. When an interaction effect is significant, the main effects involved must be retained in the equation regardless of their significance (McCullagh and Nelder 1989). The LRT can be used to test both the overall fit of the model as well as the significance of adding or deleting variables or sets of variables to the model, thus comparing two models, one with and one without the variables. When resources are uniquely defined or in the case of behaviors that are by definition uniquely defined, separate utility equations are computed for each resource and/or behavior. The LRT is used to determine whether the model supports different values for the coefficients of a given attribute in each utility equation or whether some utility equations have the same value for the coefficient of a given attribute. Using behavior as an example, time of day may have a different impact on feeding than on sleeping, but it may have the same effect for feeding and moving. The LRT is proportional to the difference in the log-likelihood of the two models in question, one with k parameters (denoted as $\mathbf{B}^k$) and the other with m parameters (denoted as $\mathbf{B}^m$) where $k > m$. This test is χ^2 distributed with $(k - m)$ degrees of freedom, and takes the form:

$$LRT = -2^*[L(\mathbf{B}^m) - L(\mathbf{B}^k)]. \tag{9.7}$$

A recent development is to use Akaike information criteria (AIC) (Burnham and Anderson 1998). The AIC value for a model with k parameters equals:

$$AIC = -2^*L(\mathbf{B}^k) + 2^*k. \tag{9.8}$$

The more parsimonious model is the one with the lower AIC value. This technique allows researchers to directly compare nonnested models, which is something neither the t-test nor the LRT can do.

Using the above techniques, the researcher will parameterize the utility equations for selecting a resource given each type of behavior. Thus, if one were exploring habitat selection and the behaviors included resting, foraging, and moving, there would be three separate analyses performed: one for the data on chosen and not-chosen sites within each choice set when the animal was resting, one for the data on chosen and not-chosen sites within each choice set when the animal was foraging, and one for the data for chosen and not-chosen sites within each choice set when the animal was moving.

Once all of the resource utility equations have been parameterized for each behavior, then the researcher calculates the inclusive values associated with each of the activities for each choice set for the entire dataset using Eq. (9.6).

Note, however, that even though we could use a subset of the choice set in estimating the resource utility equations under each behavior, the inclusive value must take into account all of the resources within the choice set. Returning to our habitat selection example, the researcher will calculate one inclusive value based on the utility equations for resting, one inclusive value based on the utility equations for foraging, and one inclusive value based on the utility equations for moving for each of the choice sets regardless of what behavior was actually chosen. To calculate the inclusive value for resting, for example, the researcher would apply the resource utility equation that was parameterized for the resting data to all of the resources within the choice set, exponentiate those values and then combine them, and then take the natural log. This value is the inclusive value of resting for that choice set. This is then performed on every choice set for the whole dataset, regardless of what behavior was chosen. The dependent variable in this discrete choice model is no longer what resource was chosen, as was the case for parameterizing the resource utility equations, but rather what behavior was performed. The inclusive value is treated as an independent variable associated with each behavior. The entire dataset is now analyzed as a discrete choice model in the identical fashion as the previous discrete choice models with behavior as the dependent variable. If the inclusive value for any behavior is statistically nonsignificant, then it can be dropped from the behavior's utility equation. The interpretation of this would be that the resources available do not influence the probability of choosing a particular behavior at any given point and time.

Readers are referred to Ben-Akiva and Lerman (1985) for the derivation of the likelihood function for nested discrete choice models. As a cautionary note, in the derivation, the marginal probability of selecting a resource given a behavior is replaced by its estimate from the conditional probability (Ben Akiva and Lerman 1985). Because of this, the estimates for the coefficients of the behavior utility equations will be consistent but not asymptotically efficient. Amemiya (1978) demonstrated that this would result in the standard errors for these coefficients being too small. McFadden (1981) developed a correction for this, but in most situations "researchers often report the erroneous standard errors, recognizing that they are somewhat too small" (Ben-Akiva and Lerman 1985:298). There are more computationally intensive techniques that allow one to get around this problem; however, they are not available in most standard statistical packages.

Once the researcher has parameterized an equation for each behavior, the probability of an individual choosing a resource can be calculated by combining Eqs. (9.4) and (9.5). The probability of selecting a resource is the product of the probability of choosing a behavior and the probability of choosing a resource given that behavior summed over all behaviors.

CASE STUDY: SUMMER, DIURNAL, AND MICROSITE RESOURCE SELECTION BY ELK IN SOUTH DAKOTA

To illustrate the importance of accounting for behavior and the use of nested discrete choice models, we consider summer, diurnal, microsite resource selection by elk (*Cervus elaphus*) in CSP from 1994 to 1997. Briefly, CSP encompasses about 29,000 ha in the southern portion of the Black Hills region of western South Dakota. Temperatures average 24°C in the summer and −4°C in winter. Coniferous forests, dominated by ponderosa pine (*Pinus ponderosa*), comprise 12,355 ha of non-fire-affected land within CSP. In 1988 and 1990, stand-replacing wildfires removed about 40% of the forested habitat in CSP. Human activities are pervasive throughout CSP; there are 34 km of roads and trails providing extensive access throughout most of the park. During our study there were about 750–1000 resident elk (Custer State Park, unpublished data). More complete descriptions of the park are available in Millspaugh et al. (1998b, 2000) and Millspaugh (1999).

Elk locations were collected from 36 (21 cows and 15 bulls) free-ranging, radio-collared, adult elk from 1994 to 1997. A location for each radio-collared elk was estimated every 28 hours using triangulation or aerial telemetry techniques (Mech 1983, White and Garrott 1990). By separating locations 28 h, observations were independent based on the examination of cumulative distances between locations during 48-h continuous monitoring of selected individuals and calculation of Schoener's ratio (Swihart and Slade 1985a,b, 1986; see also Chapter 5). Also, each day one or two elk were randomly selected for visual observation from the ground in conjunction with other study activities (Millspaugh et al. 1998b). Visual observations were obtained by "homing" on radio signals (Mech 1983) until the elk was seen. At these locations, we recorded the position (universal transverse Mercator coordinates; UTM), elk behavior (bedding, feeding or moving), and other habitat-related characteristics (e.g., hiding cover) that were used to corroborate geographic information system (GIS) data. This sampling strategy resulted in 190 (44, 48, 39, and 59 from 1994 to 1997, respectively) bedding site locations, 201 (50, 55, 47, and 49 from 1994 to 1997, respectively) feeding site locations, and 38 (9, 10, 7, and 12 from 1994 to 1997, respectively) moving site locations along with the location of the elk 28 h prior. To document resource use at each observed microsite location, the UTM coordinates were imported into the park GIS and overlaid on the habitat patch map described below.

Defining Habitat Patches

CSP was divided into discrete habitat patches by grouping areas of similar habitat characteristics. This was accomplished by joining the following 11 map layers in the CSP GIS: stem density (stems/acre); primary (paved roads open to the public), secondary (gravel roads open to the public), and tertiary (unimproved service or fire roads not open to the public) roads, hiking trails, horse trails, slope, aspect (north = 337.6–22.5°, northeast = 22.6–67.5°, east = 67.6–112.5°, southeast = 112.6–157.5°, south = Z 157.6–202.5°, southwest = 202.6–247.5°, west = 247.6–292.5°, and northwest = 292.6–337.5°), average diameter of trees at breast height (dbh), percent canopy closure, and dominant vegetation type (meadow, fire, all else assumed ponderosa pine habitat type). Categorical distance measures (to roads and trails) were converted to continuous variables by taking the midpoint of the category (e.g., 100 m for distance category 0–200, at 200-m increments, up to 1300 m for distance > 1300 m) (Cooper and Millspaugh 1999). Because these values can be treated as continuous variables in the analysis, there is only one variable for distance from each type of road or trail as opposed to seven categorical distance variables for each category. A similar approach was taken for calculating dbh (3 in., 7 in., 11 in., 15 in.), canopy closure (25%, 55%, 85%), and slope (2.5%, 7.5%, 12.5%, 17.5%, 22.5%). Meadow and fire-killed habitats were classified using indicator (0/1) variables. Aspect was classified by seven indicator variables with northwest aspects being represented as zeros for all other aspect variables (McCullagh and Nelder 1989). To account for annual changes in habitats due to timber harvest or other management activities, we used annually updated GIS layers.

Defining the Choice Sets

Choice sets were defined similar to Arthur et al. (1996). During the summer, the average linear distance between locations separated by 28 h for radio-collared elk in CSP was 2 km (J. J. Millspaugh, unpublished data). Rather than centering the choice set on the elk's initial radiotelemetry location, the choice set was centered 500 m, one-quarter the average linear distance between locations, in the direction of the chosen microsite. This approximated our assumption that elk have no *a priori* information regarding the chosen bedding, feeding, or moving sites (Fig. 9.2). In the GIS, a circle of radius 1 km was circumscribed about this point (Fig. 9.2). We assumed that all habitat patches within this circle could have been sampled and chosen by an elk for bedding, feeding, or moving given the time between the radio-location and the observed microsite selection.

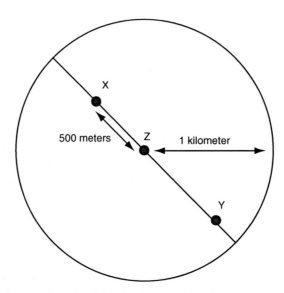

FIGURE 9.2   The method used to define the choice set for each microsite event. Point X marks the radiotelemetry location of the elk 28 h prior to the microsite event; point Y marks the observed microsite used for bedding, feeding, or moving; and point Z represents the center of the choice set. The boundary of the choice set is defined by a circle, centered at point Z with radius 1 km.

With the choice set defined, two more continuous variables were added to the attribute set of each patch within a choice set. The area of each patch (m^2) within each choice set was calculated, as well as the distance between the radiotelemetry location and the patch's closest boundary. The distance variable represents the shortest straight-line distance an elk must traverse to reach each potential choice. This entire process was repeated for each of the microsite/radio-location pairs, with between 47 and 910 potential choices contained within each choice set, depending on the heterogeneity of the landscape given the time and place of the elk's locations.

ESTIMATING THE DISCRETE CHOICE MODELS

We fit a total of five discrete choice models: one each for choosing a patch given the behavior of bedding, feeding, and moving (the lower nests in Fig. 9.1); one for choosing a behavior (the upper nest in Fig. 9.1); and a "naïve model," which used all the data but ignored behavior. Habitat patches were defined as areas that were homogeneous with respect to all the variables listed above.

Each model was estimated with the program SST (Statistical Software Tools, Dubin-Rivers Associates, Riverside, CA, 1992). Data were separated using the three behaviors mentioned above. Each of the behavior-specific habitat choice models used the chosen habitat patch and a subset of five nonchosen patches selected randomly without replacement from the individual's choice set for each microsite selection event (McFadden 1978). The full main-effects model plus all non-aspect-related interactions was fit to the data and the addition of aspect-related interactions were tested individually. Once all aspect-related interactions were tested, the remaining variables were tested using a step-wise process where the least significant variable was dropped. For the behavior choice model, the variables that were explored included time of day (in hours), sex, age, and Julian date, along with the inclusive values for each activity given the habitat patches available. For each of the above models, all pair-wise interactions were tested for significance before any main-effect variable was removed. Interaction effects were tested for significance at the 0.10 level, and the main effects were tested for significance at the 0.05 level (Sokal and Rohlf 1995). Likelihood ratio tests were used to test overall model fit and the significance of adding or deleting variables (McCullagh and Nelder 1989, Neter et al. 1990).

RESULTS

BEDDING MICROSITE SELECTION

The final model's coefficients, along with their respective standard errors, t-statistics, and P-values are given in Table 9.1. The LRT results are presented in Table 9.2. Percent canopy closure, stem density, dbh, slope, distance from tertiary roads, distance from horse trails, north/northeast/southwest aspects, and distance from radio-location all have significant effects on the probability of an elk selecting a site in which to bed. Significant interaction effects were found between north-facing aspects and slope, northeast-facing aspects and slope, patch size and stem density, and distance from secondary roads and stem density. When there is a significant interaction effect between two main effects, the main effects remain in the model regardless of their significance (McCullagh and Nelder 1989). Consequently, patch size and secondary roads remain in the final model. Given the parameterization shown in Table 9.1, the equation for the utility/satisfaction derived from selecting patch i as a bed site equals:

TABLE 9.1 Parameterization and Output for the Final Discrete Choice Model for Summer, Diurnal, Bed Site Selection by Elk in Custer State Park, South Dakota, 1994–1997

General variables	Coefficient	1 SE	t	P
Canopy closure	4.5723	1.5229	3.5225	0.0006
Stem density	0.0016	0.0015	2.0956	0.0379
Patch size	−0.4371	0.3817	−1.1752	0.2419
Slope	9.4575	5.7801	1.9059	0.0587
Dbh	−0.2848	0.1632	−1.8977	0.0597
Distance from				
Radio-location	0.0028	0.0008	3.9760	0.0001
Secondary roads	−0.0023	0.0024	−0.7091	0.4794
Tertiary roads	−0.0029	0.0014	−2.9335	0.0039
Horse trails	−0.0053	0.0017	−2.5633	0.0114
Aspect				
North	5.0159	1.4629	3.5738	0.0005
Northeast	3.9562	1.4117	2.8912	0.0044
Southwest	−3.4421	1.2118	−2.8546	0.0049
Interactions				
North-aspect-slope	−47.7662	13.4697	−3.3745	0.0009
Northeast aspect-slope	−33.4798	13.1568	−2.5479	0.0118
Stem density-patch size	0.2645	0.0691	3.8704	0.0002
Stem density-sec. rds.	0.0002	0.0001	2.4199	0.0168

$$
\begin{aligned}
U_i^{\text{Bedding}} =\ & 4.5723(\text{Canopy closure}_i) + 0.0016(\text{Stem(density}_i) \\
& - 0.4371(\text{Patch size}_i) + 9.4575(\text{Slope}_i) - 0.2848(\text{Dbh}_i) \\
& + 0.0028(\text{Distance from radio-location}_i) - 0.0023 \\
& (\text{Distance from secondary road}_i) - 0.0029(\text{Distance from} \\
& \text{tertiary road}_i) - 0.0053(\text{Distance from horse trail}_i) \\
& + 5.0159(\text{North aspect}_i) + 3.9562(\text{Northeast aspect}_i) \\
& - 3.4421(\text{Southwest aspect}_i) - 47.7662(\text{North aspect}_i \\
& \times \text{Slope}_i) - 33.4798(\text{Northeast aspect}_i \times \text{Slope}_i) + 0.2645 \\
& (\text{Stem density}_i \times \text{Patch size}_i) + 0.0002(\text{Stem density}_i \times \\
& \text{Distance from secondary road}_i).
\end{aligned}
\tag{9.9}
$$

TABLE 9.2   Model Fit Statistics for Elk Bedding, Feeding, Moving, and Overall Microsite Resource Selection in Custer State Park, South Dakota, 1994–1997

Behavior	Final model	Null model[a]	LRT[b]	df	P
Bedding	−57.61	−247.59	379.96	15	<0.0001
Feeding	−64.95	−224.44	318.98	12	<0.0001
Moving	−99.20	−273.56	348.72	13	<0.0001
Naïve	−125.12	−284.76	319.28	14	<0.0001

[a]Null model = all coefficients equal 0.
[b]LRT = likelihood ratio test = $-2 * [L(\mathbf{B}^m) - L(\mathbf{B}^k)]$.

The effect of each habitat-related parameter on the utility of choosing a patch as a bed site for elk is evaluated as in any regression model (McCullagh and Nelder 1989, Neter et al. 1990). The magnitude of the influence of each habitat-related variable on elk bed site selection is evaluated using the parameter coefficient and range of values for that parameter. For example, in Eq. (9.9), as canopy closure increases, the probability that an elk will select that site as a bed site increases. North and northeast aspects increase the probability of elk use, whereas southwest aspects decrease the probability of use. However, as the slope of the patch increases, the benefit of the northeast aspect decreases.

FEEDING MICROSITE SELECTION

The final model's coefficients, along with their respective standard errors, t-statistics, and P-values are in Table 9.3. The LRT results are presented in Table 9.2. Percent canopy closure, meadow and fire habitat types, distance from primary, secondary, and tertiary roads, distance from horse trails, northeast aspects, and distance from radio-location all significantly affect the probability of an elk selecting a site for feeding. Significant interaction effects were noted between primary roads and fire-killed habitats, primary roads and meadow habitats, and secondary roads and meadow habitats. Given the parameterization shown in Table 9.3, the equation for the utility/satisfaction derived from selecting patch i as a feeding site equals:

TABLE 9.3 Parameterization and Output for the Final Discrete Choice Model for Summer, Diurnal, Feeding Site Selection by Elk in Custer State Park, South Dakota, 1994–1997

General variables	Coefficient	1 SE	t	P
Canopy closure	−2.4794	0.6542	−2.7552	0.0069
Meadow	0.9752	0.3158	2.3597	0.0201
Fire	0.3255	0.1489	1.9344	0.0507
Stem density	−0.0042	0.0018	−2.4768	0.0144
Distance from				
Radio-location	0.0075	0.0016	3.8327	0.0002
Primary roads	−0.0016	0.0002	−2.6639	0.0089
Secondary roads	−0.0049	0.0026	−2.3497	0.0206
Tertiary roads	−0.0027	0.0011	−2.2501	0.0265
Horse trails	−0.0099	0.0020	−2.9885	0.0035
Aspect				
Northeast	1.5592	0.4845	3.1521	0.0021
Interactions				
Primary roads — fire	0.4068	0.0398	3.1165	0.0024
Primary roads — meadow	0.7933	0.1332	2.9778	0.0036
Secondary roads — meadow	0.6465	0.1966	3.7603	0.0003

$$
\begin{aligned}
U_i^{\text{Feeding}} = {} & -2.4794(\text{Canopy closure}_i) + 0.9752(\text{Meadow habitats}_i) \\
& + 0.3255(\text{Fire} - \text{killed habitats}_i) - 0.0042(\text{Stem density}_i) \\
& + 0.0075(\text{Distance from radio} - \text{location}_i) - 0.0016 \\
& (\text{Distance from primary roads}_i) - 0.0049(\text{Distance} \\
& \text{from secondary roads}_i) - 0.0027(\text{Distance from tertiary} \\
& \text{roads}_i) - 0.0099(\text{Distance from horse trails}_i) + 1.5592 \\
& (\text{Northeast aspects}_i) + 0.4068(\text{Distance from primary} \\
& \text{roads}_i \times \text{Fire-killed habitats}_i) + 0.7933(\text{Distance from} \\
& \text{primary roads}_i \times \text{Meadow habitats}_i) + 0.6465(\text{Distance} \\
& \text{from secondary roads}_i \times \text{Meadow habitats}_i).
\end{aligned}
\tag{9.10}
$$

MOVING MICROSITE SELECTION

The final model's coefficients, along with their respective standard errors, t-statistics, and P-values are given in Table 9.4. The LRT results are presented in

TABLE 9.4 Parameterization and Output for the Final Discrete Choice Model for Summer, Diurnal, Moving Site Selection by Elk in Custer State Park, South Dakota, 1994–1997

General variables	Coefficient	1 SE	t	P
Canopy closure	1.2366	0.0319	1.9780	0.0504
Fire	−0.3552	0.1157	−2.8549	0.0051
Patch size	−0.4982	0.2267	−1.0844	0.2805
Slope	15.9391	6.9542	2.3571	0.0202
Stem density	0.0117	0.0231	2.0235	0.0454
Distance from				
Radio-location	0.0059	0.0006	3.2566	0.0015
Primary roads	0.0009	0.0004	2.5680	0.0115
Secondary roads	0.0219	0.0089	1.2765	0.2044
Tertiary roads	0.0029	0.0007	1.9822	0.0499
Horse trails	0.0097	0.0041	2.3552	0.0203
Interactions				
Primary roads — fire	3.2791	1.4490	3.4719	0.0007
Secondary roads — fire	2.9533	1.1376	2.1052	0.0375
Tertiary roads — fire	2.4850	1.0484	3.5745	0.0005
Stem density — patch size	0.6731	0.2655	2.4731	0.0149

Table 9.2. Percent canopy closure, fire habitat types, slope, stem density, distance from primary and tertiary roads, distance from horse trails, northeast, and distance from radio-location all significantly affect the probability of an elk selecting a site for moving. Significant interaction effects were noted between primary, secondary, and tertiary roads and fire-killed habitats, and stem density and size. Given the parameterization shown in Table 9.4, the equation for the utility/satisfaction derived from selecting patch i as a moving site equals:

$$U_i^{\text{Moving}} = 1.2366(\text{Canopy closure}_i) - 0.3552(\text{Fire-killed}$$
$$\text{habitats}_i) - 0.4982(\text{Patch size}_i) + 15.9391(\text{Slope}_i)$$
$$+ 0.0117(\text{Stem density}_i) + 0.0059(\text{Distance from}$$
$$\text{radio} - \text{location}_i) + 0.0009(\text{Distance from primary}$$
$$\text{road}_i) + 0.0219(\text{Distance from secondary road}_i)$$
$$+ 0.0029(\text{Distance from tertiary road}_i) + 0.0097$$
$$(\text{Distance from horse trails}_i) + 3.2791(\text{Distance from}$$

primary road$_i$ × Fire − killed habitats$_i$) + 2.9533(Distance
from secondary road$_i$ × Fire − killed habitats$_i$) + 2.4850
(Distance from tertiary roads$_i$ × Fire-killed habitats$_i$)   (9.11)
+ 0.6731(Stem density$_i$ × Patch size$_i$).

Behavior Selection

In the final behavior selection model we obtain three utility equations with
moving behavior set as our baseline. This means that the utility of foraging and
bedding is relative to moving; utility is not an absolute term. Time of day,
measured in hours, was the only significant factor ($P < 0.001$) and did not
differ between foraging and bedding ($P > 0.05$). Nonlinear functions of time of
day were nonsignificant ($P > 0.05$). All data were collected between the hours
of 0900 and 1900. Interestingly, the inclusive values for each of the behaviors
was nonsignificant (all $P > 0.05$), indicating that the availability of resources
does not influence the probability of choosing a particular behavior given the
distribution of resources within CSP. The final model equals:

$$U_{Bedding} = 1.6419 + 0.0866(\text{Time of day});   \quad (9.12)$$

$$U_{Feeding} = 1.6419 + 0.0866(\text{Time of day});   \quad (9.13)$$

$$U_{Moving} = 0.   \quad (9.14)$$

The implied probabilities for choosing a given behavior are calculated using
Eq. (9.4) (remembering that choosing a behavior is based on the same
theoretical model as choosing a habitat patch) and are shown in Fig. 9.3.
Given that bedding and feeding have the same utility equations, they also have
an equal probability of being chosen.

Naïve Model That Ignores Behavior

The final naïve model's coefficients, along with their respective standard
errors, t-statistics, and P-values, are given in Table 9.5. The LRT results are
presented in Table 9.2. Percent canopy closure, stem density, slope, distance
from secondary and tertiary roads, distance from horse trails, north/northeast/
southwest aspects, and distance from radio-location all significantly affect the
probability of an elk selecting a microsite during the day in summer. Significant
interaction effects were noted between north aspect and slope, northeast

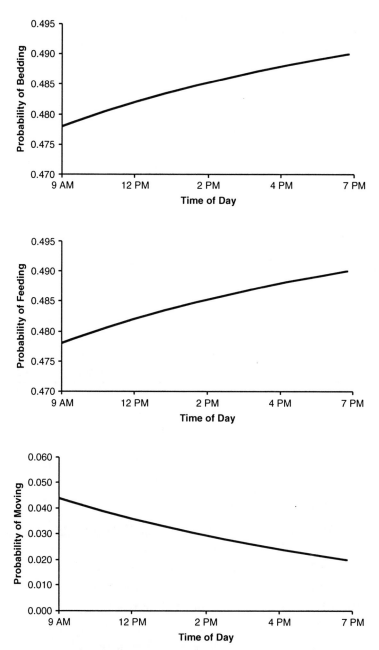

FIGURE 9.3   Probability of bedding, feeding, and moving as a function of time of day as described by Eqs. (9.12), (9.13), and (9.14) applied to Eq. (9.4).

TABLE 9.5 Parameterization and Output for the Final Discrete Choice Model for Summer, Diurnal Naïve Microsite Selection by Elk in Custer State Park, South Dakota, 1994–1997

General variables	Coefficient	1 SE	t	P
Canopy closure	3.2745	1.2220	3.1197	0.0019
Stem density	0.0393	0.1209	2.4950	0.0130
Slope	7.5611	3.2426	2.0906	0.0372
Meadow	0.1023	0.0297	1.1517	0.2479
Distance from				
Radio-location	0.0021	0.0006	3.7531	0.0002
Primary roads	0.0037	0.0016	1.7201	0.0862
Secondary roads	0.0008	0.0003	2.3796	0.0178
Tertiary roads	0.0021	0.0013	2.6551	0.0083
Horse trails	0.0009	0.0002	2.3540	0.0191
Aspect				
North	1.5412	0.0677	2.9423	0.0035
Northeast	0.7791	0.3173	2.6162	0.0092
Southwest	−1.4958	0.6750	−2.9519	0.0033
Interactions				
North aspect—slope	21.8521	9.6381	3.1185	0.0019
Northeast aspect—slope	11.4329	3.5183	3.0728	0.0023
Primary roads—meadow	0.4749	0.1890	3.0236	0.0027

aspect and slope, and primary roads and meadows. Given the parameterization shown in Table 9.5, the equation for the utility/satisfaction derived from selecting patch i during the day in the summer is:

$$
\begin{aligned}
U_i^{\text{Naive}} = {}& 3.2745(\text{Canopy closure}_i) + 0.0393(\text{Stem density}_i) \\
& + 7.5611(\text{Slope}_i) + 0.1023(\text{Meadow habitats}_i) + 0.0021 \\
& (\text{Distance from radio-location}_i) + 0.0037(\text{Distance from} \\
& \text{primary roads}_i) + 0.0008(\text{Distance from secondary roads}_i) \\
& + 0.0021(\text{Distance from tertiary roads}_i) + 0.0009(\text{Distance} \\
& \text{from horse trails}_i) + 1.5412(\text{North aspects}_i) + 0.7791 \\
& (\text{Northeast aspects}_i) - 1.4958(\text{Southwest aspects}_i) \\
& + 21.8521(\text{North aspect}_i \times \text{Slope}_i) + 11.4329(\text{Northeast} \\
& \text{aspect}_i \times \text{Slope}_i) + 0.4749(\text{Distance to primary road}_i \\
& \times \text{Meadow habitats}_i).
\end{aligned}
\tag{9.15}
$$

DISCUSSION

By applying Eqs. (9.9), (9.10), and (9.11) to Eq. (9.4) and linking this with a GIS, habitat suitability maps can be produced for bedding, feeding, and moving, respectively. An overall suitability map can be created by applying Eqs. (9.9) through (9.14) to Eq. (9.5), where each component of Eq. (9.5) is modeled as in Eq. (9.4). Because the distance from the elk's radio-location to the edge of the patch is significant in each equation, this map can only be generated within a choice set. Had this variable not been significant, utility would be independent of the elk's initial location, and a utility surface map could be generated for the entire park. The significance of this variable may be due to the fact that elk have a tendency not to stay in the same place over a 28-h period. However, as evidenced by the low probability of moving (Fig. 9.3), elk tend to remain in the same place throughout the daylight hours. The utility equations could be used to explore changes in habitat quality for each elk behavior as affected by habitat management. As with all models, the discrete choice model cannot accurately predict utility or resource selection in cases when the value of the attributes lies outside the range of attribute values used to parameterize the model.

In our case study, the discrete choice models gave insight into the factors affecting the feeding, bedding, and moving selection process of elk in CSP. Meadow and fire-killed habitats and sites containing low canopy closure offered the greatest utility as feeding sites for elk in CSP (Table 9.3). Unlike other literature (Edge et al. 1987), the utility of a nonmeadow or non-fire-killed site for daytime feeding decreased as distance from primary, secondary, and tertiary roads and horse trails increased. This pattern is similar to resource selection patterns of CSP elk at night during the summer when they tend to choose roadside habitats for foraging (Millspaugh 1999). However, meadows and fire-killed habitats follow a similar pattern, described by Edge et al. (1987), with utility increasing with increasing distance from roads and trails (see Interactions, Table 9.3). Given the large magnitude of the coefficients for the interaction terms and the scale of the measures for distance to roads (ranging from 100 to 1300 m), fire-killed and meadow habitats give by far the highest utility, and the increasing trend in utility with increasing distance from roads far exceeds the small, negative coefficients on the distance-to-roads main effects.

Sites on steep north and northeast aspects containing high levels of canopy closure and stem density offered the greatest utility as a bed site for elk in CSP (Table 9.1). These factors may reflect both thermal and hiding needs of elk during the day in summer (Zahn 1985, Millspaugh et al. 1998b). Unlike previous research, utility as a bed site decreased with increasing distance from secondary and tertiary roads and horse trails, yet increased in the presence of high stem densities. Given that Millspaugh (1999) noted elk tend to choose roadside habitats for foraging at night during the summer and that this pattern continues for nonmeadow and non-fire-killed sites during the day, we hypothesize that other factors, such as proximity to preferred forage sites, especially at night, may help explain summer, diurnal, bed site selection of CSP elk.

Moving habitat was similar to bedding habitat in that steep areas containing high stem density and canopy closure provided the greatest utility (Table 9.4). As distance from primary and tertiary roads and horse trails increased, the utility of a site for moving improved significantly, especially in the case of fire-killed habitats. These results indicate that elk may be attracted to sites offering high security, presumably related to their need for hiding cover (Lyon and Ward 1982) when moving during the day in summer.

The two key findings of the behavior model are that behavior is independent of the available resources but dependent on the time of day, and that the probabilities of choosing to feed or bed are about equal. Green and Bear (1990) found that the daily activity patterns of cow elk in Colorado were affected by season and time of day. In the summer, two major feeding bouts coincided with sunrise and sunset, although elk fed periodically throughout the daylight hours (Green and Bear 1990). The observations in our study occurred between the hours of 0900 and 1900 during the summer, which is after sunrise and before sunset, which may explain why our results do not indicate specific feeding bouts but rather feeding throughout the daylight hours. In light of this, it is important to remember that the model presented here applies only to behavior and resource selection during daylight hours in the summer and may not accurately represent the other time periods or seasons. The equality in the probability of these two behaviors is not surprising because 44.3% of the total observations were of bedding animals and 46.8% were of feeding animals. The low probability of choosing to move is not surprising because only 9.9% of the observations were of moving individuals. However, this does not mean that individuals move only infrequently during the day but rather that daytime movements are likely to be of short duration relative to the time spent foraging and bedding.

The danger of ignoring behavior becomes apparent when one compares the naïve model with the behavior-specific resource selection models. For instance, the naïve model indicates that increasing canopy closure has a positive impact on the utility of a site, but the feeding-specific model indicates that increasing canopy closure has a negative impact on the utility of a site. Without accounting for behavior, one would assume that greater canopy closure is better for elk when in fact that is not always the case. Similarly, the naïve model predicts that utility of a site increases with distance from roads and trails, whereas that is not necessarily the case with regard to feeding and bedding—the two dominant behaviors in our dataset. In some situations, elk seem to prefer to be closer to roads and trails. This may have something to do with the park's practice of thinning forests along the roads, which has been shown to increase the forage quality and quantity along roadsides (Millspaugh 1999). It is important to note that the dangers of ignoring behavior do not apply solely to discrete choice models but to all resource selection techniques.

Aebischer et al. (1993) listed four problems affecting the validity of analytical techniques used in resource selection studies. These problems include inappropriate level of sampling and inadequate sample size, the unit-sum constraint, differential habitat use by groups of individuals, and arbitrary definition of habitat availability. When properly designed, the multinomial logit form of the discrete choice model circumvents most of these shortcomings. Sampling occurs at the scale of the individual, though individuals could be grouped if their choices were not independent from one another in a biological sense, as in a mother-offspring situation. Work is currently under way to fully incorporate the repeated-measures design when resighting an individual more than once. Regarding the second issue raised by Aebischer et al. (1993), the multinomial logit formulation is affected by the unit-sum constraint. An increase in the probability of choosing a given resource will lead directly to a decrease in the probability of choosing any other resource. However, the discrete choice model avoids the problems raised by Aebischer et al. (1993) because no inferences are made about "selection" or "avoidance" of the patch in general. A resource may have a low probability of being selected given the other available resources based on the attributes of these resources, but this is a relative statement. It is possible that the resource with low selection probability could be one of the animal's preferred resources yet less preferable than other available resources. In a discrete choice framework, there are no good and bad options, there are only better and worse options. Also, the unit-sum constraint only applies within one choice set, and by integrating over all choice sets, the unit-sum constraint does not apply in the context as described by Aebischer et al. (1993). The third issue of Aebischer et al. (1993) relates to the fact that animals of different ages or sexes may use habitat differently. As previously mentioned, it is quite straightforward in the multinomial logit framework to test and incorporate the effect of an individual's characteristics on their choices. The fourth problem raised by Aebischer et al. (1993) is dealt with directly because researchers must carefully define the choice set for each individual observation such that this cannot be a hidden assumption in the analysis. However, as with other techniques, inappropriate or incorrect definitions of availability will undoubtedly bias the results of the study.

There are other advantages offered by discrete choice models. For instance, in analyses involving categorical habitat types (e.g., mature woodland, shrubland) changes in the characteristics within these habitat types (e.g., stem density) cannot be accounted for in models based strictly on the frequency of use vs. abundance of these types (Friedman 1937, Neu et al. 1974, Quade 1979, Johnson 1980). If land use practices change the tree density in one of the habitat types, one might expect that these changes could influence the selectivity of that habitat type. Without redoing the entire study, a categorical habitat model (i.e., Friedman 1937, Neu et al. 1974, Quade 1979, Johnson 1980) cannot predict what effect this change might have on resource selection. Discrete choice models circumvent this by directly modeling the attributes rather than forcing the researcher to lump patches into unique categories.

The above case study is one example of an application of nested discrete choice models to the field of wildlife resource selection. Any behavior whereby an individual chooses between discrete resources, whether they are bed sites, foraging areas, prey items, or nest sites, can potentially be analyzed using discrete choice models. Other factors such as territoriality can be incorporated into the analysis by including variables such as distance from, or density of, territorial individuals. Discrete choice models can be applied to both observational field data and controlled experiments. Even though simple utility functions are most easily interpreted and incorporated into management, complex utility equations like the ones in the case study can also assist researchers by highlighting the factors that should be controlled in future perturbation experiments.

SUMMARY

Originally developed by econometricians more than 25 years ago, discrete choice models give insight into how the characteristics of the resources in question and of the individuals selecting those resources affect the probability of a particular resource being chosen. Such resources may be a patch of habitat or a type of prey or forage item. The data required for these models are similar to those necessary for logistic regression, and the characteristics of the resources and individuals may be either continuous or categorical variables. However, unlike logistic regression, the definition of availability may differ for each selection event, thus allowing differences to be both spatial and temporal. By nesting multiple discrete choice models, it is possible to analyze resource selection under a variety of behaviors (e.g., the characteristics of habitat patches chosen for sleeping, foraging, or moving). The model is estimated using maximum likelihood techniques and produces a series of equations that describe the value an individual places on each resource as a function of both the individual's and that resource's characteristics given what activity is being undertaken. These equations can then be combined to calculate the probability of a particular resource being chosen given all of the resources that are available at that time and place. These points were explained using a case study of elk in CSP, South Dakota.

CHAPTER 10

Using Euclidean Distances to Assess Nonrandom Habitat Use

L. MIKE CONNER AND BRUCE W. PLOWMAN

Statistical tools for evaluating habitat use are important for understanding habitat ecology of wildlife species. Habitat use metrics generally rely on some type of comparison between habitat use and habitat availability (see Chapter 8), and many approaches are multinomial (i.e., animal locations are estimated to occur in one of several habitat categories) (Neu et al. 1974, Marcum and Loftsgaarden 1980, Aebischer et al. 1993).

By definition, locational errors cause incorrect conclusions concerning the exact geographic positions of animals on the landscape (see Chapter 3). When using a multinomial habitat use procedure (Neu et al. 1974, Marcum and Loftsgaarden 1980, Aebischer et al. 1993), telemetry errors can cause locations to be misclassified, resulting in a loss of power and a misclassification bias (White and Garrott 1986, Samuel and Kenow 1992, Pendleton et al. 1998).

Radio Tracking and Animal Populations
Copyright © 2001 by Academic Press. All rights of reproduction in any form reserved.

Intensified sampling may increase statistical power but may not reduce bias. Therefore, if telemetry error exists, efforts should be made to reduce bias (Samuel and Kenow 1992).

Samuel and Kenow (1992) proposed a technique for evaluating habitat use with estimated triangulation error. The technique relies on generating random points from the error distribution associated with each estimated telemetry location. Rather than determining habitat use directly from estimated telemetry locations, habitat use is derived from the sample of points taken from the error distributions associated with each location estimate. The technique reduces bias (Samuel and Kenow 1992), but it does not appear to be widely used.

Herein we discuss desirable characteristics of a habitat metric and suggest a simple method for analysis of habitat use data that incorporates these desirable characteristics. The procedure is suitable whenever a multinomial approach would normally be employed. However, much like compositional analysis (Aebischer et al. 1993), the procedure is not solely restricted to multinomial data, and other attributes (i.e., covariates) can easily be incorporated into the analysis. In addition, hypothesis tests are based on summary statistics that provide transparent evaluation of effect size and biological relevance. Furthermore, the technique is easy to perform and interpret within a geographic information system (GIS). Finally, although telemetry error will reduce statistical power as in any habitat analysis, explicitly modeling telemetry error to reduce misclassification bias is not required.

DESIRABLE CHARACTERISTICS OF A HABITAT ANALYSIS TOOL

Aebischer et al. (1993) outlined desirable characteristics of a habitat assessment tool. A tool for assessing habitat use should (1) use the animal as the sampling unit; (2) permit hypothesis testing among meaningful groups; (3) work at multiple spatial scales; and (4) allow for the nonindependence of habitat proportions (i.e., the unit-sum constraint). Although Aebischer et al. (1993) provide a good starting point for evaluating a habitat-use metric, we think there are at least two additional criteria that should be considered when selecting a habitat analysis tool to be used with radiotelemetry data.

Habitat use metrics should be robust to telemetry error. As location accuracy decreases, statistical power decreases (White and Garrott 1986) and bias increases (Samuel and Kenow 1992). Intuitively, the problem of bias becomes more pronounced as the area bounding estimated telemetry locations (e.g., error polygon or error ellipse) increases relative to the size of the average habitat patch. Thus, in patchy landscapes, a habitat use technique must be robust to telemetry error, or a bias-correcting mechanism (e.g., Samuel and Kenow 1992) must be used to ensure the validity of the analysis.

A habitat analysis procedure also should provide summary statistics for evaluation of effect size if a statistical difference is detected (Pendleton et al. 1998). Many commonly used methods for assessing habitat use do not have readily available summary statistics; rather, habitat use indices must be used to estimate strength of animal-habitat associations. Because these indices are not generally used in the analysis (i.e., in the hypothesis test), they may not adequately reflect the effect size associated with statistical tests.

HABITAT ANALYSIS
WITH EUCLIDEAN DISTANCES

Many researchers are aware of the usefulness of Euclidean distances in habitat studies. However, the use of Euclidean distances appears to be focused on distances between animal locations and linear (e.g., creeks, roads, edges) or point (e.g., trees, coarse woody debris) features (Boal and Mannan 1998, J.D. Clark et al. 1993, 1999, Jorgensen et al. 1998, McKee et al. 1998, Ormsbee and McComb 1998). Use of Euclidean distances to areal features (e.g., habitat types) has been used much less often (Conner and Leopold 1998). Clearly, Euclidean distances are not alien to habitat studies, but their use in habitat analyses has unnecessarily been limited to point and line features. We suggest that Euclidean distances can be used to evaluate habitat use relative to any spatial habitat feature.

Our Euclidean distance approach to evaluating habitat use relies on simulated points from a uniform random distribution to estimate expected distances to habitat types, or other spatial features, of interest. If habitat use occurs at random, the distances between animal locations and each habitat type should equal the distances between random points and each habitat. Furthermore, the ratio of used distances to random distances should equal 1.0. If habitat use is nonrandom, the ratios can be used to determine which habitats are used disproportionately (i.e., habitat use is more or less than expected). If the use/random distance ratio is low (< 1.0), the animal is associated with the habitat more than expected. Similarly, if the use/random ratio is high (> 1.0), the animal is associated with the habitat less than expected. Because habitat use/availability ratios are commonly used by wildlife researchers, it is intuitive to construct similar ratios based on distances.

For our description of the Euclidean distance procedure, we assume that the only interest is in determining if habitat use is nonrandom and, if so, identifying habitats that were used disproportionately. Furthermore, the following steps assume that habitat within an individual's home range represents available habitat and habitat use is estimated from telemetry locations [i.e., Johnson's (1980) third-order habitat selection]. Following the description of the technique, we provide an example that evaluates habitat use at a scale similar to that investigated by Neu et al. (1974). In the discussion, we will explain how the procedure can be extended to a variety of situations using a general linear modeling approach.

Testing for Nonrandom Habitat Use

Here we outline the basic steps necessary to test for nonrandom habitat use.

1. Simulate numerous locations from a uniform random distribution within the home range of each animal. The stability of the elements in r_i (see step 3 below) should be used to define "numerous." Because each point is a subsample, generating too many points will only increase processing time and will add little to the quality of the analysis.

2. For each animal (i), calculate the distance from each random point to the nearest representative of each habitat type.

3. Calculate the average distance from random points to each habitat type. This creates a vector of mean distances (r_i) for each animal. This vector represents the vector of expected distances for the ith animal.

4. Repeat steps 2 and 3 with estimated animal locations to establish (u_i) for each animal. This vector represents the average distance from the ith animal to each habitat type.

5. Create a vector of ratios (d_i) for each animal by dividing each element in u_i by the corresponding element in r_i. If habitat use occurs at random, the expected value of each element in the d_i is 1.0.

6. Determine if the mean vector (ρ), calculated as the mean of the d_i, differs from a vector of 1's using multivariate analysis of variance (MANOVA). If ρ significantly differs from a vector of 1's, there is evidence that nonrandom habitat use is occurring. If the vector does not differ from a vector of 1's, then habitat use does not appear to differ from random and further analysis should not occur.

7. To determine which habitat types were used disproportionately, test each element within ρ to determine if it differs from 1 using a t-test or non-parametric equivalent. A significant difference indicates that the habitat was used disproportionately. The magnitude of the difference between the element in ρ and 1 is an indicator of effect size. Intuitively, if the element in ρ is < 1 (i.e., $u < r$), then the habitat was used more than expected. Alternatively, if

the element in $\rho > 1$ (i.e., $u > r$), then the habitat was used less than expected.

RANKING AND COMPARING HABITATS

The elements in ρ provide a ranking of habitat use relative to habitat availability. The habitat with the lowest value was used most relative to availability (i.e., animal locations are closest to this habitat relative to random points), whereas the element with the largest value was used least relative to availability. However, this ranking says nothing about whether a particular habitat type is proportionately used (i.e., habitat use after correcting for habitat availability) more (i.e., significantly) than other habitat types. Fortunately, a pair-wise test comparing proportionate use of habitat types can be performed using standard hypothesis testing procedures. For example, to determine if habitat A is proportionately used more than habitat B, the appropriate null hypothesis to test is $\rho_A - \rho_B = 0$. A significant test statistic indicates that one habitat was proportionately used more than the other habitat. The habitat with the lowest value in ρ was proportionately used more than the other habitat. A ranking matrix analogous to that presented by Aebischer et al. (1993) aids in interpreting the relationships among habitats.

Recently, much emphasis has been placed on estimation as opposed to hypothesis testing (Cherry 1998, Johnson 1999; see also Chapter 8). The difference between a given element in ρ and 1.0 is an estimate of disproportional use of that habitat. Precision associated with estimated elements in ρ is estimated using standard statistical procedures (e.g., standard deviation, standard error). Likewise, estimates of the difference between two elements in ρ and the precision associated with this estimate are also obtained in a straightforward manner.

EXAMPLE OF THE PROCEDURE APPLIED TO FOX SQUIRREL DATA

METHODS

To illustrate the Euclidean approach to habitat assessment, we used the method to compare habitat associated with fox squirrel locations to habitat available on the study area. We chose a sample of 17 fox squirrels (*Sciurus niger*) to illustrate the procedure. Because of the low number of animals in our example and the fact that the animal is the sample unit, our analysis was expected to have relatively low statistical power. However, our small sample size was chosen to provide a dataset that could be used by readers desiring to duplicate the analysis.

Squirrels were located using triangulation from known reference points within the study area during 1998–1999. Each squirrel used in the analysis had more than 50 location estimates.

We delineated four habitat types: (1) agriculture, (2) pine, (3) hardwood, and (4) mixed pine-hardwood (30% < hardwood < 70%) (Fig. 10.1). Habitat types were photo-interpreted and digitized into ARC/INFO (ESRI 1997).

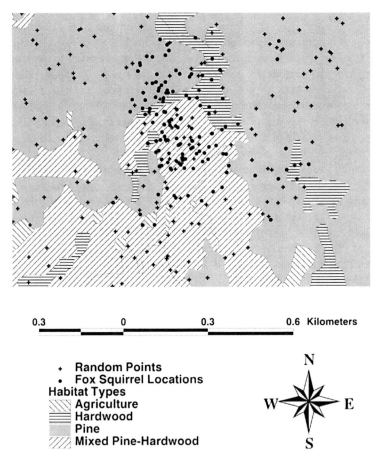

FIGURE 10.1 A portion of the study area showing animal locations, random locations, and habitat types.

We calculated the distance between each habitat type and each telemetry location using the ARC/INFO DISTANCE command (ESRI 1997). These distances were averaged for each animal (i) used in the analysis to create the u_i. Similarly, random points were generated to fall within the study area, and distances between each habitat type and each point were calculated and averaged to create r. Because we compared animal locations to random points throughout the study area, there was only one vector r for comparison with each u_i. We calculated ratios for each animal (d_i) by dividing distances associated with squirrel locations (u_i) by distances derived from random locations (r). We determined if the mean of the ratio vectors (ρ) differed from a vector of 1's using a MANOVA. Univariate t-tests were used to test the hypothesis that individual ratios did not differ from 1 for each habitat type. To rank habitat types and provide summary statistics for evaluating effect sizes, we performed all possible pair-wise habitat comparisons and constructed a ranking matrix of t-statistics (Aebischer et al. 1993). All statistical analyses were performed in SAS (SAS Institute, 1999). SAS code for performing the analysis is provided in Appendix 10.1.

RESULTS

The study area was 21% agriculture, 15% hardwood stands, 30% mixed stands, and 34% pine stands. The average distances from 2500 random points to agriculture, hardwood, mixed, and pine habitats were 67.67 m, 147.14 m, 85.97 m, and 60.71 m, respectively. The average distance of squirrel locations to each habitat varied among squirrels (Table 10.1).

The analysis of distance ratios (Table 10.2) indicated that fox squirrel locations differed from random locations ($F_{4, 13} = 16.39, P < 0.001$). Examination of distances to habitat types in univariate space indicated that fox squirrels were found closer to hardwood ($\rho_{hardwoods} = 0.43 \pm 0.07 (\bar{x} \pm SE)$, $t_{16} = -8.16$, $P < 0.001$) than expected. There were no differences between fox squirrel and random points with regard to distance to agriculture ($\rho_{agriculture} = 1.497 \pm 0.315$, $t_{16} = 1.58$, $P = 0.134$), mixed, ($\rho_{mixed} = 1.263 \pm 0.467$, $t_{16} = 0.56$, $P = 0.581$), and pine ($\rho_{pine} = 1.02 \pm 0.235$, $t_{16} = 0.08$, $P = 0.934$) habitats.

A ranking of habitats based on the values of the elements in ρ indicated hardwoods were proportionately used most followed by pine, mixed, and agriculture habitats. Pair-wise comparisons of distance ratios associated with habitat types (i.e., the elements in ρ) indicated that squirrels were found significantly closer to hardwood than to agriculture and pine habitats. There were no differences in proportional habitat use among other pair-wise comparisons (Table 10.3).

TABLE 10.1   Average Distance (m) from Fox Squirrel Telemetry Locations to Habitat Types on a Southwest Georgia Study Area, 1997–1998

Animal	Agriculture	Hardwood	Mixed[a]	Pine
F002	94.8	113.2	57.5	14.2
F014	34.5	60.7	414.2	7.6
F015	44.1	25.6	448.7	11.2
F016	30.3	41.0	92.8	10.6
F018	181.1	77.0	5.4	63.4
F019	29.1	30.4	83.0	57.2
F027	240.3	101.8	7.7	21.3
F028	98.8	63.6	485.6	15.0
F030	250.6	58.7	28.1	30.8
F031	64.4	187.3	4.7	157.4
F032	172.2	73.2	12.0	44.0
F033	35.4	23.2	30.7	167.3
F034	48.2	21.2	24.0	152.3
F035	25.7	82.9	31.4	73.5
F036	278.4	42.2	73.1	17.3
F037	23.3	26.2	40.2	51.7
F038	70.8	49.4	6.7	157.8

[a] 30% < hardwood < 70%.

COMPARISON WITH OTHER TECHNIQUES

Perhaps the most commonly used method for assessing wildlife habitat is the procedure described by Neu et al. (1974). This technique compares habitat use, based on animal locations, to habitat availability using a χ^2 test. If habitat use is nonrandom, Bonferroni confidence intervals are calculated for percent use to determine which habitats are used more or less than expected. Although this technique is easy to use and interpret, Aebischer et al. (1993) point out several problems with this approach. (1) The procedure does not use the animal as the sample unit for analysis. Therefore, animals that are monitored more intensively have greater weight in the analysis. (2) A test for differences in habitat use among meaningful groups (e.g., sex or age classification) is not available. (3) Because the technique uses animal locations in the analysis, it is difficult to apply at multiple spatial scales (e.g., comparing habitat within the home range to habitat available on the study area). In addition to the problems pointed out by Aebischer et al.

TABLE 10.2   Average Distance between Fox Squirrel Locations and a Given Habitat Divided by the Average Random Distance to the Same Habitat

Animal	Agriculture	Hardwood	Mixed[a]	Pine
F002	1.401	0.780	0.669	0.234
F014	0.510	0.413	4.818	0.126
F015	0.651	0.174	5.219	0.184
F016	0.449	0.279	1.079	0.174
F018	2.676	0.523	0.063	1.044
F019	0.431	0.206	0.965	0.941
F027	3.551	0.692	0.089	0.350
F028	1.460	0.432	5.649	0.247
F030	3.704	0.399	0.327	0.508
F031	0.952	1.273	0.055	2.593
F032	2.545	0.497	0.139	0.724
F033	0.523	0.158	0.357	2.756
F034	0.713	0.144	0.279	2.508
F035	0.379	0.563	0.365	1.210
F036	4.114	0.287	0.850	0.286
F037	0.345	0.178	0.468	0.852
F038	1.046	0.336	0.078	2.600
Mean	1.497	0.432	1.263	1.020

[a] 30% < hardwood < 70%.

TABLE 10.3   Ranking Matrix of Fox Squirrel Habitat Use[a]

	Agriculture	Hardwood	Mixed	Pine
Agriculture		3.44(0.003)	0.38(0.71)	1.06(0.305)
Hardwood	−3.44 (0.003)		− 1.71(0.11)	− 2.46(0.026)
Mixed	−0.38(0.71)	1.71(0.11)		−0.40(0.698)
Pine	−1.06(0.305)	2.46(0.026)	0.40(0.698)	

[a] Numbers are t-statistics (P-values) associated with pairwise comparisons of corrected distances to habitat[b].

[b] The t-statistic testing the null hypothesis that (mean squirrel distance to habitat A/mean random distance to habitat A) − (mean squirrel distance to habitat B/mean random distance to habitat B) = 0.

(1993), the Neu et al. (1974) technique is incapable of incorporating covariates (e.g., proximity to roads or creeks) into the analysis, and telemetry error must be explicitly addressed if misclassification bias is to be reduced. In contrast, the Euclidean distance approach does not suffer from these problems.

Recently, compositional analysis (Aebischer et al. 1993) has become popular for analyzing habitat use data. This technique correctly uses the animal as the sample unit, allows for meaningful comparisons among groups, permits analysis at multiple scales, and corrects for the unit-sum constraint. However, the analysis can be difficult to interpret because habitat use is assessed by scaling all habitats to a common habitat (i.e., the habitat proportion used in the denominator of the log-ratio differences). Furthermore, there is concern over the treatment of habitats that were not used or available at a given spatial scale, i.e., 0% availability or 0% use results in an undefined value, log(0). The Euclidean distance approach is virtually identical to compositional analysis regarding the multivariate nature of the analysis (e.g., use of MANOVA and ranking matrices). However, the Euclidean distance approach does not assess habitat use relative to a common habitat because the unit-sum constraint does not apply to distance data. Furthermore, if a habitat is not available within the home range of an animal, researchers must use an arbitrarily low value in the analysis to avoid undefined values, i.e., log(0), in compositional analysis. However, the Euclidean distance approach to habitat analysis requires no such arbitrary treatment of the data. Lastly, if estimated animal locations are used to derive habitat use trajectories, then some form of explicit error handling (e.g., modeling the error distribution associated with each location estimate) must be incorporated into the compositional analysis. Error modeling is not required with the Euclidean distance approach.

Logistic regression is commonly used to develop models of wildlife habitat selection (Brennan et al. 1986, Conner and Leopold 1998; see also Chapter 8). In many cases, the construction of logistic regression models can significantly increase our understanding of a species' habitat needs. In other cases a researcher may be interested in comparing habitat use, after correcting for habitat availability, among meaningful groups (e.g., sex or age classes). Although a logistic regression model can be constructed using two or more groups as dependent variables, the resulting model will only describe how habitat variables might be used to predict group membership. The model will not predict whether animals within the groups used habitat differently relative to availability. For example, males may use hardwood habitats more than females. The logistic regression model may be successful at discriminating between male and female habitat based on the prevalence of hardwood habitats. However, if females used areas with few hardwoods and males used areas with many hardwoods, the n females might actually use hardwoods more than expected, whereas males might use hardwoods less than expected. Under this scenario, important biological information concerning sex-specific habitat use could be overlooked.

BENEFITS OF THE EUCLIDEAN DISTANCE APPROACH

The Euclidean distance approach to assessing habitat use has several desirable attributes. The animal is correctly used as the sampling unit; thus, individual variation among animals is not assumed to be constant. Furthermore, because the animal is the sample unit, unequal sampling of individual animals does not affect the overall analysis. The only concern regarding sampling intensity associated with individual animals is that the intensity be sufficient to derive accurate estimates of mean distances to each habitat of interest.

Because the dependent data are continuous and do not suffer from the unit-sum constraint, analysis using MANOVA is intuitive. Addition of covariates or grouping variables can be accomplished within the MANOVA framework. For example, to determine if habitat use relative to availability is sex specific, a class variable for sex can be used within MANOVA to test this hypothesis. Furthermore, additional habitat features (e.g., creeks or roads) can be incorporated into the analysis. For example, to determine if distance to creeks affects habitat use (i.e., if creeks interact with habitat features to alter proportional habitat use), the mean distance to creeks could be included in a MANOVA as a covariate. Alternatively, creeks could be treated as any other habitat variable and use of creeks assessed relative to expected.

The analysis can be adapted to any spatial scale. For the procedure description, we chose to use an example that represents Johnson's (1980) third-order selection (i.e., use of habitat patches within the home range). For our example, we chose a commonly used spatial scale of analysis (Neu et al. 1974) that does not adhere to one of the hierarchies described in Johnson (1980). Other spatial scales can be used. For example, to compare habitat within an animal's home range to habitat available on the study area (i.e., Johnson's second-order habitat selection), the distances to habitats based on random points within the home range can be compared to distances calculated for random points placed throughout the study area.

Although telemetry error will reduce statistical power, the Euclidean distance approach to assessing habitat use requires no explicit error modeling. For example, assume that we have a checkerboard arrangement of four different habitats. Further assume that the animal spends virtually all of its time in one habitat. If we estimate habitat use for the animal, telemetry error may cause estimated locations to fall within wrong habitats, but the distance to the preferred habitat will be much lower than expected. Conversely, if each location must be classified into a single habitat, a misclassification bias will occur if explicit error modeling is not incorporated into the analysis (Fig. 10.2). Notably, no approach to analysis of habitat use can overcome biases associated with severe telemetry error (e.g., if error distributions are larger than habitat patches) because telemetry error of this magnitude will cause estimated animal locations to occur randomly on the study area.

The Euclidean distance approach provides summary statistics that are readily interpreted. Researchers and managers have little difficulty understanding use/availability ratios, or their derivatives, that are commonly used to index selection (Manly et al. 1993:10). The approach described herein is merely an extension of these common indices into the realm of Euclidean distances, with one notable advantage: the ratio used to index selection is the same ratio used in constructing hypothesis tests.

As an additional benefit, there is no need for all habitat types to be used or available for use by each individual animal. Some habitat use procedures require that all habitat types of interest be available to each animal studied (e.g., compositional analysis). If a given habitat is not available for use by an individual, then portions of the analysis are undefined [e.g., log (0); Alldredge et al. 1998]. Using the Euclidean distance approach negates this requirement. For example, if habitat A is not available to an individual, then the mean distance to habitat A in both u_i and r_i will be large, and d_i will approach 1.

RESEARCH NEEDS

There is a need for evaluating the type I error distribution of the Euclidean distance method under varying levels of telemetry error. Of particular interest would be a comparison of the Euclidean distance approach to other techniques (e.g., compositional analysis) with known levels of telemetry error. Furthermore, many animals use edges between two or more habitat types, use rare habitats, or use habitats that exist in narrow corridors (e.g., riparian areas). When animals are associated with rare habitats, edges, or narrow corridors, the influence of telemetry error may be high, and the Euclidean distance approach should have higher power than other techniques for assessing nonrandom habitat use. Specific simulations to assess statistical power when dealing with edge species are needed.

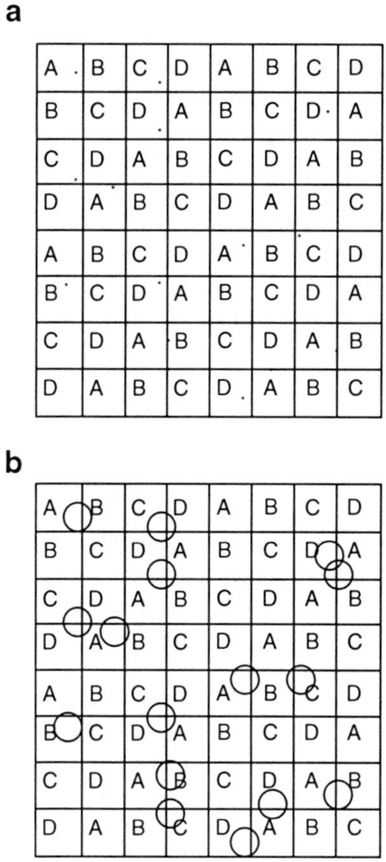

a

A	B	C	D	A	B	C	D
B	C	D	A	B	C	D	A
C	D	A	B	C	D	A	B
D	A	B	C	D	A	B	C
A	B	C	D	A	B	C	D
B	C	D	A	B	C	D	A
C	D	A	B	C	D	A	B
D	A	B	C	D	A	B	C

b

A	B	C	D	A	B	C	D
B	C	D	A	B	C	D	A
C	D	A	B	C	D	A	B
D	A	B	C	D	A	B	C
A	B	C	D	A	B	C	D
B	C	D	A	B	C	D	A
C	D	A	B	C	D	A	B
D	A	B	C	D	A	B	C

FIGURE 10.2 Example showing that explicit modeling of error associated with location estimates is unnecessary when using Euclidean distances to assess habitat use. Assume animals only occurred in habitat A but were estimated (*) with error (a). Assume location error is approximately half the width of a habitat block and is best modeled using a circle (b). Sixteen estimated locations were obtained. If each location is classified into one habitat type, as in Fig. 10.2a, habitat use appears to be random (i.e., four locations fell within each habitat), and an incorrect conclusion regarding habitat use is reached.

If the Euclidean distance approach is used, distance to habitat A is 0 in four cases, and less than half of a block width (telemetry error) separates remaining locations from habitat A. Because this relationship does not hold for other habitats (i.e., distances are often more than a half block width from remaining habitats), the distance to A will be less than for other habitats, indicating that points are associated with habitat A.

In Fig. 10.2b, error has been modeled using error circles. If proportions of each habitat type within each error circle are used to estimate habitat use, habitat A is also correctly identified as used > available. Using the Euclidean distance procedure, however, requires no explicit modeling of the error distribution to arrive at the conclusion that A is used more than expected.

FUTURE DIRECTIONS

We think that the benefits of using Euclidean distances for habitat analysis will have increased importance in future habitat studies. Biologists have long recognized the influence of linear features (e.g., edges or rivers) on wildlife habitat use, and Euclidean distances have been used to model habitat as a function of distance to linear features (Clark et al. 1993, Conner and Leopold 1998, McKee et al. 1998). Scenarios in which habitat features interact to create suitable habitat are envisioned. For example, perhaps a marginal habitat becomes a good habitat based on proximity to edge. We think that interaction among habitat features should become a primary focus of habitat use studies. Euclidean distances have the potential to elucidate the interactions between different habitat features.

We also believe that Euclidean distances will eventually prove useful for modeling spatial correlation of habitat features. Habitat features seldom exist at random within a landscape. Numerous natural and anthropogenic factors often interact to influence habitat composition of an area (e.g., hardwood forests are often associated with creeks within a managed pine matrix). When these interactions occur, spatial correlation of habitats also occurs. Methods for investigating spatial correlation of habitat features and their impact on wildlife may be developed using Euclidean distance–based approaches. If so, the spatial correlation of various habitat features will likely prove interesting as we seek to understand why an animal uses a particular geographic area.

In addition to interactions among habitat features, animal–animal interactions may influence habitat use, especially among territorial or social species. Future habitat analysis techniques should allow animal–animal interactions to be explicitly investigated as predictors of habitat use. To incorporate animal–animal interactions into an analysis of habitat use will require the merger of spatial and temporal analyses (e.g., animal use of habitats when they are near other animals). Such approaches will require intensive monitoring of tagged animals. As transmitter technology increases, investigations of habitat use as a function of animal–animal interactions will broaden our understanding of mechanisms that influence animal use of habitat.

SUMMARY

In this chapter, we present a multivariate technique for evaluating habitat use based on Euclidean distances. The technique uses a multivariate analysis of variance to compare Euclidean distances between animal locations and habitats of interest to expected distances derived from random locations. The procedure uses the animal as the sampling unit, does not suffer from a unit-sum constraint, allows for comparison among meaningful groups, and works at multiple spatial scales; thus, the test is similar to compositional analysis (Aebischer et al. 1993). However, habitat types that are not used by the animal (e.g., percent use = 0) do not result in undefined values [i.e., $\log(0)$], as in compositional analysis. Furthermore, although telemetry error may decrease statistical power of the procedure, there is no need to explicitly model telemetry error in an attempt to reduce misclassification bias. We present an example using fox squirrels.

APPENDIX 10.A   SAS Code for Analyzing Euclidean Distance Data[a]

```
data eucdist;
/*Note:input data are in Table 10.2*/
infile 'table2.txt';
input sq $ ag hwd mix pine;

/*Set up all possible habitat comparisons for ranking matrix */
dagmix=ag-mix;
daghwd=ag-hwd;
dagpine=ag-pine;
dhwdmix=hwd-mix;
dhwdpine=hwd-pine;
dmixpine=mix-pine;

/*Modify data for overall test of null hypothesis that mean ratios are a vector of 1's
(actually a vector of 0's because 1 is being subtracted from each element of the vector).*/

ag=ag-1;
hwd=hwd-1;
mix=mix-1;
pine=pine-1;
run;

/*Overall test (distances=expectation) */
proc glm;
class sq;
model ag hwd mix pine=/nouni;
manova h=intercept/printe printh;
title 'Test for used distances = random distances';
run;

/*Test if habitat use = availability: univariate test*/
proc means data=summ mean stderr t prt;
var ag hwd mix pine;
title 'Test that each habitat ratio = 1.0, (e.g., ag-1=0)';
run;

/* This section constructs the ranking matrix (Table 10.3) */
proc means data=summ mean stderr t prt;
var daghwd dagmix dagpine dhwdmix dhwdpine dmixpine;
title 'Pairwise comparison of distance da – db';
run
```

[a]Code performs analysis on data in Table 10.2.

CHAPTER 11

Effect of Sample Size on the Performance of Resource Selection Analyses

FREDERICK A. LEBAN, MICHAEL J. WISDOM, EDWARD O. GARTON, BRUCE K. JOHNSON, AND JOHN G. KIE

Assessing resource selection is one of the main objectives of many wildlife radiotelemetry studies (Haney and Solow 1992, Dixon 1995, Copeland 1996, Tufto et al. 1996). Methods for determining resource selection vary from the widely used χ^2 analysis (Neu et al. 1974, Byers et al. 1984, Jelinski 1991) for categorical data to complex statistical modeling using continuous data (see

Chapters 2, 8, and 9). How data were obtained in part determines what statistical analyses can be performed (i.e., which of these methods of resource selection is appropriate for the data). Manly et al. (1993) described the application of resource selection methods that are commonly used to analyze habitat data. Recently, Alldredge et al. (1998) surveyed and compared many of these methods. The "performance" of such methods also was evaluated by Alldredge and Ratti (1986, 1992), Cherry (1996), and McClean et al. (1998). However, the choice of the most appropriate method remains problematic, even though many authors (e.g., Alldredge and Ratti 1986, 1992, Thomas and Taylor 1990, White and Garrott 1990, Aebischer et al. 1993, Manly et al. 1993, Alldredge et al. 1998) have offered their insight.

Alldredge and Ratti (1986) compared four statistical methods of analyzing resource selection; computer simulations of field data were used to estimate rates of type I and type II errors. Alldredge and Ratti (1986) also presented the hypotheses, assumptions, and data requirements associated with use of the Neu et al. (1974), Johnson (1980), Friedman (1937), and Quade (1979) methods. For their simulations, Alldredge and Ratti (1986) considered the number of habitats (4, 7, 10, 15), the number of animals (10, 20, 50), and the number of observations per animal (15, 50). They found that the type II error rate for the omnibus test decreased as the number of animals or the number of observations per animal increased. The exception was the Johnson (1980) method, where in some cases the number of errors increased as the number of observations per animal increased. For multiple comparisons, Alldredge and Ratti (1986) reported that in two-thirds of cases the actual error rate was lower than the nominal ($\alpha = 0.05$) type I error rate. In another one-third of cases (four-habitat case), the error rate was much larger than 5%. The Friedman (1937) method had a large average type I error rate of 21.3% for the four-habitat case. Frequency of type II errors for their multiple comparisons depended on the same factors as the overall test, including the number of habitats and the magnitude of differences.

Cherry (1996) used simulation parameters similar to those of Alldredge and Ratti (1986) to compare several methods of constructing confidence intervals. He studied three alternatives: the usual method with a continuity correction factor, and two methods based on multinomial proportions. He reported poor performance of binomial confidence intervals with and without continuity correction, even with large sample sizes. The binomial confidence interval without continuity correction was used in the Neu et al. (1974) and Byers et al. (1984) methods. Cherry (1996) recommended use of the Bailey intervals, which provided the best combination of low error rates and interval length.

McClean et al. (1998) compared six methods (Neu et al. 1974; Johnson 1980; Friedman 1937; Quade 1979; compositional analysis, Aebischer et al.

1993; and multiresponse permutation procedures, Mielke 1986) using data from radio-marked Merriam's wild turkey (*Meleagris gallopavo merriami*) hens with poults in the Black Hills, South Dakota. These authors chose hens with poults for comparing use-availability analyses because habitats that meet the dietary requirements of poults are well documented. Habitat availability was defined by three circular buffers (100–, 200–, and 400–m radii) around individual hen locations, as well as the total study area. They found that only the Neu et al. (1974) method identified selection patterns consistent with known requirements of poults at all levels of availability. Compositional analysis was the only method that did not reject the omnibus null hypothesis when availability was defined as the whole study area. The authors recommended defining availability at the study area level, which was generally superior to use of circular buffers, regardless of the statistical method used.

New radiotelemetry technologies [e.g., Argos satellite-based methods, global positioning system (GPS) transmitters; see Chapter 4] facilitate collection of an enormous number of animal locations, which can be combined with GIS data of resource variables based on remotely sensed descriptors of vegetation, topography, and human activity (see Chapter 8). These new technologies make it necessary to reassess the performance of resource selection methods in relation to the number of animals vs. the number of observations per animal.

Accordingly, we used an extensive database of elk (*Cervus elaphus*) observations from the Starkey Experimental Forest and Range (referred to hereafter as Starkey) (Rowland et al. 1997, 1998) to evaluate sample size requirements for the Neu et al. (1974) (Neu et al. method) and compositional analysis (Aebischer et al. 1993) methods of estimating resource selection. A sample of 42 elk, each monitored intensively during spring 1994, was used as a case example for our analysis. Our main objective was to evaluate the issue of sampling allocation (i.e., to sample more animals less intensively or fewer animals more intensively, in relation to accuracy of the two methods).

STUDY AREA AND TECHNOLOGIES

Starkey is located 35 km southwest of La Grande in northeast Oregon (Rowland et al. 1997). An ungulate-proof fence encloses the 10,102–ha area (Bryant et al. 1993), which contains four separately fenced areas (Rowland et al. 1997). The main (7762 ha) and northeast (1453 ha) study areas contain known, controlled populations of mule deer (*Odocoileus hemionus*) and elk, and are used for telemetry studies during spring through fall (Rowland et al. 1997).

Movements of a sample of elk, deer, and cattle were monitored with an automated telemetry system (ATS) (Rowland et al. 1997, Johnson et al. 2000). Animals were systematically located approximately once every 3–4 h with the ATS (Wisdom 1998). Location accuracy of the ATS averaged ± 53 m (Findholt et al. 1996).

METHODS

SELECTING ELK AND RESOURCE DATA FOR ANALYSIS

We selected data from 42 radio-collared elk for our analysis. Each animal had at least 300 locations that were collected during spring (mid-April to mid-June) 1994. We subsequently limited this dataset to the daytime period (1 h after sunrise to 1 h before sunset) to reduce the potential confounding effects of time of day on use patterns, based on results of an earlier, separate analysis on time-of-day use patterns (Leban 1999).

Resource variables were selected based on their assumed importance to elk (Table 11.1), per rationale and literature review of Johnson et al. (2000). Resource availability was fixed for all animals because elk were contained by a game fence surrounding the main study area of Starkey. Availability was calculated as the proportion of each resource type in the study area. Use was determined from resource variables that were associated with each animal location. Results were accumulated by resource types for each radio-collared animal and summarized for all animals.

ESTABLISHING THE BASELINE PATTERN OF RESOURCE SELECTION

We used the full set of elk locations for the 42 radio-collared elk to establish the baseline pattern of resource selection. We used this baseline pattern to approximate the true pattern of resource selection for these 42 elk. We then used this approximation to test the accuracy of various levels of sample size as estimators of resource selection (Table 11.2) under compositional analysis and the Neu et al. method. Using our baseline pattern to approximate the true pattern of resource selection was reasonable because (1) resource variables were derived as a census of all resource values taken from all 30 × 30 m pixels in our study area (Rowland et al. 1998); (2) exploratory analysis of location data on radio-collared elk in our study area indicated that estimates of resource use reached asymptotic properties at or before collection of 100 locations per

TABLE 11.1 Resource Variables[a] Used to Assess Performance of Neu et al. and Compositional Analysis Methods for Assessing Resource Selection

Description	Variable name	No. of classes	Categories[a]
Distance to nearest open road	DistOpen	4	$0 - \frac{1}{2}, \frac{1}{2} - 1, 1 - 1\frac{1}{2}, > 1\frac{1}{2}$ km
Aspect	Aspect	4	0–90, 91–180, 181–270, & 271–359 degrees
Percent canopy closure of trees > 12cm dbh	Sz > 3Can	3	0–39, 40–69, > 70% canopy closure
Distance to forage[b]	DistForg	3	0–100, 101–200, > 200 m
Distance to cover[c]	DistCovr	4	0–100, 101–200, 201–300, > 300 m
Vegetation (plant community types)	Eco13	13	1—Buildings, structures, roads 2—Ponderosa pine—Douglas fir 3—Juniper 4—Lodgepole pine 5—Ponderosa pine—grass 6—White fir—mixed conifer 7—Bunchgrass 8—Dry meadow 9—Moist meadow 10—Wet meadow 11—Nonvegetated areas 12—Low sagebrush 13—Water-covered areas

[a]Based on literature search by Johnson et al. (2000).
[b]Less than or equal to 40% canopy closure of trees > 12cm dbh.
[c]Greater than 40% canopy closure of trees > 12cm dbh.

animal (M. J. Wisdom and A. A. Ager, unpublished data); by contrast, mean
(± SE) number of locations per animal for the 42 radio-collared elk in our
study was substantially greater than 100 (912 ± 40) (Rowland et al. 2000);
and (3) the large number of radio-collared animals represented a large percentage (approximately 15%) of the total population of elk in the study area.
Moreover, the number of locations per animal and the number of radio-
collared animals that composed our dataset far exceeded the typical sample
sizes used in most radiotelemetry studies of resource selection.

SIMULATING DESIRED LEVELS OF SAMPLE SIZE

We used resampling procedures to simulate varying levels of sample size in
relation to accuracy of resource selection estimates (Table 11.2). Resampling
procedures were designed to simulate five levels of the number of radio-
collared animals (5, 10, 20, 30, and 42 animals) in combination with six levels
of the number of locations collected per animal (10, 20, 30, 50, 100, and all
observations). As a preliminary step in resampling elk and elk locations, the
42 elk were ordered and numbered from 1 to n. Similarly, telemetry locations
for each elk were ordered by date and time to ensure that locations were in the
proper sequence.

Resampling of elk was accomplished by drawing a random number from 1
to n. Using a random starting point where the first elk location was in the first
x location, we sampled every x (interval) location. The sampling interval was
determined as the total number of locations for each elk divided by the
number of locations required to achieve each level of sampling. This resampling approach corresponded to a common procedure researchers apply by
locating animals systematically every other day or every week or at some other
regular interval (see Chapter 2).

TABLE 11.2 Sampling Scheme for Evaluating Resource Selection Methods

Factors	Levels
Sampling design	Systematic
Season	Spring
Time of day	Day
Number of radio-collared animals	5, 10, 20, 30, 42 (all)
Locations per animal	10, 20, 30, 50, 100, and all observations
Habitat types	6 (fixed) ranging in 3–13 categories each (see Table 11.1)

Elk were resampled randomly without replacement for each simulation. We then calculated the percentage of correct conclusions per 1000 simulations as our measure of accuracy at each level of sample size, as described in the next section.

EVALUATING ACCURACY OF RESOURCE SELECTION ESTIMATES

We analyzed locations under the various levels of resampling to determine if we would have reached the same conclusions as those reached for the true resource selection patterns. We then tallied the frequency of correct conclusions that was reached for each method, which composed our assessment of accuracy. The specific procedures for evaluating performance of each resource selection method in relation to sample size are described below.

Neu et al.

Consider the habitat variable aspect with four categories. The difference between the true value and the estimated value for the first category in the first sample might be 0.01 when resampling 20 locations per animal. For each resampling at this level (20 locations per animal), the difference between the true and estimated value was obtained to produce a distribution of differences. This process was repeated for increasing numbers of locations per animal. Similarly, the distribution of differences was calculated for the other three aspect categories. These distributions of differences can be described by computing the MSE, i.e., [(proportion of use for each sample − proportion available) − (proportion of use for all animals − proportion available)]^2/number of runs. Plotting the number of locations against the square root of MSE would give a measure of the performance of the Neu et al. method.

Consider again the aspect variable with four categories. Presuming that the omnibus test resulted in a rejection of the null hypothesis (that the resource categories were used in proportion to their availability), the three possible conclusions were that each category was selected, neither selected nor avoided (i.e., no difference), or avoided. Assuming that the true conclusion from all locations on all animals was that the first category was significantly selected, resampling would result in either a correct conclusion (significant selection) or an error (no difference or significant avoidance). The relation of the percentage of errors made and the sample size indexes how well this method performed.

Compositional Analysis

The MSE for compositional analysis was computed using the λ values from the true population and the resampled data (λ is the ratio of the determinant of the mean-corrected sum of squares and cross-products and the raw sum of squares and cross-products). The percentage of errors in habitat rankings was calculated by tallying the number of incorrect rankings from resampled data.

RESULTS

OMNIBUS TEST USING ALL OBSERVATIONS

Analysis of all data showed significant selection for all resource variables under both the Neu et al. and compositional analysis methods (Table 11.3). However, compositional analysis could not be used to evaluate resource selection of vegetation type due to singularity of the data matrix.

PERCENTAGE OF CORRECT CONCLUSIONS

Neu et al. Method

With the Neu method, the percentage of correct conclusions typically increased with increasing number of animals and increasing observations per animal for all variables (Fig. 11.1). Generally, the percent correct was low (<60%) with few (10) observations even when all 42 animals were simulated. Doubling

TABLE 11.3 Baseline Patterns of Resource Selection Based on Applying Two Analysis Methods for All Locations of 42 Elk during Spring Daytime Period, 1994, Main Study Area, Starkey Experimental Forest and Range, Northeast Oregon

Habitat types	Method	
	Neu et al.	Compositional analysis
Vegetation	1	−1
Distance to open roads	1	1
Aspect	1	1
% canopy closure	1	1
Distance to forage	1	1
Distance to cover	1	1

Note: 0 = no selection (P > 0.05); 1 = there was selection (P ≤ 0.05); −1 = analysis cannot be performed because of singular matrix.

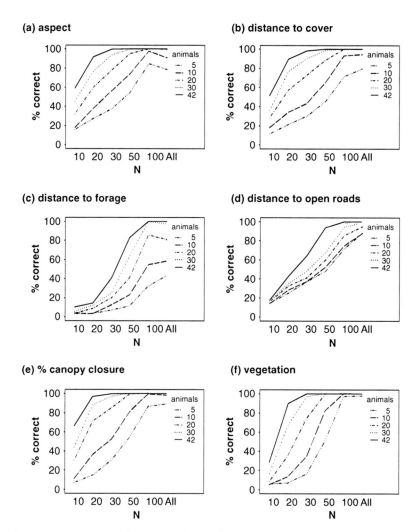

FIGURE 11.1 Percentage of correct conclusions for overall test using the Neu et al. method on six habitat variables, based on resampling of radiotelemetry data at 5 levels of number of radio-collared animals (legend) in combination with six levels of number of locations per animals (N), using data from 42 female elk monitored during day, spring 1994, main study area, Starkey Experimental Forest and Range, northeast Oregon.

the observations per animal from 10 to 20 dramatically increased the percent correct to 90% or better in some cases. Similarly, doubling the number of animals (e.g., from 10 to 20) with 20 observations each improved the percent correct by as much as 100% for some variables (e.g., % canopy closure) (Fig. 11.1).

Compositional Analysis

The percentage of correct conclusions for the overall test for compositional analysis did not vary much with increasing observations per animal but was determined primarily by the number of animals sampled (Fig. 11.2). The only

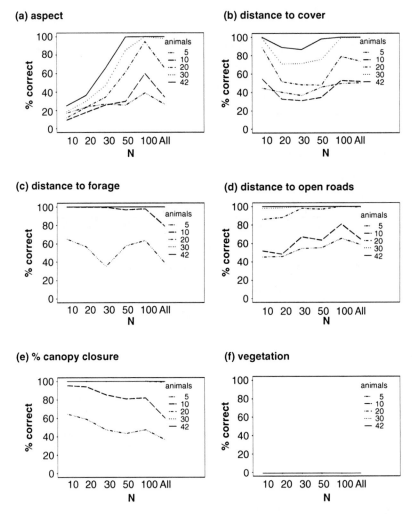

FIGURE 11.2 Percentage of correct conclusions for overall test using compositional analysis on six habitat variables, based on resampling of radiotelemetry data at 5 levels of number of radio-collared animals (legend) in combination with six levels of number of locations per animal (N), using data from 42 female elk monitored during day, spring 1994, main study area, Starkey Experimental Forest and Range, northeast Oregon.

exception was the aspect variable whereby the percent correct increased with increasing number of animals and observations. Generally, the percent correct was low (<60%) for few (five) animals but was very high (100%) for all 42 and even 30 animals. The percent correct was also high even for moderately large numbers of animals (10 and 20) with as few as 10 observations in some cases (e.g., distance to forage, % canopy closure) (Fig. 11.2).

Multiple Comparison Tests of Compositional Analysis

Multiple comparison tests are possible under compositional analysis; these tests evaluate the relative strength of resource selection among all resource types. Under our simulations, the percentage of correct conclusions reached for multiple comparisons for compositional analysis generally was low (51%) for small sample sizes (5 animals with 10 observations) (Fig. 11.3). Moreover, even for 42 animals with 10 observations the percent correct was only as high as 62%, except for percent canopy closure, where it was as high as 86%. The percent correct increased with sample sizes for some cases (e.g., aspect, distance to forage) while remaining fairly constant for others (e.g., percent canopy closure, distance to open roads). The percent correct for the 42–animal case was generally higher than other cases. It required about 50 observations for the 30– and 42–animal cases to attain a high percentage of correct conclusions for pairwise comparisons (Fig. 11.3).

DISCUSSION

SUGGESTED SAMPLE SIZE

The percentage of correct conclusions for the overall test of resource selection increased with sample sizes (number of animals and number of observations) for both the Neu et al. and compositional analysis methods; this was consistent with findings of Alldredge and Ratti (1986) for the Neu et al. method. We would have expected the results from compositional analysis to follow this same pattern but the number of observations had negligible effect in some cases. This result from compositional analysis suggests that researchers should radio-mark as many animals as possible while collecting a minimum number of observations per animal, in contrast to more intensive collection of locations on fewer animals.

Overall, our results suggest that a minimum sample of 20 animals with 50 observations per animal is needed to adequately determine resource selection for a population during a season at one time of day. This recommended sample size is identical to that of Alldredge and Ratti (1986) for simulated data.

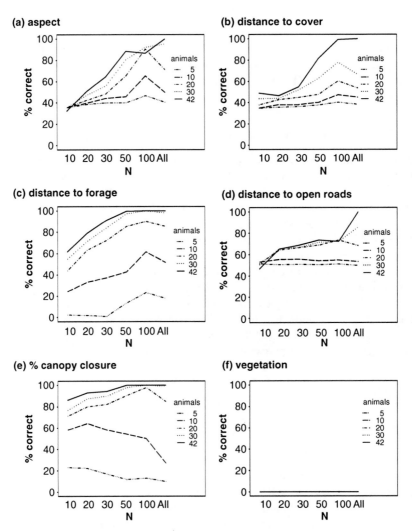

FIGURE 11.3   Percentage of correct conclusions for multiple comparisons using compositional analysis on six habitat variables, based on resampling of radiotelemetry data at 5 levels of number of radio-collared animals (legend) in combination with six levels of number of locations per animal (N), using data from 42 female elk monitored during day, spring 1994, main study area, Starkey Experimental Forest and Range, northeast Oregon.

Aebischer et al. (1993) recommended radio-marking at least 10, and preferably 30, animals for compositional analysis. When sample sizes are small, alternative statistical tests, such as randomization tests (Romesburg 1985, Manly 1997), may be employed.

When a sample of animals was taken but all observations for each of these animals were used, the percentage of correct conclusions was lower than when only 100 observations were used (e.g., Fig. 11.3a). The decrease in accuracy may be attributed to the unequal number of observations among animals. Some animals had twice as many observations as others and using all their observations unduly biased the overall estimate.

Computational Requirements of Compositional Analysis

The failure of compositional analysis to analyze resource selection of vegetation type illustrates the problem posed by this method's requirement that the use of resource data in the analysis not result in a singular matrix. That is, if data have linearly dependent rows or columns, certain mathematical operations (matrix inversion) would fail, making the calculations required for compositional analysis impossible to complete (Leban 1999). Compositional analysis was inappropriate for analyzing vegetation because some vegetation types were rare, thereby creating a singular matrix.

Unfortunately, Aebischer et al. (1993) neglected to warn prospective users of this potential problem when using compositional analysis. Even though Aebischer et al. (1993) suggested several ways of treating missing habitat types, compositional analysis still does not guarantee linearly independent rows and columns. Furthermore, these authors cautioned readers that any adjustment to the successful calculation of λ could make its value no longer independent of the habitat type chosen as denominator, and that it may depart from the standard χ^2 distribution.

Multiple Comparison Tests under Compositional Analysis

The accuracy of multiple comparisons under compositional analysis followed a similar pattern to that of the overall test, i.e., accuracy increased with increasing sample sizes. However, the accuracy of pairwise comparisons for two resource variables (distance to forage and % canopy closure; Fig. 11.3) using compositional analysis resulted in large differences between the 5– and the 42–animal cases. The large differences may be attributed to the magnitude of the proportion used and proportion available for those resources. It is widely known that a small or moderate sample is sufficient to detect large differences. However, a large sample is required to detect small differences.

Considering 10 observations per animal, the 42–animal case had more than eight times the number of observations as the 5–animal case (420 vs. 50 observations).

OTHER FACTORS OF POTENTIAL INFLUENCE ON RESULTS

A variety of factors could affect results of simulations designed to assess the accuracy of resource selection estimates in relation to sample size. In particular, the factors inherent to the methods used to collect animal locations, as well as the unique characteristics of a given species' movement and resource use, could influence the simulation results. One such example is the degree of serial correlation among locations of each animal (Swihart and Slade 1985a; see also Chapter 5). We found that the full set of animal locations for each of our 42 animals were highly correlated, based on analysis methods of Swihart and Slade (1985a; A. A. Ager, unpublished data). By contrast, randomly sampled locations at low sample sizes (e.g., 5 or 10 locations per animal) for each animal were not correlated. Relations between accuracy of resource selection estimates under varying levels of serial correlation and sample size need further investigation to assess how such differences in serial correlation might affect simulation results.

Another factor that could affect simulation results is the movement characteristics of the animals themselves. Species that move quickly over large areas and that have specific habitat preferences that vary by time of day or season may require larger sample sizes to estimate resource selection. By contrast, species with small home ranges and specialized patterns of resource selection that vary less by time of day or season may require relatively small sample sizes for accurate estimation of resource selection. Assessing the effects of such characteristics of animal movement and patterns of resource selection in relation to accuracy of resource selection estimates and sample size would be a fruitful endeavor for future simulations.

In a similar manner, simulation results could be affected by sampling that mixes locations across times of day and seasons. For example, if wide variation in resource use occurs between an animal's foraging vs. resting sites, mixing this variation together to estimate patterns of resource selection and to evaluate sample size requirements under simulation could confound the results of simulation efforts such as those described here. These relations deserve further attention in future work on the evaluation of sample size requirements.

STRENGTHS AND WEAKNESSES OF EACH METHOD

Alldredge and Ratti (1986, 1992) and McClean et al. (1998) have assured us that there is no clear choice as to which method of resource selection is best in all cases. Choosing a method is further complicated by controversy surrounding resource selection analyses (Hobbs and Haney 1990, Aebischer and Robertson 1994, Haney 1994). Consequently, "the choice of method depends ultimately on which statistical hypothesis is most closely related to the biological question of interest, on how observations and individuals are weighted, and on which assumptions are most likely to be satisfied" (Alldredge and Ratti 1992:8). A researcher has to choose among many methods.

The Neu et al. method is the most commonly used method due to its simplicity, but it has been applied erroneously (Jelinski 1991). The technique uses each observation on each individual animal as the sample unit and pools across animals, ignoring any dependencies among observations of the same animal (see Chapter 8). Each individual animal may or may not have the same selection pattern as the others. Recent findings suggest that the use of individual locations as the sample unit to compare across animals (as in the Neu et al. method) is pseudoreplication (Otis and White 1999). These authors recommended against the use of the Neu et al. technique for analysis of resource selection data.

Although compositional analysis has the potential problem of failures related to singularity of the data matrix, Aebischer et al. (1993) emphasized the method's advantages over other methods. This method does not seem to be affected much by the number of observations. This strongly suggests that a researcher should radio-mark as many animals as possible and take a moderate number of observations per animal in order to obtain a high percentage of correct conclusions. Consequently, we recommend compositional analysis as a better method for analyzing resource selection due to its modest requirement for sample sizes, capability to evaluate resource selection among identifiable animal groups (e.g., age, sex), and utility in conducting multiple comparisons among resource types and animal groups. However, a major computational requirement of compositional analysis is that resource data used for this method not result in singularity of the data's matrix structure.

SUMMARY

We evaluated the effect of sample size on results for the Neu et al. (1974) and compositional analysis (Aebischer et al. 1993) methods of evaluating resource selection. We conducted our evaluation with elk location data generated from an automated tracking system, which allowed collection of a substantially larger dataset (42 radio-collared animals containing more than 300 locations per animal for a 2–month period) than could be collected with conventional methods of radiotelemetry. We used these data to estimate the baseline patterns of resource selection for our population of elk and used resampling methods to simulate effects of varying number of animals (5, 10, 20, 30, 42) and varying number of locations per animal (10, 20, 30, 50, 100, all) on the accuracy of resource selection estimates. Specifically, we calculated the percentage of correct conclusions (accuracy) for 1000 simulations for elk selection of six resource types (aspect, distance to open roads, distance to cover, distance to forage, % canopy closure, and vegetation) under the varying levels of resampling for the two methods of resource selection. In general, accuracy of resource selection increased with increasing number of animals and increasing number of observations per animal for all resource variables. Accuracy was particularly low (<60%) for few animals (5 or 10) with only 10 observations. Our results suggest that a minimum sample of 20 animals and 50 observations per animal are needed to accurately estimate resource selection for our population and season. This finding has strong implications for resource selection studies that rely on conventional methods of radiotelemetry, which typically fail to meet these minimum levels of sampling. Moreover, when radiotelemetry resources are limited, monitoring a higher number of radio-collared animals while collecting fewer locations per animal appears more accurate than more intensive monitoring of fewer animals. Of the two methods tested, the Neu et al. method is less desirable because the animal location must be used as the sample unit, which results in pseudoreplication. By contrast, compositional analysis is more desirable due to its modest sample size requirements, capability to evaluate resource selection among identifiable animal groups (e.g., age, sex), and utility in conducting multiple comparisons among resource types and animal groups. However, a major computational requirement of compositional analysis is that resource data used for this method not result in singularity of the data's matrix structure. Regardless of method, the potentially high sample sizes needed to obtain accurate estimates of resource selection under our analysis warrant further research to better understand the relations among the number of radio-marked animals, number of locations collected per animal, and accuracy of resource selection estimates.

ACKNOWLEDGMENTS

We thank Drs. J. R. Alldredge and J. T. Ratti for their thought-provoking questions and suggestions about this manuscript. M. M. Rowland, R. J. Stussy, and A. A. Ager (all associated with the Starkey project) provided assistance during various stages of this project. We extend our gratitude to E. J. Fox for permission to use the computers in the Remote Sensing/GIS Laboratory, University of Idaho, without which the simulations would not have been possible.

CHAPTER 12

High-Tech Behavioral Ecology: Modeling the Distribution of Animal Activities to Better Understand Wildlife Space Use and Resource Selection

JOHN M. MARZLUFF, STEVEN T. KNICK, AND JOSHUA J. MILLSPAUGH

Space Use
    Using the Utilization Distribution
    Moving beyond *Where* Animals Occur
Resource Selection
    Measuring Use
    What Is Available?
    Mapping Important Habitats
    Assumptions for Resource Selection Models
    Identifying Mechanisms from Patterns
Improving Our Approach to the Study of Wildlife
    Radiotelemetry
Summary

Wildlife management has benefited tremendously from analytical and technological breakthroughs in radiotelemetry techniques that allow animals to be observed more often and over larger areas than ever before. We are able to identify what habitat features are used by animals with increasing clarity. However, despite these advances, we remain relatively ignorant about *why* animals use particular habitat features. Here we argue that detailed behavioral

Radio Tracking and Animal Populations
Copyright © 2001 by Academic Press. All rights of reproduction in any form reserved.

observations and robust models of animal home ranges, utilization distributions, and resource needs can be combined with accurate mapping in a geographic information system (GIS) environment to improve our understanding of the behavioral mechanisms driving animals to use particular resources.

Our objective is to illustrate how investigations of space use and resource selection can be integrated to increase our understanding of resources that are important to wildlife. As such, our chapter compliments the chapters by Kernohan et al. (Chapter 5) and Erickson et al. (Chapter 8). We suggest that users of radiotelemetry (1) increase their understanding of animal behavior by answering why their study animals use resources in a nonrandom pattern, (2) make greater use of the utilization distributions to quantify space use and resource selection, and (3) consider alternative resource selection techniques such as mapping important habitats by using predictive models. We develop each of these topics below, beginning with a brief review of important issues, approaches, and definitions related to animal space use and resource selection. We present new data illustrating how utilization distributions can be created for specific behaviors and how they can be used to investigate resource selection. Lastly, we discuss modeling and mapping procedures useful in studies of resource selection that were briefly introduced by Erickson et al. (Chapter 8).

SPACE USE

Ecologists have tried to measure "home range" since Burt (1943) first formally introduced the concept. First, we simply connected the outer points to define a maximum area of use, the "minimum convex polygon" (Mohr 1947). Concerned that we were not adequately sampling all movements, and in an attempt to account for use patterns within a range, a series of alternative analytical models were developed (e.g., bivariate normal, harmonic mean, peeled polygons, nearest-neighbor clustering, and adaptive and fixed kernels) to interpolate and extrapolate total space use from a sample of observations (Jennrich and Turner 1969, Dixon and Chapman 1980, Kenward 1987, Worton 1989, 1995). Our quest has been to find the best analytical technique to define first the total area used by an animal followed by subsequent analyses to look for patterns of use within that area. Recent simulation studies suggest that notable improvements have been made. Under many circumstances, including adequate sample size and appropriate user-defined options, which vary in each study, kernel estimators accurately transform point locations into the area used by an animal (Seaman and Powell 1996, Seaman et al. 1999).

In the search for accurate models of space use, a holistic measure of the pattern of space use, termed the "utilization distribution" (UD; Van Winkle 1975), also was developed. The UD is a potentially powerful though under-utilized analytical tool available to those studying radio-tagged animals. The UD is a probability density function that maps an individual's or group of individuals' relative use of space. It depicts the probability of an animal occurring at each location within its home range as a function of the number of telemetry locations (White and Garrott 1990:146; Fig. 12.1).

The UD has been used in telemetry studies to (1) provide an objective (albeit arbitrary) way to exclude location estimates perceived as "excursive" from the calculation of home range, (2) define areas of frequent or "core" use (Samuel et al. 1985), (3) incorporate geographically referenced time budgets into home range calculations (Samuel and Garton 1987), and (4) assess animal interactions (Seidel 1992, Millspaugh et al. 2000). The latter three uses contribute to our understanding of how animals use space because they help determine important areas and habitats.

We believe that use of the UD will enhance future studies of animal move-ments, species interactions, and resource selection because it focuses on animal use of the entire range. Using the UD, our understanding of animal movements would improve by describing use patterns in a probabilistic sense (e.g., low or high probability of using a particular site). Additionally, compari-son of UDs would allow for a more accurate estimate of space use overlap or species interaction (see Chapter 5). The UD may be equally advantageous for studies of resource selection because instead of focusing on used or nonused sites, we could assess the probability of using particular resources within the animal's range based on the UD.

To properly use UD estimates, certain assumptions must be met. Primary among these is the assumption that our interpolation technique accurately represents use where we did not locate animals. That is, we assume that our subjects frequent areas near areas in which we commonly observed them and avoid areas far from frequently used areas. The degree to which this assump-tion is met depends on our ability to obtain an unbiased sample of locations that represents the full range of space use (Otis and White 1999). In other words, we must systematically sample animals throughout the range of times, seasons, and activity periods likely to affect their movement (see Chapter 2). Such unbiased representation of the entire pattern of space use is paramount to the calculation of an accurate UD. A second important assumption is that we have enough data to compute the UD estimate. For example, accurate kernel-based estimates are quite data intensive (typically more than 50 loca-tions per animal; Seaman et al. 1999, Chapters 2 and 5) and will likely differ among studies. If a researcher cannot meet sample requirements, a biased estimate of the UD may be computed (Seaman et al. 1999).

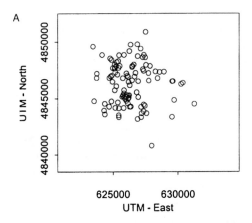

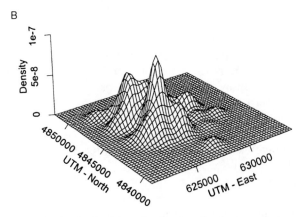

FIGURE 12.1   An example of a fixed kernel utilization distribution with least-squares cross-validation smoothing (bottom; B) for elk locations determined by radiotelemetry (top; A) in Custer State Park, South Dakota. The height of the utilization distribution provides an index of relative use. (Kernel plotting routines from Bowman and Azzalini 1997).

USING THE UTILIZATION DISTRIBUTION

The UD could become a standard metric by which an animal's total use of space is described and compared. Using the UD as a standard metric of total space use will (1) reduce the impact of telemetry error because clusters of points, rather than single points, are used to construct the estimate; (2) eliminate concerns about independence of points because a systematic sampling strategy, with short time intervals that may be statistically dependent, will provide an adequate UD estimate provided data are collected systematically throughout the period to which inferences are drawn; and (3) correctly emphasize the animal (or group of animals), not the point estimate, as the experimental unit. We illustrate how the UD can be used to improve our understanding of space use by animals with examples from our ongoing research on prairie falcons (*Falco mexicanus*) in Idaho (Marzluff et al. 1997a) and American crows (*Corvus brachyrhynchos*) in Washington.

The Idaho Army National Guard uses approximately one-third of the Snake River Birds of Prey National Conservation Area (NCA) for military activities (e.g., maneuver exercises, tank and gunnery firing practice, and missile launching) (U.S. Department of the Interior 1996). A large proportion of the world's population of prairie falcons nest in the NCA and forage in areas used for military training. To determine if foraging falcons changed their pattern of space use when firing occurred, we developed UDs using fixed kernel analyses (see Chapter 5) for falcons on days when firing occurred and days when it did not. Falcons that nested near the training area and regularly hunted within the training area changed their space use about 30% when firing occurred, compared with periods of no firing [intersection of UDs = 68%, Fig. 12.2a,b; see Eq. (5.10) for a discussion of how to calculate V.I. (volume of intersection)]. However, falcons that nested far from the training area and rarely used the area showed nearly a 50% change in space use when firing occurred (V.I. = 56%, Fig. 12.2c,d). This suggests that, contrary to expectation, falcons farthest from the training may suffer most from firing, perhaps because they are not habituated to the disturbance. We currently are conducting these comparisons for individual falcons so that individual variability in overlap values can be determined and UDs can be compared statistically.

MOVING BEYOND *WHERE* ANIMALS OCCUR

As suggested above, studies of animals' space use patterns can be improved by increasing our understanding of why animals use particular areas. We can accomplish this by using telemetry to observe animals more closely and by extending the UD concept to incorporate the behavior of an animal at a given location. That is, a UD can be constructed for each behavior. Each area within the animal's range may used for a variety of behaviors, but the relative use for any given behavior would be represented by the depth of each behavioral layer at that location (Fig. 12.3). This might allow the researcher to determine (1) if animals spatially segregate their behaviors and (2) why certain locations or habitats are important in the overall home range.

We decomposed an individual crow's use of space into four behaviorally specific UDs (Fig. 12.3). Using this approach, we determined that behaviors

A. Nested Near; Training

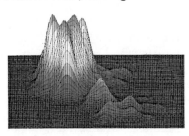

B. Nested Near; No training

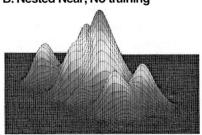

C. Nested Far; Training

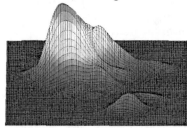

D. Nested Far; No Training

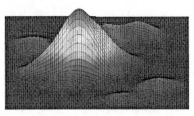

FIGURE 12.2 Space use by prairie falcons in relation to military training. Utilization distributions quantify the use (z axis) of space by populations of falcons nesting near (A, B) and far (C, D) from the Idaho Army National Guard's Orchard Training Area. Each utilization distribution is plotted on the same portion of the study area. Training occurred in the southeast corner of each plot. During periods of training (A, C) falcons used the northwest portion of the study area more than when training did not occur (B, D).

are spatially segregated. The most common activities (perching and foraging) were concentrated in different areas and maintained little overlap (V.I. = 32%). Below, we expand this approach to determine if these areas differ in resources (see Resource Selection, below).

These examples show some of the ways in which the UD can be directly analyzed, but much more can be done. Skilled biometricians could extend our quantification of the shape and structure of the UD. Perhaps by collaborating with geomorphologists we can determine how to measure the "ruggedness" of a UD (rugosity), or the dominance of a peak on the UD, and then use these measures to quantify an individual's pattern of space use. For example, an individual with a rugged UD may be using patchy resources. The shape of the UD could alert us to ways animals partition space and stimulate us to look for patchiness at a scale not entirely intuitive.

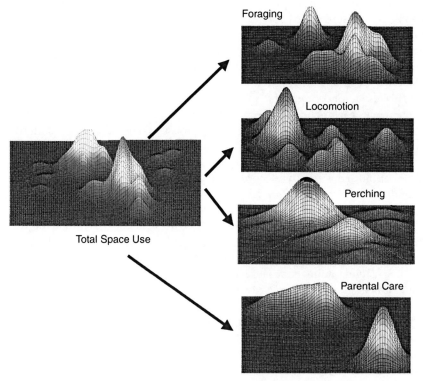

FIGURE 12.3 Decomposition of a single American crow's utilization distribution into four behaviorally specific utilization distributions. This crow foraged and perched in different areas. Most perching was done near the nest where parental care was provided.

RESOURCE SELECTION

The study of resource selection has proceeded from the development of simple metrics designed to quantify food selection (Scott 1920, Ivlev 1961) to univariate techniques that allow researchers to discriminate between resources that are "selected" or "avoided" (Neu et al. 1974, Johnson 1980, Aebischer et al. 1993) and complex multivariate models that allow researchers to predict resource use based on important resource variables (Manly et al. 1993, Cooper and Millspaugh 1999; see also Chapter 9). The early metrics were rooted in the study of animal behavior and typically were used to quantify true preference for food types when a variety of foods were offered simultaneously in controlled settings. However, despite incorporating the scale of selection and the importance of multiple resource attributes, recent models have been unable to measure actual "preference" of resources because they are descriptive rather than experimental. Specifically, they rely on the assumption that we can accurately define resource use and availability, and that statistical differences in use and availability equate to inferences about resource preference (Manly et al. 1993).

This assumption remains problematic. "Usage" (use), "selection," and "preference" each define different animal behaviors, and a distinction among these terms and their associated concepts must be maintained. These terms are not synonyms, although they are often used as such. Wiens (1989a) and others have long recognized a distinction between selection and preference, although not consistently. According to Johnson (1980), resource use is the quantity of a resource utilized by the consumer in a fixed period. If use is selective, resources are used disproportionately to their availability (Johnson 1980). Preference for a particular resource reflects the likelihood of that resource being chosen if offered on an equal basis with others (Johnson 1980). Therefore, it is nearly impossible to determine resource "preference" for free-ranging animals because resources are never randomly available in equal units.

Equally troublesome is the way some researchers define habitat availability. The availability of a resource component is its accessibility to the consumer (Johnson 1980). A key element of selection studies is what, why, and how much the biologist chooses to include as available to the animal. Selection is conditional upon availability and many factors can affect availability (Wiens 1989b, Mysterud and Ims 1998). For instance, social hierarchies and predator behavior can limit what is actually available to individuals (Mysterud and Ims 1998). Assumptions about availability can drastically affect the results and interpretation of resource use studies (Porter and Church 1987). For this reason, the choice of what is available must reflect the biological and social constraints on the individual.

The use of discrete choice modeling (Cooper and Millspaugh 1999; see also Chapter 9) narrows the scale at which we test for selection by allowing the researcher to specify a spatial and temporal period around what is considered "available." However, we remain ignorant of the temporal and spatial scales on which most animals base choices (Wiens 1989a). The central, unanswered question is, how do animals perceive their world? Do animals really only consider resources available from a recent travel trajectory when they select resources or do they perceive resources over large areas and remember encounters over long time periods when they decide what to use? Do animals even define "available" the same way for different resources? These are the types of questions we must answer to move the field of resource selection ahead.

Measuring Use

The first step in determining resource selection is documentation of resource use. The UD explicitly defines the total relative use within an animal's range. We should take advantage of this information and integrate it with the distribution of resource attributes in the range. Resources used for different behaviors may sum to use that is proportional to availability when considered in total. However, if use is considered for a specific behavior, then it may be quite different from availability (see Chapter 9). In addition, behavioral-based analyses allow easy construction of mechanistic explanations for documented resource use. We want to know why a resource was used, not simply that it was used more or less than expected.

The amount of use indicated by the UD provides a quantitative description of resource use by our study animal(s). Determining the volume of the UD above each habitat type indicates the relative amount of use each resource type received. In this way we obtain one measure of the relative importance of each resource type to the animal. This is an improvement over reporting the percentage of the used area composed of each resource type within a home range because no indication of the amount of use is incorporated in such an index. Furthermore, we consider the entire distribution of animal movements instead of focusing solely on individual sampling points. The UD allows us to quantify the amount of use and assign that to a series of resources.

The UD can be used to refine the scale of use of a particular habitat or vector of habitat features. Take the case of discrete vegetation types for illustration. The usual method of determining the likelihood of use for each vegetation type would involve collecting vegetation information at points animals used and did not use or at used and random (available) points. Logistic regression could then assign a likelihood of use to each vegetation type based on differences in used and unused or available types (see Chapter 8). However, this assumes that all used vegetation types are used to an equal degree. We can improve on this by using the UD to assign a relative amount of use to each used vegetation type. Rather than give each used location a score of 1.0, we could compute the relative use, measured by the height of the UD at each point, and add those scores by habitat type. This score ranges from 0 (an available point that was not used) to 1.0 (if all use occurred in a single habitat type). Put another way, the UD allows us to assign a value from a continuous variable (height of the UD) to each habitat type that indicates how much it was used by our study animal(s). This is similar to estimating use in each habitat using results from a logistic regression analysis, but it relates more closely to the observed pattern of space use. The UD approach is directly related to observed space use because it interpolates continuous use directly from the discrete observations of animals, regardless of where they were located. In contrast, the logistic approach first determines how discrete observations of animals correlate with habitat and then uses the association between use and habitat to interpolate a continuous measure of use.

As an example of this process consider the crow we used to define behaviorally specific UDs. We can overlay a UD on a resource map of this individual's range using a GIS. Then, each pixel in the range receives a use score (the height of the UD at that point) and a score for the resource(s) of interest. Resources are then regressed on use with standard multiple regression techniques. The standardized regression coefficients represent the relative importance of each resource to the animal (J. M. Marzluff and J. J. Millspaugh, unpublished data). Importance is determined along the continuous scale of use rather than on a dichotomous scale of use vs. availability, as is typically done with logistic regression. In addition, average use per pixel can be calculated for each resource type as a measure of selectivity. Average use ranges from 0 (no use despite availability) to 1 (exclusive use). In our crow example, forest and urban habitats were used less than 30% as often as grass, despite their abundance in the crow's range.

WHAT IS AVAILABLE?

In the lab, an investigator can provide resources and thereby determine what is truly available to an animal. However, in the field, only the animal can "tell" us what is truly available. Even with today's technology, we are limited to inferring availability from detailed behavioral observations documenting the pattern of travel and resource use on very fine time scales.

Johnson (1980), Porter and Church (1987), Wiens (1989b), and others have made us aware of the implications of assuming resources are available in our assessment of resource selection. A few points made in these papers should be emphasized. Johnson (1980) found selection analyses sensitive to *a priori* decisions concerning the inclusion and abundance of resources. In addition, changing the occurrence or availability boundaries changes the proportional availability of habitats (Johnson 1980). Porter and Church (1987) illustrated that these changes in proportional distributions were related to the pattern of dispersion of the resource cover types. If habitat types are regularly spaced throughout the study area, the amount of each habitat is not dependent on scale (Porter and Church 1987). Consequently, a comparison of use with availability provides similar results at large and small spatial scales. However, in aggregated environments, conclusions may vary substantially at different spatial scales (Porter and Church 1987).

As Porter and Church (1987) and Johnson (1980) point out, the biologist should carefully consider what resources to include as available. Furthermore, we agree with Johnson (1980) that results from several orders or spatial scales of selection aid in identifying resource selection patterns (Marzluff et al. 1997a, b). Although availability is inherently difficult to quantify, we urge wildlife biologists to carefully consider the behavior of their study animal and the temporal dynamics of the resource to more realistically study resource selection. Resource availability is much more than a buffer around an estimated location. New methods of resource selection (Arthur et al. 1996, Cooper and Millspaugh 1999) allow researchers to consider availability as a function of where the animal was located, the area it traversed, and the point of use. These techniques also incorporate temporal dynamics of resource selection and thus can be useful for assessing habitat selection. Analytical techniques allow biologists to assess the effects of including different resource types in varying amounts. The results of these new analyses will be much more informative if their interpretation is based on knowledge of the animal's behavior at the sample locations (see Chapter 9).

We recognize that there are limitations to the gathering of behavioral data. In fact, we do not advocate that every resource selection study describe and quantify animal behavior. However, we can expect biologists to refrain from making inferences about behavioral concepts and resource distributions for which they are not gathering the appropriate data. We also recognize that specific questions may be answered from "points on a map." In many cases, we may not need behavioral information for certain types of management to be implemented. Wildlife biologists, conservation biologists, and resource managers can gain useful information and apply it with knowledge of how much area an animal uses and what habitat types/vegetative cover are within its range. It is not necessarily illogical, and it is most likely practical, to assume that if we manage and conserve the resources in areas where animals have been sustained, we probably are "managing for" the animal.

MAPPING IMPORTANT HABITATS

Our understanding of animal distribution and abundance increased significantly when ecologists incorporated GIS in resource selection modeling (Verner et al. 1986, Manly et al. 1993). As a result, resource selection functions can include spatial variables, describing habitat composition and configuration, in addition to nonspatial components of the environment. When coupled with remote sensing data, GIS techniques allow a researcher to map environmental variables and predict use over large areas at a relatively fine resolution. Previously, only small-scale variables were possible. Thus, GIS technology has significantly increased our analysis of cross-scale patterns and processes (Peterson and Parker 1998).

Concurrent with advances in GIS technology, spatial models have increased in complexity from simple spatial overlays (a series of if–then statements producing a cookie-cutter effect) to fitting a mathematical function to an observed distribution of the selection response. In this section, we review the major techniques used for spatial mapping in a GIS and their mathematical and ecological assumptions. Finally, we discuss future applications of these techniques to mechanisms underlying the patterns identified by mapping techniques.

Mapping resource selection in a GIS is a form of calibration or classification (ter Braak 1995) in which we classify all cells (or pixels) in a map into either an expected number of animals or a probability of a class response from a set of predictor variables. Using remote sensing technology, we first develop a signature from a dataset of known characteristics. The known dataset is often derived from radiotelemetry observations or from geographically referenced surveys. Subsequently, we determine the value of the response in each map cell using an algorithm of the individual variables. The resulting mosaic of cell values presents a spatial distribution of the species response [Fig. 12.4; the logistic regression map of sage sparrows (*Amphispiza belli*)].

The most common statistical approaches for GIS mapping are based on a continuum of values for the response variable or a categorical response such as presence/absence. In multiple regression, the GIS maps usually predict the number of individuals at each site (ter Braak and Looman 1995). Our

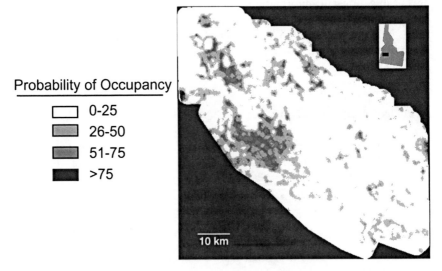

Probability of Occupancy

☐ 0-25
▨ 26-50
▨ 51-75
■ >75

10 km

FIGURE 12.4   Probability of sage sparrow presence determined from a logistic regression analysis of used and available habitats in southwestern Idaho.

development of the UD to assign a level of use to habitat and landscape features on a pixel by pixel basis can be used to extend the application of multiple regression to predict the amount of use expected by an individual at each site (J. M. Marzluff and J. J. Millspaugh, unpublished data). The logistic regression technique is used to predict a probability of occupancy from presence/absence data based on a sigmoid response curve (Manly et al. 1993, Trexler and Travis 1993). The Mahalanobis distance technique is used to predict the probability of similarity to the multivariate mean habitat vector of sites used by the species (Clark et al. 1993, Knick and Dyer 1997).

The regression methods require sampling of used and unused, or used and available habitats. (Here we define "used habitats" as the set of resource units where animal use was observed; "unused habitats" as the set of resource units where animal use was not observed; and "available habitats" as the set of resource units from which an animal can use.) The coefficients determined by these techniques measure selection relative to our definition of available habitats. Thus, they likely will change with any change in the set of available habitats. The Mahalanobis distance technique does not measure selection but rather is developed only from the set of used sites to estimate a probability of similarity to the mean habitat vector (the mean habitat vector is the multivariate mean of the set of habitats at sites used by the species). Although the problem of defining the available habitat is avoided, the Mahalanobis distance technique is extremely conservative in its predictions. The Mahalanobis distance technique also is highly sensitive when attempting to extend prediction to habitat vectors not included in the original sampling distribution (Knick and Rotenberry 1998).

ASSUMPTIONS FOR RESOURCE SELECTION MODELS

We develop models of resource selection based on a number of assumptions (see Chapter 8). First, we assume that animals fill habitats and distribute themselves optimally. Our statistical models assume that the best habitats always will have individuals and in the highest densities (Fretwell and Lucas 1969), which is not always accurate (Van Horne 1983, Hobbs and Hanley 1990, Vickery et al. 1992). We assume that the variables that we select are truly explanatory. Most often, we decide which variables to measure based on our perception of what is important to the species. We assume that the scale and selection of environmental variables are relevant to the species (Rotenberry 1986, Wiens 1989a). GIS permits analysis of multiple scales of selection. Each scale of analysis permits a different ecological question (Peterson and Parker 1998). We assume that our multivariate statistical model appropriately describes the species–habitat associations and that the shape of the statistical function represents the actual shape of the selection (ter Braak 1995). Most statistics assume a linear, curvilinear relationship, or uni- or multimodal distribution. However, some commonly determined GIS functions, such as distance to stream, road, or point, may be more appropriately described by logistic or linear distribution rather than by a relationship based on a mean and variance. Finally, we also assume that animal selection does not change during our study when extrapolating outside of our sampling period. We assume that animals are selecting for the original habitat conditions even though the habitat configuration may change because of seasonal or successional change, or because of disturbance. Indeed, predicting changes in animal distribution in response to disturbance or management action remains a most intriguing challenge (Shugart 1998).

IDENTIFYING MECHANISMS FROM PATTERNS

GIS mapping technology has greatly increased our understanding of spatial patterns of home range use and habitat selection. Using GIS, we now can describe the spatial distribution of elements in the landscape by mapping variables or indexes. Our emphasis on spatial patterns largely results from the relative ease with which spatial patterns are determined. Spatial patterns also answer a manager's need to know where in the landscape animals are likely to be found. As a result, we now can begin to manage landscapes for species based on concepts from theoretical biogeography, which has important implications for conservation biology (Urban and Shugart 1984, Temple and Cary 1988, Burkey 1989, Opdam 1991, Danielson 1992, Hansson and Angelstam 1991). For example, using maps of habitat distributions, managers can identify regions containing high probability of use and maintain areas of sufficient size to contain viable populations or identify habitat corridors for dispersal. Thus, spatial patterns are an important foundation. However, we have pursued techniques and technologies that will allow us to map patterns to the exclusion of underlying ecological mechanisms.

We must be careful to understand the assumptions we make when relying more on technology and less on direct observation of animals. We have become increasingly dependent on technology in delineating or describing resource use and availability, possibly violating key assumptions in the process. For example, one would expect resource availability (e.g., vegetation, prey) to change in quality and quantity seasonally and annually. Despite acknowledging this, many researchers assess habitat availability from a single point in time while their estimates of use span many seasons and years. A single assessment of availability must be validated or resources must be assumed to be constant over the time of study. Such validation requires spending time in the field and collecting data to determine if resource availability is different among seasons or years.

We now need to use GIS technology and spatial analyses to understand the underlying mechanism for the patterns. For example, how is productivity related to habitat characteristics and landscape configuration (Van Horne 1983, Vickery et al. 1992)? By studying productivity of individuals within a population as well as by populations within a species range, we can begin to understand how landscapes influence population change.

Identifying a set of minimal conditions necessary for a species remains a significant challenge. We need to pursue models that extend beyond the study area to the species' range (Knopf et al. 1990). Coefficients of selection models are relative to the study area in which they were determined. As such, the coefficients tell us what variables are important but do not answer the question of what a species needs. We are exploring a statistical concept, Pearson's "planes of closest fit" (Pearson 1901), that identifies a first-order, linear approximation of the most basic combination of environmental variables used by a species. If successful, range-wide mapping may be possible based on an ecological rationale of fundamental requirements (Rotenberry et al. in press).

We suggest that the technology to answer many important ecological questions already exists. The mechanisms by which a species distributes itself across a landscape in time and space might be answered by looking at differences between statistical techniques and spatial patterns, coupled with field analysis of population behavior. By coupling population dynamics with spatial mapping, we can map population sources and sinks across the landscape. By doing so, we ultimately will be able to test theoretical models of metapopulation structure such as core satellite models (Hanski and Gilpin 1997), evaluate the use of habitat corridors, or identify species requirements.

IMPROVING OUR APPROACH TO THE STUDY OF WILDLIFE RADIOTELEMETRY

Researchers using radiotelemetry rely on technology. Some might even say that we are obsessed with technology! Our increasing reliance on technology has in some cases replaced our reliance on detailed observations of animal behavior. This is especially true in studies of resource use by animals. The typical study remotely estimates animal locations with telemetry and documents habitat occurrence via aerial satellite imagery (e.g., Marzluff et al. 1997b). However, only by careful observation and experimentation can we learn about resource availability to animals, resource requirements for animals, and resource preferences of animals. We should embrace technology to improve our understanding of animal ecology by realizing that a complete

assessment of resource use, availability, and selection requires an understanding of how and why animals use resources (Wiens 1972, Partridge 1978).

Future studies of radio-tagged animals should strive to answer "why"
questions (Gavin 1991). Why do animals use certain resources? Why do
they survive at a given rate in some habitat configurations or resource avail-
abilities and at different rates in others? Go beyond documenting where
animals are and document why they are there. Obtaining a thorough answer
to the why question is a two-step process. First, the proximate cue determining why an animal is at a particular place should be understood so that a
causal or mechanistic understanding is obtained. Second, the ultimate reason
should be understood so that an evolutionary explanation for the behavior can
be postulated (Gavin 1991). The only way to answer "why" is to observe what
an animal does at a particular location. Failure to improve our understanding
of animal behavior through radiotelemetry methods will reduce our ability to
effectively manage species. We need to use telemetry to get close to our
animals, not to remain far from them if we want to understand why animals
do what they do.

We recognize and do not discount the tremendous amount of behavioral
information that has been inferred simply from remotely collected location
data. For example, the spatial behavior of territorial species is readily obtained
from radio-locations. Similarly, foraging distances (and consequently an estimate of energy expenditures) and migration routes can be interpreted using
successive locations of radio-marked individuals. Delimiting species ranges,
defining populations, and estimating dispersal distances also have contributed
significantly to conservation biology. But we want researchers to remember
that seemingly primitive uses of radiotelemetry can also assist in the approach
to our study animals. For example, "homing" to a transmitter signal (Mech
1983) and gaining proximity that allows observation, or telemetering sensor
data about behavior (Samuel and Fuller 1994), may often have significant
rewards. This simple use of telemetry often allows us to observe animals for
long periods of time in natural settings. Such observation has been the key to
major breakthroughs in our understanding of animal behavior during the last
century (Dewsbury 1985). We suggest that it will continue to be an integral
part of all future breakthroughs in our understanding of what resources are
needed for effective species conservation. Through detailed observations of
our study subjects, we will begin to understand the temporal and spatial scales
animals work on, thereby allowing us to better define selection.

We conclude by considering how we would like future reviews of our field
to characterize our studies. Given incredible progress in telemetry technology
(see Chapter 4), remote sensing abilities, and concerted efforts to follow and
observe our animals, we are sure that future studies will (1) clearly state and
consider their biases and assumptions (which includes estimating errors for

telemetry techniques, map classification schemes, and habitat delineations), (2) quantify the effects of attaching transmitters and tracking animals, (3) include large samples of animals or groups that are considered independent experimental units, (4) accurately define how the study animals perceive their world so that available resources are defined, (5) test scientific hypotheses that posit causal relationships between resources and animal space and resource use, and (6) identify important resources for animals by linking resource use to individual fitness and population viability.

Will future students of resource use rarely visit the field? We hope not. Advances in technology now allow detailed studies of wildlife resource selection once thought impossible. However, we must not lose sight of basic ecological and behavioral principles. Accurate assessments of resource selection can benefit from technological innovation, but they require careful study of animal behavior. We believe that the best way to have a reasonable idea about the more subtle aspects of whether use is nonrandom is to observe the animal's behavior and quantify the resources at the time of observation and use.

In the future, a radio-tagged animal will be used to answer multiple questions. Investigators will have integrated research programs that allow all six hallmarks stated above to flow from every study. Our field will have matured when the typical study can no longer be pigeonholed into one dealing with space use, resource selection, or demography. Instead, all studies will simultaneously investigate the interaction between these subdisciplines and answer the fundamental management question: what resources do wildlife populations need to remain viable?

SUMMARY

We remain relatively ignorant about why animals use particular habitat features despite remarkable advances in technological applications in the field of radiotelemetry. We now can translate detailed, spatially explicit behavioral observations into accurate utilization distributions amenable to overlaying on resource maps. We then can interpret those maps to improve our understanding of why animals use particular resources. Investigators should design studies using radiotelemetry that (1) increase their understanding of animal behavior by establishing why their study animals use resources in a nonrandom pattern, (2) make greater use of utilization distributions to quantify space use and resource selection, and (3) consider alternative resource selection techniques such as mapping important habitats by using predictive models.

We recommend the utilization distribution as a valuable analysis tool that will enhance future studies of animal movements, species interactions, and resource selection because it focuses on animal use of the entire range. However, to properly use UD estimates we must assume that interpolation accurately represents where we did and did not locate animals, which requires that we have enough observations. We used behaviorally specific UDs to illustrate (1) how we might determine if and how animals segregate their behaviors and (2) why certain locations or resources are important in the overall spatial distribution. However, much more can be done with UDs, and we encourage future research into quantification of the pattern, or ruggedness, of UDs.

We review the meanings of use, selection, and preference as they apply to studies using radiotelemetry. Of these, only "use" is reliably measured in the typical study of wildlife–resource relationships. We encourage researchers to focus on understanding resource use rather than selection (which is entirely dependent on our definition of available) or preference (which requires controlled laboratory conditions). Measuring overall use is best done with the UD, and this measure of use can be related to a variety of factors (such as habitat type, animal age or sex, habitat patch metrics, or distances to various resources) with multiple regression.

Resource use can be accurately mapped with current GIS technology given reliable statistical models. We discuss common approaches and their assumptions, noting that we have often pursued techniques and technologies to the exclusion of underlying ecological mechanisms. We now need to use GIS technology and spatial analyses to understand the underlying mechanism for the patterns. For example, how is productivity related to habitat characteristics and landscape configuration? Identifying a set of minimum conditions necessary for a species remains a significant challenge. If solved, range-wide mapping may be possible based on an ecological rationale of fundamental requirements. Accomplishing such breakthroughs will require that researchers strive to answer "why" questions. Going beyond documenting only where animals exist to answering why they are there will allow us to solve critical management issues about which resources wildlife populations need to remain viable.

PART VI

Population Demographics

Population Estimation with Radio-Marked Animals

GARY C. WHITE AND TANYA M. SHENK

Estimation of the size of a geographically and demographically closed but free-ranging population is a common problem encountered by wildlife biologists. The earliest approaches to this problem were developed by Petersen (1896) and later Lincoln (1930), where capture-recapture techniques were applied. Extensions to the simple two-occasion Lincoln-Petersen estimator were developed for multiple occasions (Schnabel 1938, Darroch 1958), for removal experiments (Zippin 1956, 1958), and for heterogeneity of individual animals (Burnham and Overton 1978, 1979, Chao 1988). Otis et al. (1978), White et al. (1982), and Seber (1982) provided summaries of the available methods for analysis and study design of capture-recapture surveys.

Radio-marked animals provide considerably more information about estimation of population size (N) because the location of the animal can be determined. Because an investigator knows whether the marked animal is present on the study area and hence available for resighting during a survey, the issue of movement on and off the study area can be resolved. Seber (1986) listed the use of radio-marks as a major technological advancement relevant to population estimation.

In this chapter, we discuss three uses of radio-marked animals to estimate population size. Direct mark-resight estimation of N is considered first, where the probability of observing an individual in the population is estimated from observation rates of radio-marked individuals that are present in the population. Second, we discuss the use of sightability models developed from radio-marked animals, where trials are conducted to estimate the probability of sighting given that the radio-marked animal is present on a survey unit. Covariates, such as size of the group of animals containing the radio-marked individual at the time of the survey or the percent vegetation cover of the animal's location during the survey, are used to develop models that predict the probability of sighting individuals in the population. These sightability models are then used to estimate population size in other populations with similar conditions, or in the same population at a later time, if conditions remain essentially the same. Finally, we consider how radio-marked animals can resolve the issue of how to estimate animal density from a trapping grid of size A with $\hat{N}$ individuals estimated from mark-recapture methods. By radio-marking a sample of individuals captured on the grid and tracking them shortly after the grid trapping procedure is finished, the proportion of their locations that occur on the former trapping grid can be used to correct the bias of the naive estimate of density, $\hat{D} = \hat{N}/A$.

DIRECT MARK-RESIGHT ESTIMATION

More technologically advanced approaches to estimate abundance have incorporated animals marked with radiotransmitters. The initial sample of animals is captured and marked with transmitters, but recaptures of these animals are obtained by observing them, not actually recapturing them. A limitation is that unmarked animals are not marked on subsequent occasions; therefore, the use of typical estimators for population estimation of closed populations (Otis et al. 1978, White et al. 1982) are not applicable. The advantage of using an initial sample of radio-marked animals is that resighting occasions are generally much cheaper to acquire than when the animals must be physically captured and handled on each occasion.

A critical assumption of the mark-resight methodology is that radio-marked animals have the same sightability as unmarked animals. That is, the radio-marked animals are used to estimate the sightability of unmarked animals.

As discussed below, if sightability of radio-marked individuals is somehow different from that of unmarked individuals, estimators of population size will be biased. This assumption is not the same as assuming that radio-marked animals are randomly distributed in the population during sighting surveys. Rather, the assumption requires that the sightability of radio-marked animals be the same as that of unmarked animals, so that a scenario where marked animals occur in only one portion of the study area may be feasible if the researcher is willing to assume that sightability is the same for animals everywhere in the study area. Clearly, a study area with two levels of vegetative cover, and marked animals in only one of these cover types, would violate the assumption and lead to biased estimates of population size.

A second important assumption of mark-resight estimators is closure, which is required for all closed population estimation (White et al. 1982). Closure of the population assures that there exists a population size that is estimable (i.e., that N is a fixed value during the mark-resight surveys). Demographic closure (i.e., no births or deaths) is typically assured by keeping the time period for resightings short enough so that no deaths are likely to occur, and by conducting resightings during a time of year when survival rates are high. Geographic closure, although not required for the immigration/emigration joint hypergeometric estimator (IEJHE), described below, is required for most mark-resight estimators. Geographic closure is evaluated using movements of radio-marked animals. If marked animals are a representative sample of the population and they are found to emigrate off the study area, geographic closure is violated. When geographic closure is violated, density estimates ($\hat{D} = \hat{N}/A$) are biased high for the study area size (A) because more animals are estimated to inhabit the area than actually do. To minimize the violation of geographic closure, surveys should be conducted during periods when little emigration/immigration is expected (i.e., little movement is taking place in the population, and animals show strong fidelity to home ranges).

The mark-resight procedure has been tested with known populations of mule deer (*Odocoileus hemionus*) (Bartmann et al. 1987), and used with white-tailed deer (*Odocoileus virginianus*) (Rice and Harder 1977), mountain sheep (*Ovis canadensis canadensis*) (Furlow et al. 1981, Neal et al. 1993), desert bighorn sheep (*Ovis canadensis nelsoni*) (Leslie and Douglas 1979, 1986), black (*Ursus americanus*) and grizzly (*Ursus arctos*) bears (S. D. Miller et al. 1987, S. G. Miller et al. 1997), coyotes (*Canus latrans*) (Hein and Andelt 1995), and harbor seals (*Phoca vitulina*) (Ries et al. 1998).

Early attempts to estimate population size with radio-marked animals (e.g., Leslie and Douglas 1979) used the Lincoln-Petersen estimator (cf. Le Cren 1965):

$$\hat{N} = \frac{n_1 n_2}{m_2},$$
(13.1)

where n_1 is the number of radio-marked animals in the population, n_2 is the number of animals counted on a resighting survey, and m_2 is the number of radio-marked animals observed on the resighting survey. Chapman (1951) extended this estimator to be unbiased for $n_1 + n_2 \geq N$:

$$\hat{N} = \frac{(n_1 + 1)(n_2 + 1)}{(m_2 + 1)} - 1. \tag{13.2}$$

When multiple resighting surveys are conducted, White and Garrott (1990) showed that using the mean of Chapman estimates divided by study area density as an estimator of the average density generates a biased estimate when a low proportion (<10%) of the population is marked with low resighting rates (<20%). However, the reason for not using the arithmetic mean of the Chapman estimates is that the procedure is less efficient than the joint hypergeometric maximum likelihood estimator (JHE) described below. Confidence interval coverage is not as close to the desired 95% coverage as the JHE, plus confidence interval length is generally wider than the JHE (White and Garrott 1990), reflecting the lack of efficiency of the mean estimator compared with the JHE. Furthermore, the procedure of taking the mean of the Chapman estimates does not incorporate information from surveys where no marked animals were observed, whereas the JHE does.

Eberhardt (1990) further investigated the Petersen estimator with the Chapman correction (Chapman 1951) for small population sizes where animals could move on and off the study area during the survey period. We do not discuss his estimators further because they do not contribute any advantages over the estimators presented here. None of the estimators he considered were maximum likelihood estimators for the complete set of data, nor did any of the estimators alleviate the assumption of constant capture probability of all animals on each occasion.

Garshelis (1992) also considered extending the Chapman estimator for correcting population estimates for immigration and emigration of animals from the study area. Garshelis (1992) was primarily interested in recaptures of marked animals (rather than resightings) and used radio-marked animals to estimate the proportion of time animals spent on the study area. As with the estimators proposed by Eberhardt (1990), his estimator also did not provide any advantages over the immigration-emigration estimator discussed below for the same reasons: (1) the estimator is not a maximum likelihood estimator for the complete set of data and (2) it does not alleviate the assumption of constant capture probability of all animals on each occasion.

Here we discuss in some detail four estimators of abundance: the JHE (Bartmann et al. 1987, White and Garrott 1990, Neal et al. 1993), the JHE extended to incorporate animals moving on and off the study area (Neal et al. 1993, White 1993), the Minta-Mangel estimator (Minta and Mangel

1989), and Bowden's estimator (Bowden 1993). The latter two estimators are not maximum likelihood estimators, but they do not require the assumption that each animal in the population has the same probability of resighting on a particular occasion, as the first two estimators require. All four estimators are provided in Program NOREMARK (White 1996) (see Appendix A).

NOTATION

T_i Number of marked (radio-marked) animals in the population at the time of the ith survey, $i = 1, \ldots, k$. When the number of marked animals is assumed constant across surveys, the value is denoted as T.

M_i Number of marked animals in the population that are on the area surveyed at the time of the ith sighting survey. For all M_i constant, define $M \equiv M_i$.

n_i Number of animals seen during the ith sighting survey, consisting of m_i marked animals and u_i unmarked animals, so that $n_i = m_i + u_i$.

f_i Number of times marked animal i was observed during the k surveys (sighting frequencies), $i = 1, \ldots, T$. . (Note that this is not the same use of f_j as in Otis et al. 1978, White et al. 1982).

$m.$ Total number of sightings of marked animals, so that $m. = \sum m_i = \sum f_i$.

$u.$ Total number of sightings of unmarked animals, so that $u. = \sum u_i$, where $i = 1, \ldots, N-T.$.

$\bar{f}$ Mean capture frequency of marked animals, $m./T.$.

s_f^2 Variance of the sighting frequencies of the marked animals,

$$S_f^2 = \frac{\sum_{i=1}^{T.}(f_i - \bar{f})^2}{T.}.$$

ESTIMATORS

The JHE (Bartmann et al. 1987, White and Garrott 1990, Neal 1990, Neal et al. 1993) provides the estimate ($\hat{N}$) of population size N, which maximizes the likelihood:

$$\mathcal{L}(N|M, n_i, m_i) = \prod_{i=1}^{k} \frac{\binom{M}{m_i}\binom{N-M}{n_i - m_i}}{\binom{N}{n_i}} \tag{13.3}$$

with the various terms defined for all $i = 1$ *to* k sighting occasions. $\hat{N}$ can be estimated by iterative numerical methods. Confidence intervals can be determined with the profile likelihood method (Hudson 1971, Venzon and Moolgavkar 1988). This estimator assumes that all marked animals are on the area surveyed for each survey (i.e., that the population is geographically and demographically closed), and thus N is constant across sighting occasions (i.e., no animals are removed from or added to the population). The number of marked animals (M) is the same for each survey in the above equation, although the probability of sighting animals is not assumed to be the same for each survey. An extension is to allow unmarked animals currently in the population to be marked between sighting occasions. Thus, M_i replaces M in the above equation, but the value of N is still assumed constant across occasions.

The JHE has been extended to accommodate immigration and emigration (Neal et al. 1993) through a binomial process. The immigration/emigration JHE (IEJHE) does not assume that the population is geographically closed, but the population is still assumed to be demographically closed. Assume that the total population with any chance of being observed on the study area is N^*, and that at the time of the ith sighting occasion, N_i animals occur on the study area. We are interested in estimating the mean number of animals on the study area, $\bar{N}$, and possibly N^*. At the time of the ith sighting occasion, a known number of the marked animals (M_i) are on the study area of the possible T_i animals with radiotransmitters. The probability that an individual is on the study area on the ith occasion can be estimated as M_i/T_i or, in terms of the parameters of interest, as N_i/N^*. Then the likelihood function for the model that includes temporary immigration and emigration from the study area is a product of the binomial distribution for the probability that a marked animal is on the study area times the joint hypergeometrical likelihood of Eq. (13.3):

$$(N^*, N_i | T_i, M_i, m_i, n_i) = \prod_{i=1}^{k} \binom{T_i}{M_i} \left(\frac{N_i}{N^*}\right)^{M_i} \left(1 - \frac{N_i}{N^*}\right)^{T_i - M_i} \frac{\binom{M_i}{m_i}\binom{N_i - M_i}{n_i - m_i}}{\binom{N_i}{n_i}}. \tag{13.4}$$

The parameters N^* and N_i for $i = 1$ *to* k can be estimated by numerical iteration to maximize this likelihood, with the constraints that $N_i > (M_i + u_i)$ and $N^* > N_i$ for $i = 1$ *to* k. Profile confidence intervals can be obtained for the $k + 1$ parameters. Typically biologists are interested not in the k population estimates for each sighting occasion but rather in the mean of the N_i estimates. Therefore, the likelihood can be reparameterized to estimate the total population and mean population size on the study area directly, and their profile likelihood confidence intervals. In the reparameterized likelihood, $N_i = \bar{N} + \alpha_i$, where $\sum \alpha_i = 0$.

Minta and Mangel (1989) suggested a bootstrap estimator (MM) of population size based on the sighting frequencies of the marked animals, f_i. For unmarked animals, sighting frequencies are drawn at random from the observed sighting frequencies of the marked animals until the total number of captures equals $u.$. The number of animals sampled is then an estimate of the number of unmarked animals in the population, so that M plus the number sampled is an estimator for N. Minta and Mangel (1989) took as the bootstrap sample the bootstrapped population when the cumulative number of sightings equaled or exceeded $u.$. A less biased stopping rule is when the cumulative number of sightings exactly equals $u.$, i.e., cases where the cumulative sightings exceeded $u.$ are rejected. Minta and Mangel (1989) suggested the mode of the bootstrap replicates as the population estimate. Confidence intervals were computed as probability intervals with the 2.5th and 97.5th percentiles from the bootstrapped sample of estimates. White (1993) demonstrated that the MM estimator is basically unbiased, but that the confidence interval coverage was not at the expected 95% for $\alpha = 0.05$ because the MM estimator conditions on the number of unmarked animals seen (i.e., the number of unmarked animals seen is taken as a fixed constant rather than a random variable). Although White (1993) suggested a modified procedure, coverage still was not satisfactory.

Bowden (1993) and Bowden and Kufeld (1995) suggested an estimator for the MM model where the confidence intervals on the estimate were computed based on the variance of the resighting frequencies of the marked animals. Bowden approached the problem from a survey sampling framework, where each animal in the population has the attribute f_i of the number of times it was resighted. The values of f_i are known for the marked animals, and the sum of the $f_i's$ ($= u.$) are known for the unmarked animals. Then, an unbiased estimator of the population size is

$$\hat{N} = \frac{\left(\dfrac{(u. + m.)}{\bar{\bar{f}}} + \dfrac{s_f^2}{\bar{\bar{f}}^2} \right)}{\left(1 + \dfrac{s_f^2}{T.\bar{\bar{f}}^2} \right)} \tag{13.5}$$

with variance

$$\hat{\mathrm{Var}}(\hat{N}) = \hat{N}^2 \frac{\left(\dfrac{1}{T.} - \dfrac{1}{\hat{N}} \right) \dfrac{s_f^2}{\bar{\bar{f}}^2}}{\left(1 + \dfrac{s_f^2}{T.\bar{\bar{f}}^2} \right)^2}. \tag{13.6}$$

Confidence intervals for N are computed from a log-transformation as

$$\hat{N}/\exp\left(t_{1-\frac{\alpha}{2}, T.-1}\hat{C}V(\hat{N})\right) \text{ and } \hat{N} \times \exp\left(t_{1-\frac{\alpha}{2}, T.-1}\hat{C}V(\hat{N})\right), \qquad (13.7)$$

where $\hat{C}V(\hat{N})$ is $\hat{V}ar(\hat{N})^{1/2}/\hat{N}$ and $t_{1-\frac{\alpha}{2}, T.-1}$ is a t distribution with $T. - 1$ degrees of freedom.

A common problem is that marked animals are observed but their individual identity is not known. An extension of Bowden's estimator allows some radio-marked animals to be unidentified. Simulations with NOREMARK (White 1996) have shown that as long as 90% or more of the marked animal resightings identify the individual, no bias and little loss in efficiency results.

Madsen (1995) developed an immigration/emigration version of Bowden's estimator by assuming that two or more teams of observers survey the study area simultaneously during a single occasion when the population is assumed to be closed and each radio-marked animal is identified as being on or off the study area. Later, after allowing animals to move on and off the study area, the teams resurvey the site, repeating the process for r surveys. Madsen (1995) shows that the mean of the Bowden estimates [$\hat{N}_i$ of Eq. (13.5)] from the r surveys across time:

$$\frac{1}{r}\sum_{i=1}^{r} \hat{N}_i \qquad (13.8)$$

provides an approximately unbiased estimator of the mean number of animals on the area $(\bar{N})$. Note that the strategy used by Madsen (1995) can be used generally, i.e., if animals are moving on and off the study area, a "quick" survey with multiple observation teams will provide an estimate of $\hat{N}_i$ for time i.

DETERMINING WHICH ESTIMATOR IS APPROPRIATE

As described above, the mark-resight estimators require that marked and unmarked animals have the same sightability. Given that assumption, each of the four estimators has different assumptions (Table 13.1) and thus works best in certain situations.

The JHE assumes a geographically closed population and equal sighting probabilities among animals on a particular occasion. In addition, animals are assumed to be sampled without replacement (i.e., no animal is observed or counted twice during the survey for a particular occasion). The IEJHE extends this model to the situation where the study area is no longer geographically closed but each animal in the population still has the same sighting probability on a particular occasion and is sampled without replacement.

TABLE 13.1   Assumptions Required for Unbiased Estimates from Four Mark-Resight Population Estimators When Radio-Marked Animals Are Used

Estimator	Demographic closure	Geographic closure	Sampling with replacement	Individual heterogeneity
Joint hypergeometric estimator (JHE)	Required	Required	Not allowed	Not allowed
Immigration/emigration JHE	Required	Not required	Not allowed	Not allowed
Bowden's	Required	Required	Allowed	Allowed
Minta-Mangel	Required	Required	Allowed	Allowed

Bowden's estimator allows animals sighting probabilities to differ, and sampling can be with or without replacement. Sighting heterogeneity allows for study areas not geographically closed, in that some animals can be off the study area for a particular occasion(s) and hence have a zero sighting probability. The resulting estimate is the total population using the study area, which is not the same as the average density of animals on the study area. Thus, Bowden's estimate may not be exactly what the researcher envisioned if the study area is not geographically closed.

The MM estimator requires the same assumptions as Bowden's estimator, although the MM estimator is derived under the assumption of sampling with replacement. However, using this estimator for situations where sampling is done without replacement is not necessarily inappropriate. In this case, the estimator would not be particularly biased and confidence interval length would be only slightly larger than the estimator derived under the assumption of sampling with replacement. We do not recommend using the MM estimator because of the poor performance of its confidence interval coverage (White 1993) and suggest use of Bowden's estimator if heterogeneity of sighting probabilities is serious or sampling is performed with replacement. The extensions to the MM estimator suggested by Gardner and Mangel (1996) do not provide any advantages over Bowden's estimator.

If there is little or no heterogeneity in sighting probabilities, the JHE should generate slightly smaller confidence intervals than Bowden's estimator because stronger assumptions are made. Neal et al. (1993) showed that confidence interval coverage drops to 80% for an expected 95% interval when sighting heterogeneity was simulated from a beta distribution ($\alpha = 3$, $\beta = 3$) with $\bar{x} = 0.5$ and SD = 0.19, although the JHE was still unbiased. The choice of Bowden's estimator over the JHE will depend on the degree of heterogeneity in sighting probabilities, and whether sampling is conducted with or without replacement.

One additional criterion is useful in deciding between the JHE and Bowden's estimator. Often, resighting surveys are not conducted as discrete occasions (i.e., animals may be counted multiple times during one "occasion"). Examples include photographing bears at bait sites with motion-sensing cameras, where the same bear visits the same bait site multiple times in a few hours. Besides this survey approach having heterogeneity of sighting probabilities, structure of the survey basically precludes use of the JHE. Such surveys should be considered as sampling with replacement.

The issue of closure on the study area can be viewed from two perspectives. If there is a closed population, N^*, that uses the study area, then Bowden's estimator should give about the same estimate of N^* as the IEJHE. That is, the study area only contains a portion of the population of interest during any survey. Immigration and/or emigration from the study area creates individual heterogeneity of sighting probabilities, which is addressed by Bowden's estimator. IEJHE addresses spatial heterogeneity by modeling it with the binomial distribution, i.e., each animal has a probability of being on the study area during the survey. In this case, JHE may be inappropriate depending on the amount of individual heterogeneity caused by animal movements. As shown by Neal (1990), JHE is somewhat robust to individual heterogeneity, but sufficient individual heterogeneity will result in poor confidence interval coverage. Furthermore, if the movement is large enough so that the concept of a constant $N = \bar{N}$ on the study area is inappropriate, then JHE is definitely inappropriate.

The concept of a closed population that uses the study area may be irrelevant. In this case, Bowden's estimator will also be inappropriate because it estimates the indefinite quantity N^*, something not of interest in this situation. Likewise, IEJHE will also likely provide an unrealistic estimate of N^*, but its estimate of $\bar{N}$ should be reasonable and be of biological interest.

Selection of the appropriate estimator to a large degree depends on the biology of the population and the survey techniques used. Issues such as whether animals are sampled with or without replacement should be addressed before data collection through survey protocols. Violation of this assumption can be assessed by determining if some marked individuals were seen multiple times during a single survey, although from a practical point of view, most surveys would have a small chance of detecting much effect of sampling with replacement, and the best strategy to avoid sampling with replacement is careful survey design.

The assumption of no individual heterogeneity required by the JHE and IEJHE is obviously never completely true because animals are genetically and phenotypically different. The real question is whether there is enough individual heterogeneity present in the data to justify the use of a less efficient estimator. To be conservative, some researchers may decide to always use

Bowden's estimator and avoid the issue entirely. A more liberal approach is to evaluate whether the capture frequencies of individuals are overdispersed in comparison with a binomial distribution using logistic regression (Neal et al. 1993). A more sophisticated approach for performing this kind of test is to fit a beta-binomial distribution to the observed frequencies, such as was done to survival rates by Unsworth et al. (1999). However, all of these procedures will undoubtedly lack power because of sample size limitations, and thus provide no guarantee that even considerable individual heterogeneity will be detected.

Design of Studies: Field Methods

Researchers have been very innovative in obtaining sightings of marked and unmarked animals. As described above, a key assumption of the mark-resight approach is that sighting probabilities are identical for marked and unmarked animals, or, more generally for Bowden's estimator, the distribution of sighting probabilities must be the same for marked and unmarked animals. Careful choice of field techniques is necessary to insure this assumption is approximately met.

As an example, an unwise practice would be to radio-mark deer via helicopter net-gun capture and then obtain resightings from a helicopter. The first bias caused by this approach is that the more visible and hence more resightable animals will be captured and marked. Second, animals are more likely to hide from the helicopter during resighting efforts, given their initial experience of capture from the helicopter. To meet the assumption of the mark-resight method, a researcher could capture animals for radio-marking with a different capture method than would be used for resighting efforts, such as drop nets or Clover traps.

An innovative approach to obtain resightings of grizzly bears in forested habitats could involve use of cameras (e.g., Mace et al. 1994). However, consideration of potential biases caused by similar capture and resighting methods must still be made. Capture of bears using attractants to traps may mean that radio-marked bears then avoid similar attractants. Consequently, use of similar baits to attract bears to camera resighting sites potentially could result in biased estimates of population size. As a general rule, capture methods to mark animals should be as different as possible from methods used to resight animals so that the potential for bias is minimized.

Bowden's estimator requires that the identity of marked animals be determined. Typically, individually identifiable collars are placed on animals so that animals can be identified visually. Use of the radio signal to identify the animal is a questionable practice because observers could easily misidentify the animal. It is more likely that the observer will discover another marked animal in the vicinity and be inclined to include it in the observed sample when in fact the animal was only resighted because a radio signal was heard. Rather than risk such a bias, the visual identification of each animal should be considered during the design of the study so that visual observations nearly always provide the animal's unique identify.

DESIGN OF STUDIES: STATISTICAL METHODS

Design of mark-resight surveys can be optimized by considering the cost of marking animals vs. the cost of resighting them. A design that minimizes the confidence interval length on $\hat{N}$ for a given cost is desired, i.e., what is the allocation of effort (i.e., dollars) to capture versus the effort placed on resighting animals that produces the shortest possible confidence interval. How can one do this?

The program NOREMARK (White 1996) provides two methods for optimizing the efficiency of mark-resight surveys. Simulation results from Neal (1990) have been included with the program so that users can interpolate expected confidence interval length for the JHE for a value of N, given the proportion of the population that is marked, the probability of resighting on an occasion, and the number of resighting occasions. In addition, simulation procedures have been included in NOREMARK for all four estimators, so that design of studies outside the range of values in the interpolation tables is possible. Simulation takes longer than the table look-up procedure but is well worth the small price in computer time given the cost of most mark-resight studies.

For both of these procedures, predicted confidence interval length from surveys with identical costs are examined. The goal is to find the optimal trade-off between marking animals and increasing resighting probabilities (i.e., to find the design that for the chosen cost minimizes the variation of the population estimate).

EXAMPLES

To demonstrate the JHE, IEJHE, and Bowden's estimator, an example of a mountain sheep study from Neal et al. (1993) will be used. Twenty-five adult ewes were marked with radio-collars using drop nets (15), net guns (3), and capture guns (7), all from the ground (i.e., no helicopter captures were conducted). Fourteen resighting occasions were conducted with a helicopter. The area surveyed was not the entire winter range of this population, so not all animals were on the study area for each survey. Summary statistics to compute the JHE and IEJHE estimators are provided in Table 13.2, with frequency distributions necessary for Bowden's estimator provided in Table 13.3. Results are given in Table 13.4.

Approximately 81% of the marked animals were on the study area during surveys. Thus, we would expect differences between N^* and $\bar{N}$ of the IEJHE, with N^* being more comparable to $\hat{N}$ of the JHE and Bowden's estimator. Interestingly in this example, the JHE and Bowden's estimates are similar. However, JHE confidence intervals are considerably shorter than Bowden's estimator. As shown by Neal et al. (1993), individual heterogeneity causes little bias in the JHE estimate, but confidence interval coverage is less than 95%. Thus, we would expect the confidence intervals from Bowden's estimator to be wider and provide closer to 95% coverage. These results are consistent with the knowledge of the sheep on the study area. The limited winter range in the area dictated a closed population, N^*, of which only a fraction were on the study area during each survey. Thus, all three estimators generate similar estimates in this example, although IEJHE provides the most information on the population because both the total population size and the study area density are provided. Because Neal et al. (1993) did not find strong evidence of individual heterogeneity with a test based on logistic regression, we believe

TABLE 13.2 Summary Data for the Joint Hypergeometric Estimator (JHE) and Immigration/Emigration JHE from a Mark-Resight Survey of Mountain Sheep from Neal et al. (1993)

Occasion	Marked	Marked on study area	Animals seen	Marked seen
1	25	22	40	9
2	25	21	63	11
3	25	21	66	14
4	25	20	57	11
5	25	20	52	10
6	25	20	61	9
7	25	22	87	19
8	25	18	46	8
9	25	23	63	13
10	25	18	30	8
11	25	20	69	14
12	25	17	35	6
13	25	20	51	13
14	25	21	57	17

TABLE 13.3 Summary Data for
Bowden's Estimator from a Mark-
Resight Survey of Mountain Sheep
from Neal et al. (1993)[a]

Animal No.	No. of resightings
1	10
2	9
3	3
4	6
5	7
6	5
7	5
8	12
9	2
10	5
11	8
12	9
13	8
14	3
15	12
16	10
17	4
18	9
19	4
20	3
21	6
22	6
23	6
24	3
25	7

[a]Total number of unmarked animals
sighted was 615.

that use of this estimator is appropriate for the data even though IEJHE
assumes no individual heterogeneity. Further, IEJHE provides an estimate of
mean density in the presence of immigration and emigration from the study
area, and thus acknowledges the biological realities of the survey.

Table 13.4 Results of Three Mark-Resight Estimators for Mountain Sheep Data from Neal et al. (1993)

Estimator	Parameter	Estimate	95% CI
Joint hypergeometric estimator (JHE)	$\hat{N}$	120	110–132
Immigration/emigration JHE	N^*	120	109–133
	$\bar{N}$	96.1	88.7–105.5
Bowden's	$\hat{N}$	119	101–140

EXTENSIONS OF DIRECT MARK-RESIGHT ESTIMATION

Arnason et al. (1991) considered the estimation of closed population size when the number of marked animals in the population is unknown, but the marks are individually identifiable. The estimator does not require radio-marked animals. However, the approach may be useful in circumstances where a survey has been designed to use radio-marked animals but transmitter failure precludes use of the estimators described above.

An innovative estimator that uses the transmitter signals of radio-marked animals to find and count unmarked animals is described by Rivest et al. (1998) for caribou (*Rangifer tarandus*) in northern Labrador and northeastern Québec. A two-phase Horvitz–Thompson estimator is developed based on the probability that a group of animals contains one or more radio-marked animals and on the probability that a transmitter signal from a group of caribou containing one or more radio-marked animals is found. A critical assumption of the estimator is that radio-marked animals are distributed randomly among the groups, with a test of this assumption provided by Rivest et al. (1998).

As noted above, the usual mark-resight estimators do not require that radio-marked animals are randomly distributed in the population, but only that the sighting probabilities of radio-marked animals are representative of unmarked animals. In contrast, the estimator developed by Rivest et al. (1998) requires that radio-marked animals be randomly distributed with unmarked animals because transmitter signals are used to find unmarked animals. Therefore, careful planning of how transmitters will be placed in the population is required for Rivest's estimator to provide useful results.

SIGHTABILITY MODELS

Most aerial surveys of wildlife populations use a two-stage sampling technique. The first stage entails dividing the entire survey area into smaller units (quadrats), which are called the primary sampling units. During the second stage, a selected sample of these quadrats is surveyed, with the goal to observe (count) every animal on each of the surveyed quadrats. Estimates from these surveys contain two sources of error, namely, errors due to sampling variation in both stages. The first source of error may be controlled by the choice of a suitable sampling plan for the first stage. The second error is usually termed "visibility bias" because only a portion of the animals on the sampled quadrats are counted. Sightability models attempt to correct for visibility bias.

The use of sightability models developed from radio-marked animals has been widely applied to population estimation of elk (*Cervus elaphus*) (Samuel et al. 1987, Unsworth et al. 1990, Otten et al. 1993), mule deer (Ackerman 1988, J. W. Unsworth, personal communication), bighorn sheep (*Ovis canadensis*) (Bodie et al. 1995), moose (*Alces alces*) (Anderson and Lindzey 1996), and waterfowl (Smith et al. 1995). Sightability models have also been applied to correcting age and sex ratios for visibility bias (Samuel et al. 1992).

Sightability models are developed from sighting trials on radio-marked animals. The marked animals are located, and a search area surrounding the animal is identified. The area is then searched by individuals lacking knowledge of the marked animal's location. If the marked animal is sighted, the covariates being considered for inclusion in the sightability model, such as group size, vegetation cover, observer, etc., are recorded. If the marked animal is not located after the search area has been completely covered, then a radio receiver is used to find the marked animal and the covariate values are collected.

From these sighting trials, logistic regression models are constructed to predict the probability of sighting an animal (p) given the values of the covariates used in the model. For example, Samuel et al. (1987) suggested the logistic regression model

$$\hat{p} = \frac{\exp[1.22 + 1.55\log_e(\text{group size}) + -0.05(\% \text{ vegetative cover})]}{1 + \exp[1.22 + 1.55\log_e(\text{group size}) + -0.05(\% \text{ vegetative cover})]}, \quad (13.9)$$

where the covariates group size and % vegetative cover are used to estimate p for an elk population in Idaho. As an example, a group of three elk at a location with 10% vegetative cover would have 0.919 probability of being sighted. If the vegetative cover increased to 90%, the probability of sighting would drop to 0.171.

This model is then used to remove sightability bias by correcting the observed counts. For example, if three elk at a location with 10% vegetative cover have a 0.919 probability of being seen, then they represent a total of $3/0.919 = 3.26$ elk. Had the three elk been standing in thick vegetation with

90% cover, then $3/0.171 = 17.54$ elk would be the corrected count. The sightability correction is applied to each observed group to obtain a less biased (hopefully unbiased) count.

Radio-marked animals are not directly included in the sampling design. A key feature of the application of sightability models has been the use of a sampling frame of quadrats to survey, with a random sample surveyed. Between-quadrat variation is incorporated into the population estimation process, with the sightability model derived from radio-marked animals correcting the count for each quadrat (Steinhorst and Samuel 1989).

Several important assumptions must be made when applying sightability models. First, the conditions under which the survey is performed must be encompassed within the sightability model. That is, you would not want to apply a sightability model developed in an area with only low vegetative cover (0–30%) to an area of much higher vegetative cover (50–80%). In this case, the data used to build the model do not apply to the new area. Other abuses of the method would include the development of a model with one type of aircraft, then use of the model with data collected from a different type of aircraft. The point is that sightability models can only be applied in areas with the same conditions as the area in which data were collected to build the model. Applicability of a sightability model to an area different from where the model was built, or to the same area but under conditions different from those in which the model was built (e.g., prefire model, postfire application), is a judgment call, and there is no way to assess the validity of the judgment short of building a new model. Ideally, new models would be built from a large sample of radio-marked animals on each new study area, and each year, but obviously costs preclude this extreme.

Extrapolation of sightability models to new areas has often not been as critical as implied above because the sighting probabilities are close to 1, with only small corrections derived from the model. In particular, this has been the situation with elk, but much less so with deer and sheep. Even with deer and sheep, the majority of the variation of the population estimate derives from the spatial heterogeneity of the quadrat survey, and not from the visibility correction supplied by the sightability model.

A second critical assumption when using sightability models is that the covariates predicting sighting probability are measured without error. For example, Cogan and Diefenbach (1998) found that observers undercounted group sizes of elk, and the percentages of elk missed increased as percent canopy cover increased. Simulations demonstrated that undercounting group sizes of elk could result in negative biases of 20%. As an example, we will reconsider the model of Samuel et al. (1987) presented above. Assume a group of 10 elk with 10% canopy cover. If all 10 elk were observed, a sighting probability of $\hat{p} = 0.986$ would result. However, suppose only 8 are observed because 2 stood still while the other 8 ran and the observers missed recording the 2 standing individuals. This scenario results in a sighting probability of $\hat{p} = 0.981$. The corrected counts for the two scenarios are then $10/0.986 = 10.142$ and $8/0.981 = 8.155$, giving a 19.6% negative bias. In effect, a sightability model is needed to correct the group size of elk sighted! The result of this undercount of group size is to bias the population estimate low.

Additional work on the design and estimation of population size with sightability models has been done by Wong (1996). In particular, she derived an unbiased version of the variance estimator proposed by Steinhorst and Samuel (1989) and explored the performance of the various estimators with simulation.

CORRECTING BIAS OF GRID TRAPPING ESTIMATES

Trapping grids have been a standard technique for estimating population density of small mammals (Otis et al. 1978, White et al. 1982, reviewed in Thompson et al. 1998). Grids of traps are run for multiple occasions to estimate a population size ($\hat{N}$) for the grid. For estimating density, the trapping grid is assumed to be demographically and geographically closed, so that density $\hat{D} = \hat{N}/A$ where A is the area of the trapping grid. The issue of demographic closure is not difficult to meet because the trapping interval can be short enough to preclude much, if any, mortality. Juveniles recruited to the population can be recognized and excluded from the sample. However, geographic closure is more difficult to handle. Animals at the edge of the trapping grid, with only a portion of their home range included in the grid, cause uncertainty when defining the exact grid area (A). The effective value of A is greater than the actual trapping grid area. Nested grids (Otis et al. 1978, White et al. 1982) have been used to remove this bias, but such attempts have not been totally successful (Wilson and Anderson 1985).

Radio-marked animals provide an innovative approach to obtaining unbiased estimates of density. A sample of the M_{t+1} animals captured on the grid are radio-marked. After trapping of the grid is complete, so that the grid with its baited traps no longer provides an attraction, these animal's locations are monitored to determine the proportion of time they spend on the trapping grid. Thus, $p_i = g_i/G_i$, where p_i is the probability of a location being on the grid for animal i, estimated by the number of total locations (G_i) divided into the number of locations found on the grid (g_i). The mean ($\bar{p}$) and its variance [$\hat{\text{Var}}(\bar{p})$] of the p_i's can be used to correct the population estimate ($\hat{N}$) and its variance [$\hat{\text{Var}}(\hat{N})$] to obtain an unbiased density estimate as

$$\hat{D} = \frac{\hat{N}\bar{p}}{A}, \tag{13.10}$$

with variance estimated as

$$\hat{Var}(\hat{D}) = \frac{\hat{N}^2\hat{Var}(\bar{p}) + \bar{p}^2\hat{Var}(\hat{N})}{A^2}, \tag{13.11}$$

assuming that $\hat{N}$ and $\bar{p}$ are independent.

An alternative estimator could estimate $\bar{p}$ as the proportion of the radio-marked animals that spent at least 50% of their time on the trapping grid. However, we can envision a scenario where none of the animals whose home range overlaps a portion of the trapping grid have more than 50% of their home range on the trapping grid. Thus, the estimate of $\bar{p}$ would be zero, resulting in $\hat{D} = 0$. However, the trapping grid still has a density, consisting of the fractions of the home range of all the animals occupying the trapping grid. Therefore, we believe that the best estimator of $\bar{p}$ is the mean of the proportion of locations on the trapping grid of each of the radio-marked animals. A mean of the p_i weighted by the number of locations for each animal can be used for situations where there is considerable variation in the number of locations obtained for each mouse in the radio-marked sample. We would not recommend weighting the estimates of p_i by their variance because $\hat{var}(\hat{p}_i)$ and $\hat{p}_i$ are not independent, so a biased estimate of $\bar{p}$ will result.

The proposed estimator may be slightly biased for situations where animals are not attracted to the grid. This bias occurs because animals spending little time on the grid are less likely to be radio-marked than animals that spend most of their time on the grid, resulting in a size-biased sample. The resulting estimate of $\bar{p}$ is too high, so that $\hat{D}$ is biased high. For situations where animals are attracted to the grid, we suspect that this bias disappears because animals that would normally not use the grid are now captured and radio-marked, resulting in a less biased estimate of $\bar{p}$ for animals trapped on the grid. Simulation studies of animal movement within defined home ranges patterned after those done by Wilson and Anderson (1985) could be used to assess this bias.

EXAMPLE

Locations of radio-collared Preble's meadow jumping mice (PMJM, *Zapus hudsonius preblei*) were used to provide an unbiased estimate of mouse density per kilometer of stream stretch. Population size (N) was estimated using mark-recapture techniques. Parallel trap lines, with traps placed every 5 m, were set on either side of three different streams. During consecutive 8-day trapping sessions, all captured PMJM were individually marked with PIT (passive integrated transponder) tags (Table 13.5). Capture histories of the PIT-tagged mice were used to estimate N with programs CAPTURE (White et al. 1978, 1982, Rexstad and Burnham 1991) and MARK (White and Burnham 1999) at each site for both trapping sessions. An unadjusted density (PMJM per km stream stretch) was calculated as $\hat{D}_U = \hat{N}/A$ where A is length of the trap line. This unadjusted density estimate in mice per kilometer of stream is, however, positively biased by the mice captured on the trap line that spent very little time along that stretch of stream covered by the trap line.

A subset of captured PMJM at each of the three sites were also radio-marked to facilitate study of movement patterns of the mice (Table 13.5). Once traps were removed, radio-collared mice were tracked every other night for approximately 4 weeks (the duration of the battery life of the transmitters), with locations of each mouse recorded every hour during 8-hour tracking sessions. To correct the bias in the density estimates of PMJM, the proportion of locations within the trap lines (p_i) was estimated for each radio-collared mouse. Mean of the p_i's ($\bar{p}$) per site and session was then used to estimate an adjusted density, $D_A = \hat{N} \times \bar{p}/A$, with $\hat{Var}(\hat{D}_A)$ calculated as in Eq. (13.11) (Table 13.5).

In this example, correction of the positive bias in each of the density estimates resulted in an increase in variance of the estimate (Table 13.5). Larger sample sizes of both the number of animals radio-collared and in locations per animal would reduce the error associated with estimates of $\bar{p}$ and thus minimize variance associated with the adjustment.

FUTURE DEVELOPMENTS

Immediate improvements of the methods described in this chapter will develop with more technologically advanced methods of marking and resighting animals rather than improvements to the statistical estimators. Further miniaturization of radiotransmitters with a reduction in cost will make the trapping grid estimator an effective technique. Use of PIT (Fagerstone and Johns 1987, Schooley et al. 1993) or radio-tags combined with camera systems and PIT tag readers will allow more efficient and economical methods of remotely obtaining counts of individually marked and unmarked animals. DNA markers (Palsbøll et al. 1997, Reed et al. 1997, Wasser et al. 1997, Foran et al. 1997) may also provide new technologies that will lead to resourceful field procedures. As these technological advancements provide more flexible sampling schemes, modifications of the current statistical estimators will be adapted to accommodate the improved field sampling techniques.

TABLE 13.5 Stream Reach Abundance ($\hat{N}$) and Density Estimates for Preble's Meadow Jumping Mouse (PMJM) from Three Sites in Douglas County, Colorado for Two Trapping Sessions[a]

Site	Trapping session	No. PIT-tagged	No. radio-collared	$\hat{N}$	$\hat{SE}(\hat{N})$	Trap line length (m)	Unadjusted density ($\hat{D}_U$) (PMJM/km)	$\bar{p}$	$\hat{SE}(\bar{p})$	Adjusted density ($\hat{D}_A$) (PMJM/km)	$\hat{SE}(\hat{D}_A)$
Maytag	July	28	16	31	11.4	608	51.0	0.68	0.09	34.61	13.53
Maytag	September	42	18	44	1.8	494	89.1	0.95	0.03	84.58	4.39
Pine Cliff	July	16	13	17	0.9	504	33.7	0.80	0.08	27.15	2.91
Pine Cliff	September	46	16	51	3.3	504	101.2	0.74	0.10	75.16	11.11
Woodhouse	July	14	12	20	5.2	516	38.8	0.64	0.10	24.84	7.50
Woodhouse	September	18	13	20	3.1	574	34.8	0.86	0.06	29.96	5.03

[a]Numbers of mice PIT-tagged and radio-collared at each site during each session used to estimate N and p are presented. Both adjusted and unadjusted density estimates (PMJM per km of stream stretch) are reported. Adjustments were made to the density estimates to account for the positive bias introduced when animals are attracted to a trap line.

Statistical methodology will also be expanded, with incorporation of co-variates into mark-resight models. Estimators developed by Alho (1990) and Huggins (1989, 1991) are available for such analyses in program MARK (White and Burnham 1999). We also expect to see additional innovative uses of the Horvitz–Thompson estimator, such as that of Rivest et al. (1998).

SUMMARY

Three uses of radio-marked animals in population estimation methods are reviewed: direct mark-resight techniques, sightability models, and bias correction of grid trapping estimates. A suite of models to handle a variety of assumptions for estimating population size from direct mark-resight population estimation, including individual heterogeneity, immigration/emigration, and unequal effort across sampling occasions, are available. A key assumption of the direct mark-resight or mark-recapture methodologies, needed to obtain unbiased estimates of population size, is that sighting or recapture probabilities of marked animals are representative of unmarked animals. This assumption is often met by radio-tagging animals with a capture method different than the method used for resighting.

Sightability models are developed from surveys of radio-marked animals and then extrapolated to populations with no marked animals. A key assumption needed for unbiased estimates is that the sightability model correctly predicts sighting probabilities for the population to which it is applied.

Density estimates computed from trapping grids divide the population size on the grid by the area of the grid. Because trapping grids often attract animals from outside the grid (e.g., small mammals), past attempts at estimating density corrected the area of the grid for the region of attraction. However, radio-marked animals trapped on the grid can be used to estimate the proportion of time actually spent on the grid and thus correct the population estimate prior to computing density.

The use of radio-marked animals provides innovative new ways to estimate wildlife population parameters. Most importantly, bias of population size estimates can be reduced because radiotransmitters allow correction for immigration and emigration from the study area and provide efficient ways to correct other biases inherent in traditional population size estimators. Whether the reduction in bias is worth the additional cost of marking animals with radiotransmitters is decided by the problem to be solved.

CHAPTER 14

Analysis of Survival Data from Radiotelemetry Studies

SCOTT R. WINTERSTEIN, KENNETH H. POLLOCK, AND CHRISTINE M. BUNCK

Radiotelemetry is a well-established and accepted methodology for studying survival in wild animal populations. An animal is captured, fitted with a radiotransmitter, and released. From release, the animal's unique radio signal is monitored to determine the animal's fate at more or less regular intervals. For each animal, the investigator must know (1) the date it was radio-marked and released, (2) the date it was last located, and (3) its status when last located. The date on which an animal entered the study (t_0) is known exactly, as is the date on which an animal was last observed (t_i). At each location time, the status of each animal is recorded as alive, dead, or missing. Missing animals are considered censored, meaning that the event of interest (death)

Radio Tracking and Animal Populations
Copyright © 2001 by Academic Press. All rights of reproduction in any form reserved.

cannot be observed. This type of censoring is generally referred to as "right censoring" and is caused by such factors as radio failure, topography that inhibits signal reception, and permanent or temporary emigration (White and Garrott 1990, Tsai et al. 1999a). Technically, at the end of the study (the last location time), all live animals are considered censored, again because we never observe the death. However, in a practical sense we differentiate between animals that we know are alive when the study ends and those which we lose track of during the course of the study. Throughout this chapter, we will consider only those individuals that are missing during the course of the study to be censored.

As with all research projects, we hope that prior to capturing any animals the researchers have specified the goals of the study and have outlined the estimation and model selection procedures and any hypothesis tests (see Chapter 2). The purpose of this chapter is to examine how survival data from radiotelemetry studies are analyzed. We make no attempt to present a comprehensive review of all of the analytical techniques available to researchers. White and Garrott (1990) and the references contained therein (as well as the references we supply herein) provide an excellent guide. We will present a brief overview of the three techniques most commonly used during the past 10 years and the assumptions associated with each method. Next we will pinpoint a few areas of concern in an attempt to review what has been done correctly and what has not been done, but probably should be. Many of the available analytical techniques have been broadly implemented (e.g., staggered entry Kaplan–Meier analysis), but others have not gained general use (e.g., risk ratio analysis).

APPROACHES FOR ESTIMATING SURVIVAL

What are the primary techniques used during the past 10 years to analyze survival data from radiotelemetry studies? We reviewed each paper from *The Journal of Wildlife Management* from 1989 to 1999 that used radiotelemetry and estimated survival probabilities (Fig. 14.1). While, in general, Kaplan–Meier analysis has been the method of choice since 1993, there has been increased use of the Mayfield-type method. Techniques in the *Other* category have changed over time. In the early years, it was primarily Apparent Percent Success (APS), but also rarely included numerical estimation (White 1983) or the capture-recapture (or resighting) methods outlined by Lebreton et al. (1992). In later years, a mixture of methods dominates the *Other* category (e.g., logistic regression, maximum likelihood estimates; see, for example, Bro et al. 1999).

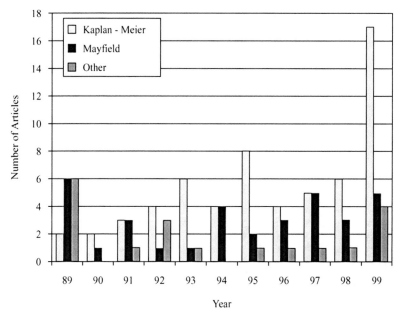

FIGURE 14.1 Number of articles in *The Journal of Wildlife Management* (1989–1999) that used radiotelemetry and estimated survival probabilities using the Kaplan–Meier, Mayfield, and *Other* methodologies.

APPARENT PERCENT SUCCESS

The APS or Simplistic estimator is easy to calculate and requires few assumptions, but it makes inefficient use of the data. We need only the initial number of radio-collared animals (number at risk) and the number that died during the study (number of deaths). As with all survival estimates, we assume that a random sample has been obtained (Table 14.1). The estimator does not allow for censored observations (those individuals that are missing during the study). The fate of each animal in a study can be viewed like an independent coin flip; the animal dies (heads) or is alive at the end of the study (tails). In this framework, we can estimate the final survival rate ($\hat{S}$) as

$$\hat{S} = 1 - \frac{\text{number of deaths}}{\text{number at risk}}. \tag{14.1}$$

Survival rates calculated in this manner assume that the entire sample of animals was marked at the beginning of the period of interest and therefore each animal has the same t_0 (Heisey and Fuller 1985, Hensler 1985). The final survival rate is the weighted average of the survival rates of each mutually

TABLE 14.1   Assumptions for the Apparent Percent Success, Mayfield Method, and Kaplan–Meier Survival Estimators

Assumption	Apparent percent success	Mayfield method	Kaplan–Meier estimate
Random sample	X	X	X
Experimental units independent	X	X	X
Observation periods[a] independent		X	X
Working radios always located	X	X^b	X^b
Random censoring allowed		X^c	X
Radios do not impact survival	X	X	X
Constant survival		X	
All experimental units have same t_0	X		

[a]Observation period = time between observations.
[b]Bunck et al. (1995) indicate that this assumption can be relaxed and present the necessary formula modifications.
[c]Strictly speaking, the Mayfield method yields biased survival estimates if censoring occurs. However, the bias is generally small if censoring is negligible. See text and Vangilder and Sheriff (1990).

exclusive group (e.g., age class) and therefore will be biased toward the group having the largest sample size.

Use of APS precludes generating a survival curve unless the researcher assumes that survival is constant throughout the study. Perhaps the biggest drawback to APS is that censoring is not allowed. There is no standard adjustment to use should animals disappear from the study. The APS estimator [Eq. (14.1)] assumes that no animals disappear from the study, so that censored animals are counted as alive and the estimate is positively biased. If all censored animals are treated as dead, the estimate will be negatively biased. The only other alternative is to remove censored animals from the initial set of animals at risk thus decreasing the sample size by assuming they never existed in the sample.

It is difficult to imagine many cases in which APS is the most appropriate analytical technique. It is rare when a radiotelemetry study does not have some animals disappear. It is even more unusual when the fate of an animal is not known on a regular basis.

MAYFIELD METHOD

The basic Mayfield methodology was first presented by Mayfield (1961, 1975) for studies of nesting success. It was generalized to wildlife and survival studies by Trent and Rongstad (1974). Heisey and Fuller (1985) provided a framework and computer program (MICROMORT) to examine cause-specific mortality rates from telemetry data. Strictly for convenience, we will refer to this collection of techniques as the Mayfield Method.

For each individual, we must know t_0 and the last date that an individual was observed (t_i), which will be a time of death, a censoring time, or the end of the study for animals that are still alive. Over the period of interest (L) the survival rate is assumed to be constant, and an estimate of the daily survival rate ($\hat{S}_d$) is

$$\hat{S}_d = 1 - \frac{\text{number of deaths}}{\text{total exposure days}} \tag{14.2}$$

and the period survival rate ($\hat{S}_p$) is

$$\hat{S}_p = \hat{S}_d^L. \tag{14.3}$$

The total number of exposure days for each animal is calculated as the number of days from t_0 to t_i. The sum of the individual animal's exposure days yields the *total exposure days* for all animals at risk.

Survival need not be calculated on a daily basis and should correspond to the observation interval. For example, if animals are checked weekly, then a weekly survival rate is appropriate. If the animal is observed daily, t_i for any individual is obvious. If the interval between observations is greater than a day, t_i must be estimated if the animal dies or is censored. For these animals, t_i can be estimated as halfway between the last known live sighting and the date on which the animal was first known to be dead or censored (Mayfield 1961, 1975). Miller and Johnson (1978) suggest that if the interval between observations exceeds about 15 days, t_i should be estimated as 40% of the visitation interval (see also Johnson 1979). Bart and Robson (1982) provide formulas for $\hat{S}_d$, variance estimates, and confidence intervals when an animal's status is determined at irregular intervals.

Unbiased estimates require no censoring (Table 14.1). However, if censoring is light, the associated bias is negligible (Vangilder and Sheriff 1990). If there are censored observations, they must be random and should be included in the analysis up to the point at which the animals are censored (Vangilder and Sheriff 1990). Bunck and Pollock (1993) suggest that if the fate of each animal is determined daily and animals disappear permanently from the study site, $\hat{S}_d$ should be modified slightly so that

$$\hat{S}_d = 1 - \frac{\text{number of deaths}}{(\text{total exposure days}) - (\text{number of censored animals})}. \tag{14.4}$$

The Mayfield Method does not require that all individuals enter the study at the same or nearly the same time. Entry can be staggered (see Pollock et al. 1989b), as long as the survival rate is constant for the duration of the study. In long-term studies, the assumption of constant survival may be too restrictive. To circumvent this problem, the study period can be divided into shorter intervals, each having constant survival, and the overall survival rate taken as the product of the rate for each interval (Johnson 1979, Bart and Robson 1982, Heisey and Fuller 1985).

KAPLAN–MEIER ANALYSIS

The Kaplan–Meier or product limit estimator was developed by Kaplan and Meier (1958) and has been used extensively in medical, engineering, and product reliability studies (Kalbfleisch and Prentice 1980, Cox and Oakes 1984, Lee 1992). Muenchow (1986) and Pyke and Thompson (1986) used Kaplan–Meier estimates, in the form of failure time analysis, in general ecological studies. Among others, Krauss et al. (1987), Conroy et al. (1989), and Pollock et al. (1989a,b) used Kaplan–Meier analysis techniques (henceforth referred to simply as Kaplan–Meier) to analyze survival data from radiotelemetry studies on wildlife.

The survival function $(\hat{S}_t)$ is the probability of an arbitrary animal in the population surviving t units of time from the beginning of the study. If we know t_0 and t_i (or can reasonably estimate it) for each individual and the fate of that individual, we can estimate the survival function as

$$\hat{S}_t = \prod \left[1 - \frac{\text{number of deaths at time } j}{\text{number at risk at time } j} \right]. \tag{14.5}$$

The calculation is conducted when a death is observed (see Pollock et al. 1989b, Lee 1992, or White and Garrott 1990, for detailed examples of the calculations). An animal is at risk at time j if alive when the time period started. For example, if animals are monitored daily, all animals alive at the beginning of a new observation day (e.g., midnight) are at risk for that day. All animals that die within a particular time period or are censored during that time period are at risk for that time period.

Kaplan–Meier has no underlying assumption of constant survival (Table 14.1), and the method provides unbiased estimates even when there are censored observations. As with the Mayfield Method, Kaplan–Meier does not require that all individuals enter the study at the same time (Pollock et al. 1989b), but any newly radio-tagged animals are assumed to have the same survival function as previously radio-tagged animals.

AREAS OF CONCERN

In the majority of radiotelemetry-based survival studies, researchers generally:

(1) Estimate survival probabilities, sometimes with confidence intervals,
(2) Generate survival curves,
(3) Compare survival curves (e.g., between sexes or among years), almost always with some form of the log-rank test, and
(4) Test for the impact of covariates (e.g., weight, condition) using the Cox proportional hazards model.

While most of the published analyses are likely done correctly, we are concerned that too often the necessary steps are overlooked. Researchers need to make sure that they:

(1) Meet the assumptions that underlie the analyses,
(2) Use the proper estimators, with sufficient sample sizes and the appropriate t_0,
(3) Do what they *should* do, not what the software makes it easy to do, and
(4) Use the appropriate tests to compare survival curves and test for the impact of covariates.

ASSUMPTIONS

Regardless of study objectives or type of data collected, all radiotelemetry survival studies share a number of assumptions. We will limit our discussion to the assumptions associated with the Mayfield Method and Kaplan–Meier (Table 14.1) and the problems that ensue when these assumptions are violated. Where possible, we offer solutions and suggestions. Pollock et al. (1989b), White and Garrott (1990), and Tsai et al. (1999a) also discuss problems associated with violation of these assumptions.

Random Sample

All survival studies assume that a random sample has been obtained. This sample is random for any particular classification made by the investigator (i.e., within a given age or sex class). Suppose in a study of overwintering white-tailed deer (*Odocoileus virginianus*) survival, lighter weight deer are more likely to be captured. If lighter deer also have lower survival rates, a negative bias in the survival rate for the population will be introduced.

It is virtually impossible for a researcher to obtain a truly random sample of a wild population. Therefore, every effort must be made to use unbiased capture and collaring techniques. If not, the biases must be recognized or the researcher is in danger of misidentifying the population under study. For example, older (≥ 2.5 years) white-tailed deer bucks rarely enter Clover traps in northern Michigan (S. Winterstein, unpublished data). If all deer in a study are caught in Clover traps, the population under investigation is not all white-tailed deer in the area. An insufficient number of older bucks would have been captured and the sample would not be representative of the entire population.

Experimental Units Are Independent

The first critical aspect of this assumption is that the experimental units are correctly identified. In the overwhelming majority of cases, each individually radio-collared animal will be an experimental unit. However, there are situations in which this would not be true (see Eberhardt et al. 1989, Winterstein 1992). Take, for example, a study designed to examine the impact of hen age on reproductive success of ruffed grouse (*Bonasa umbellus*). One way to design the study would be to select a sample of yearling hens and a sample of 2-year-old hens. When a hen's brood hatches, we place a radio on each hatchling. In this case, the hen, or more correctly her brood, is the experimental unit, *not* each individual hatchling. In this case, the proper calculation of $\hat{S}$ may not be obvious because the status of the experimental unit (brood) may not be clear. If four of the seven brood members die, is the brood alive or is it dead? Using methods based on cluster sampling, Flint et al. (1995) provide an appropriate method for obtaining $\hat{S}$ in this situation.

The second aspect is that survival rates are assumed to be independent for different animals. This means that whether one individual survives or not has no impact on the probability of any other individual surviving. Such an assumption may not hold for social animals, mated pairs, or parents and young. Take, for example, a female mammal (e.g., black bear, *Ursus americanus*) caring for her young. If the mother and young are all included in the study, the death of the mother could influence cub survival. Violation of this assumption should not cause bias, but it will make the estimates appear more precise. The obvious solution is to avoid radio-collaring of multiple members of a family or any other social group (Millspaugh et al. 1998a; see also Chapter 2).

Observation Periods Are Independent

The observation period is the time between consecutive attempts to locate an animal. In most radiotelemetry studies, the time that elapses between observations is dictated by such factors as number of experimental units, size of the study site, number of field assistants, technology available (e.g., continuously operating towers vs. aircraft vs. ground vehicles), weather, and "acts of God" (e.g., forest fires, flat tires, over-sleeping, breaking the on/off switch on both the primary and back-up receivers on the same day). In spite of these constraints, every effort should be made to locate each radio-marked animal often and regularly.

It is common in survival studies to discuss the daily survival rate, regardless of the actual interval between observations. Daily survival rates have an intuitive feel and allow us to standardize comparisons among intervals of differing lengths. Is the probability of surviving 1 day of the 15-day hunting season different from the probability of surviving 1 day of the 45-day spring migratory period? Also, if our observation period is 1 week and an animal is alive for consecutive locations, we "know" it was alive on each day between the locations. However, filling in the intermediate location intervals can lead to biased estimators (Bart and Robson 1982, Bunck et al. 1995). Survival rates are best analyzed and reported in terms of the length of the actual observation period (Tsai et al. 1999a).

Regardless of observation period length, a broad interpretation of this assumption is that what happens to an experimental unit on one day has no bearing on its fate the next day. So the fact that a radio-collared animal was wounded on Monday has absolutely no impact on the probability of its surviving Tuesday. At the extreme, this means that on each day of our study we should have a completely new sample of radio-marked animals. This, of course, is impossible, and little can be done to avoid violating this assumption. If the researcher is aware of a wounding and subsequent death, the death date could be adjusted back to the wounding date or some midpoint. However, it is unlikely that such information will be available.

Working Radios Are Always Located

Under this assumption, if an animal is on the study site and has a working radio, the animal will be located. Therefore, one assumes that animals never wander off the study site or crawl into burrows or small valleys or move great distances between observation periods. Usually, if an animal cannot be located in one period but is found in a subsequent period we fill in the missing data. In many cases this does not create a large problem. Suppose, however, that the purpose of the study is to examine survival relative to a particular treatment (e.g., reduced hunting pressure or a silvicultural practice) on a portion of the study site. If an animal leaves the study site, is it still in the study? If the animal is gone for only a few hours, no problem. But what if the animal is gone for several days or weeks? Also, how do we differentiate between an animal that is off the study site and one that is on the study site but not located?

If an animal is temporarily not located, it should only be considered at risk for those observation periods when it was located. Bunck et al. (1995) demonstrate that when the Kaplan–Meier and Mayfield Method estimators are modified to use data on animals only when they are known to be at risk, they each provide unbiased survival estimates. The authors show that when relocation probabilities are less than about 80%, the traditional calculation formulas [Eqs. (14.2), (14.3), and (14.5)] yield positively biased estimates.

Pollock et al. (1995) provide a generalization of the Kaplan–Meier estimator that allows for relocation probabilities that are less than 1. They show that survival data from a radiotelemetry study are virtually identical in form to those obtained in an open capture-recapture study. However, in contrast to what occurs in a capture-recapture study, dead animals are located in a radiotelemetry study. This allows modification of the standard Jolly–Seber formula for estimating the number of marked individuals at any observation period (Pollock et al. 1990) to yield an estimate of the number of working radiocollars at any observation period. An estimate of the relocation probability at any observation period is then the number of animals located divided by the estimated number of working radios.

Random Censoring

Censoring is not unique to telemetry studies. In all marking and banding studies, animals disappear and are usually considered dead. In telemetry studies, we assume that all animals with functional radiotransmitters present on the study area will be found (but see above). Therefore, in a perfect world all deaths are known. This further assumes that the censoring mechanism is random. Any tendency for healthier animals, for example, to emigrate or for predators to destroy radios while taking prey would violate this assumption.

It is worth noting that *censoring during the course of a study is not a good thing*. When the fate of an animal is unknown, information is lost. Just because there are analytical techniques that provide unbiased estimates when censoring is present does not make it desirable. Every effort should be made to minimize censoring by keeping the interval between observation periods as short as possible, purchasing the best radio-collars one can afford from reliable companies, designing a study that ends before radio-collar batteries are scheduled to fail, and ensuring that all field personnel are adequately trained.

The question of how much censoring can be tolerated in a study is difficult to answer. It is a function of survival rate, length of the study, and pattern of censoring. Clearly, if a majority of the animals are censored, even though survival estimates would be unbiased (Kaplan–Meier) or nearly so (Mayfield Method), the researcher should question the accuracy of the estimates. In

Kaplan–Meier analysis, precision is a function of the number of animals at risk (Cox and Oakes 1984, Lee 1992). In the Mayfield Method analysis, precision is generally a function of the number of exposure days (Johnson 1979, Hensler and Nichols 1981). Censoring during the course of the study results in a decrease in precision for both methods. In Kaplan–Meier, censoring also indirectly impacts the power of tests to compare survival distributions (George 1983). When censoring is present, power becomes a function of the number of deaths, not the number of individuals that started the study. As censoring increases, power decreases (George 1983). Based on our experience, in wild-life studies where survival is being estimated for more than 90 days, more than 10–15% censoring is cause for concern unless the initial sample size is large (>50 animals; see below).

The assumption that censoring is random is extremely important. We assume that survival time and censoring are independent. That is, censoring must be independent of death. Any animal not located must be as likely to be alive as dead. Furthermore, there must be an equal likelihood of locating dead and live animals. Possible violations could result from a predator killing an animal and subsequently destroying the radiotransmitter or an animal emigrating because it is more (or less) healthy than its companions.

Tsai et al. (1999a) describe the impact of nonrandom censoring on Kaplan–Meier estimates. They recognized that censoring could be a function of being on the study site (probability of staying = δ if alive and δ^* if dead) and being located (probability of relocation = p if alive and p^* if dead). They found that if $(\delta p)/(\delta * p*) = 1$, there is very little bias in the survival estimate. However, as the ratio becomes >1 (indicating that there is nonrandom censoring because live animals are more likely to stay on the study and/or are more likely to be located than dead animals) the bias increases, particularly as the actual survival rate decreases (Tsai et al. 1999a).

Tsai et al. (1999a) provide a modification of the Kaplan–Meier estimator to account for differences in staying and relocation probabilities. The modification requires estimates for δ, δ^*, p, and p^*, which are not likely available to most researchers. If, however, censoring is high the researcher must either assume that $(\delta p) = (\delta * p*)$ or design a study to obtain the estimates. The researcher could also assume that $\delta = \delta^*$ and $p = p^*$ and that all four values are close to 1, but this makes no sense if censoring is high.

In some cases, it may be reasonable to assume that either emigration or radio failure (but not both) is 0. The timing of the censoring could provide useful information. For example, in a study of winter survival of waterfowl with reliable radios ($p = p^*$), the censoring times would primarily reflect emigration. Estimation of this emigration time distribution could be informative to the biologist, especially if related to covariates reflecting weather severity.

If the researcher cannot estimate or assume values for δ, δ^*, p, and p^*, it is possible to put "bounds" on the survival curve, which allows censoring to take two extreme forms. We can obtain a lower bound by assuming that every censored observation was really a death and an upper bound by assuming that every censored animal survived to the end of the study.

Radios Do Not Impact Survival

We must assume that capturing and radio-marking an animal does not influence its future survival. Clearly, failure of this assumption will cause a negative bias on the survival estimates (see Chapter 3). As radio-tags become more sophisticated and smaller, this should be less of a problem. However, long-term radiotelemetry studies increasingly suggest that radio-collars impact survival (see Chapter 3). As researchers estimate population size or expected productivity using age-specific survival rates, we find that survival estimates do not make sense relative to the biological realities we observe. An age-structured population dynamics model was constructed for eastern wild turkeys (*Meleagris gallopavo*) in the Virginias (Alpizar-Jara et al. in press). This model uses age-specific survival probabilities obtained from radiotelemetry studies (Pack et al. 1999). The authors found that annual growth rates predicted by their model did not correspond well with regional population growth rates based on spring harvest data. They chose to calibrate their model by adjusting recruitment rates but conceded that the survival estimates could be negatively biased. One of us (SRW) encountered a similar situation with ruffed grouse in Michigan. Radiotelemetry was used to estimate chick survival, juvenile survival, and adult survival. The final model estimated population size too low to match the increases in population size known to have occurred.

While the two examples above do not prove that radio-collars impacted survival, they do suggest that researchers should carefully consider the impact of radio-marking their study animal. To offset these problems we need to develop and use minimum size requirements to include an animal in the study. We also need an acclimation period for each individual. An animal can not be included in the analysis until there is evidence that it has recovered from trapping stress and adjusted to the radio-collar. Short-term effects could be eliminated by using a conditioning period of, for example, 2 weeks after tagging. In this case, an animal's survival time is not considered until it has survived for 2 weeks. This acclimation period will be species specific and should be determined by evidence of normal activities. For example, daily movement patterns must be observed. Derleth and Sepik (1990) excluded American woodcocks (*Scolopax minor*) from their analysis of summer–fall survival if the birds died without moving from the area in which they were trapped. Often no suitable behavior exists or can be easily monitored, and a best guess time frame is used. The acclimation period was set at 5 days in a study evaluating the impact of hunting on ruffed grouse in northern Michigan (Gormley 1996). This acclimation period was determined empirically. Dead radio-collared grouse were necropsied to determine cause of death. Necropsy results indicated that deaths attributable to trapping stress were no longer detected past 5 days post trapping. *All* grouse that died within 5 days of trapping (regardless of the cause) were removed from the study and not considered in the survival analyses.

Having a well-defined acclimation period requires additional experimental units because the usable sample size is generally not the total number of animals captured. We also recommend that when determining the necessary sample size (either analytically using prior data or using some best guess) you also plan for censored observations (see above). In the grouse study mentioned above, there was no analytical basis for calculating necessary sample sizes, so the researchers used a rule of thumb of 50 birds per treatment (see below). They targeted 60 birds per treatment, giving a cushion of 10 that could be lost to acclimation and censoring.

STAGGERED ENTRY AND t_0

One of the most important considerations when estimating survival from radiotelemetry data is determining the appropriate point (t_0) at which to begin measuring survival. In medical studies, the natural time origin is generally the time treatment began. In radiotelemetry there is no such natural time origin. In studies where all of the animals are captured at or near the same time, the obvious time origin might be the date when the last animals were captured. In many studies involving wild populations, animals may be introduced into the study gradually over a long period. This is known as *staggered entry* (Pollock et al. 1989b) and is usually equivalent to left truncation in medical studies (Cox and Oakes 1984). Staggered entry could be due to practical problems associated with capturing all animals at once or due to the biologist's wanting to introduce more animals into the study to increase precision after some of the animals have died. It should be emphasized that any newly tagged animals are assumed to have the same survival function as the previously tagged animals. Kaplan–Meier and Mayfield Method analyses are easily generalized to account for staggered entry. In Kaplan–Meier, the number of animals at risk at any time j [Eq. (14.5)] is adjusted to reflect the addition of new animals (Pollock et al. 1989b). In the Mayfield Method, no adjustment is needed because the total exposure days [Eq. (14.2)] is the sum of the individual animal's exposure days, which is determined in the same manner regardless of when an animal enters the study (see Bunck and Pollock 1993, Bunck et al. 1995, Tsai et al. 1999a).

Despite the fact that both methods can mathematically accommodate staggered entry, there must be a time origin that makes biological sense (and if necessary accounts for an acclimation period as described above). Both methods can generate survival estimates and curves from the date on which the first animal enters the study. However, using the time at which the first animal was captured as day 0 may not be the best choice. Suppose we have two similar study scenarios. In scenario 1 we radio-collar 50 ducks one day (day 0) and 5 individuals die on day 2. In scenario 2, 10 ducks are captured and radio-collared on day 0 and 40 more are radio-collared on day 4. On day 2, 5 individuals die. After day 4, the morality rates are identical in both scenarios. When should we start estimating survival? From day 0? From day 4? The decision has less impact if the Mayfield Method is used to analyze the data than if Kaplan–Meier is used. Using the Mayfield Method, scenario 1 had 1560 exposure days and 27 deaths, yielding an $\hat{S}_d$ of 0.9827 and, for $L = 63$, an $\hat{S}_p$ of 0.3329. Scenario 2 had 1400 exposure days and 27 deaths, yielding an $\hat{S}_d$ of 0.9807 and, for $L = 63$, an $\hat{S}_p$ of 0.2932. The difference is negligible when L is short, but the disparity increases as L increases (Fig. 14.2). Using Kaplan–Meier, scenario 1 yields a survival estimate of 0.39 after 63 days and scenario 2 yields an estimate of 0.22 after 63 days (Fig. 14.3) or 0.43 if the estimate is

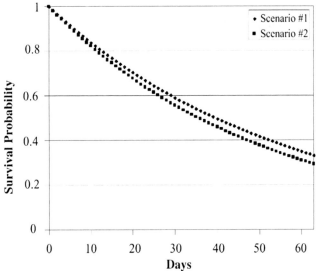

FIGURE 14.2 Results of Mayfield Method analyses for two different scenarios (see text for scenario details).

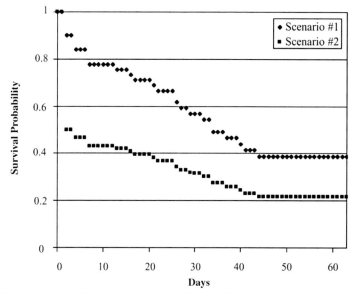

FIGURE 14.3 Results of Kaplan–Meier analyses for two different scenarios (see text for scenario details).

calculated from day 4 to day 63. The loss of 5 ducks on day 2 when the sample in scenario 2 was extremely small (10 ducks) has an unduly large influence on the final estimate and, because of the manner in which the final survival estimate is calculated, essentially initializes the survival probability at 0.50, not 1.0.

In the real world, a researcher will not have two competing scenarios from which to choose; there will only be data from the project that was actually conducted. Often animals will enter the study in small numbers over an extended period of time (e.g., three or four animals each week for 3 months), not simply in two pulses as in scenario 2. The researcher must decide when the sample size is adequate to obtain a reasonable estimate of survival. If radio-collaring is protracted, it is likely that some animals will die before they can become part of the survival analysis. Researchers must recognize the potential problems associated with beginning an analysis before an adequate number of animals are radio-collared.

SOFTWARE

Many statistical programs exist that estimate survival (see Appendix A). All of the major commercially available statistical software packages (e.g., SAS,

SPSS, BMDP, NCSS) execute some form of Kaplan–Meier. There are web-based programs that also do Kaplan–Meier (e.g., PROPHET; see Appendix A). PROPHET also provides a number of excellent analytical tools for testing model assumptions. More specialized programs are also available. KAPLAN (Missouri Department of Conservation) provides Kaplan–Meier estimates for staggered entry data. A SAS macro written by Dennis Heisey allows the researcher to test for the impact of covariates on survival when there is staggered entry. MICROMORT (Heisey and Fuller 1985) conducts Mayfield Method analyses and calculates cause-specific mortality rates. MARK (Gary White, Colorado State University; see Appendix A) is an extremely flexible, and therefore powerful, program that provides a framework to, among other things, conduct Kaplan–Meier analyses and examine the impact of multiple classes of covariates. In addition, the subroutines in MARK can be used to conduct Kaplan–Meier analyses incorporating the modifications suggested by Pollock et al. (1995) and Tsai et al. (1999a).

To be used, programs must be user friendly. Many specialized programs are not simple to use. There is generally a trade-off; simplicity of use almost by definition requires a reduction in control, meaning that the more "canned" simple programs lack flexibility. The reverse is also true: the more flexible the program, the harder it is to use. For example, many of the commercially available programs apparently allow for staggered entry. They allow input of a starting date and ending date for each individual (e.g., PHREG in SAS). This gives the illusion that the risk set is augmented with the addition of each new animal, when in reality all animals are given a common t_0. For example, the staggered data presented in Table 14.2 were analyzed using KAPLAN and PHREG. We radio-collared 10 animals on each of 5 days, assumed an acclimation period of 7 days for entry into the study, and estimated survival from September 1, 1999. KAPLAN yielded a survival estimate of 0.2070 at 196 days. We verified this value by conducting the analysis on a spreadsheet. PHREG produced a survival estimate of 0.2200 at 196 days. Another SAS procedure, LIFETEST, which requires that all animals have a common t_0 by using the number of days an animal was observed as the input, also produced an estimate of 0.2200 at 196 days. For most of these same programs (e.g., SAS) it is possible to conduct a staggered entry analysis, but to do so requires a thorough understanding of the software [e.g., for SAS users, we recommend Allison (1995)], not simply a knowledge of which pull-down menu to select.

Researchers also need to be aware of other subtle differences in the way software packages carry out calculations. For example, in the analysis of the data in Table 14.2, KAPLAN considers animals to be at risk on the day they are radio-collared (date of radio-collaring = day 1), whereas PHREG and LIFETEST consider animals to first be at risk on the day after they are radio-collared (date of radio-collaring = day 0). This difference impacts the results

TABLE 14.2   Fictitious Staggered Entry Data to Demonstrate the Potential Impact of the Analysis Software

Date radio-collared	Last date observed	Number of days observed[a]	Fate[b]
9/1/99	9/13/99	12	1
9/1/99	9/23/99	22	1
9/1/99	10/13/99	42	1
9/1/99	10/13/99	42	1
9/1/99	10/29/99	58	1
9/1/99	12/11/99	101	1
9/1/99	12/13/99	103	1
9/1/99	3/15/00	196	0
9/1/99	3/15/00	196	0
9/1/99	3/15/00	196	0
9/8/99	9/26/99	18	1
9/8/99	9/28/99	20	1
9/8/99	11/4/99	57	1
9/8/99	11/6/99	59	1
9/8/99	11/13/99	66	1
9/8/99	12/6/99	89	1
9/8/99	12/27/99	110	1
9/8/99	1/23/00	137	1
9/8/99	2/9/00	154	1
9/8/99	3/15/00	189	0
9/15/99	9/24/99	9	1
9/15/99	9/28/99	13	1
9/15/99	11/10/99	56	1
9/15/99	11/13/99	59	1
9/15/99	11/22/99	68	1
9/15/99	11/26/99	72	1
9/15/99	3/15/00	182	0
9/15/99	3/15/00	182	0
9/15/99	3/15/00	182	0
9/15/99	3/15/00	182	0
9/22/99	10/13/99	21	1
9/22/99	11/2/99	41	1
9/22/99	12/9/99	78	1
9/22/99	12/14/99	83	1

(*continues*)

TABLE 14.2 (*continued*)

Date radio-collared	Last date observed	Number of days observed[a]	Fate[b]
9/22/99	1/2/00	102	1
9/22/99	1/3/00	103	1
9/22/99	1/21/00	121	1
9/22/99	2/27/00	158	1
9/22/99	3/15/00	175	0
9/22/99	3/15/00	175	0
9/29/99	11/5/99	37	1
9/29/99	11/23/99	55	1
9/29/99	12/6/99	68	1
9/29/99	1/5/00	98	1
9/29/99	1/22/00	115	1
9/29/99	2/7/00	131	1
9/29/99	2/23/00	147	1
9/29/99	2/27/00	151	1
9/29/99	3/6/00	159	1
9/29/99	3/15/00	168	0

[a]Date radio-collared = day 0.
[b]1, mortality; 0, censored.

only if some animals die on the same day that other animals enter the study. We suggest that prior to relying on any software package, researchers calculate survival estimates for a small subsample of their data (e.g., 10 observations) with a calculator or spreadsheet. These values can be tested against the "canned" program results to ensure accuracy.

The actual analytical technique used by some researchers has become overly dependent on the availability of software. Unfortunately, the availability of software drives the analysis. In certain cases, this results in improper analyses. In other cases, the analyses are not necessarily wrong, but they also may not be as complete as they could be. For example, prior to about 1992 (Fig. 14.1), Mayfield Method analyses were commonly used to estimate overall survival rates. After that date, almost all of the analyses shown in Figure 14.1 were Heisey–Fuller analyses to look at cause-specific mortality rates (e.g., hunting deaths vs. nonhunting deaths). Such analyses do not require the Mayfield Method, but MICROMORT exists, so it is used. No simple-to-use program that has a labeled category or pull-down menu called "cause-specific mortality analysis" exists for the Kaplan–Meier programs. It is not uncommon to find published articles that use both Kaplan–Meier and

MICROMORT to analyze the same data: Kaplan–Meier to get the overall survival estimate and MICROMORT to get cause-specific mortality rates. Although it is relatively straightforward to conduct a cause-specific mortality analysis with Kaplan–Meier, it is rarely done. The marginal (cause-specific) survival curve can be obtained for a particular cause of death by treating deaths from any other cause as censored observations (Allison 1995). In some studies, there may be multiple identifiable causes of death. For example, suppose that we are interested in survival from early September to the following mid-March (Table 14.3) and we have two categories of deaths: hunting mortalities and natural mortalities. For simplicity, all of the animals in Table 14.3 entered the study on the same day (September 1, 1999) and all survived the 7-day acclimation period. Hunting season started on September 8, 1999 (day 7) and continued until December 31, 1999 (day 121). In addition to the overall survival curve, the biologist might be interested in the survival curve for natural mortality in the absence of hunting mortality. That is, for the survival curve for natural mortality, all deaths attributable to hunting would be classified as censored (Fig. 14.4). No deaths attributable to hunting occurred after the hunting season ended, but animals did die of natural causes during the hunting season. This approach to examining cause-specific survival assumes that the different causes of death are strictly independent. Violation of this assumption can produce misleading results.

HYPOTHESIS TESTING

Johnson (1979), Hensler and Nichols (1981), Bart and Robson (1982), and Heisey and Fuller (1985) provide tests to compare between survival curves (e.g., males vs. females) when the Mayfield Method is used. Here we restrict ourselves to a brief discussion of some issues related to testing between curves when Kaplan–Meier is used.

Hazard Function

An understanding of the hazard function is critical to properly testing hypotheses about survival curves. The hazard function or conditional mortality rate is the probability that an animal dies in the next small interval given the animal is alive at the beginning of the interval. Virtually every statistical package that calculates the Kaplan–Meier survival estimate also computes the hazard function. But it can be estimated fairly easily (if the relocations are regular or nearly regular) as the number of animals dying per unit time in the interval divided by the number of animals alive at the beginning of the

TABLE 14.3   Fictitious Dataset Created to Demonstrate Cause-Specific Survival and Risk Ratios

Date radio-collared	Final date observed	Number of days observed[a]	Fate[b]	Cause of death[c]	Migratory status[d]
9/1/99	9/13/99	12	1	H	1
9/1/99	9/23/99	22	1	N	1
9/1/99	9/24/99	23	1	H	1
9/1/99	9/26/99	25	1	H	1
9/1/99	9/28/99	27	1	H	1
9/1/99	9/28/99	27	1	H	1
9/1/99	10/13/99	42	1	H	1
9/1/99	10/13/99	42	1	H	1
9/1/99	10/13/99	42	1	H	1
9/1/99	10/29/99	58	1	H	1
9/1/99	11/2/99	62	1	H	1
9/1/99	11/4/99	64	1	H	1
9/1/99	11/5/99	65	1	H	1
9/1/99	11/6/99	66	1	N	1
9/1/99	11/10/99	70	1	H	1
9/1/99	11/13/99	73	1	N	1
9/1/99	11/13/99	73	1	N	0
9/1/99	11/22/99	82	1	H	1
9/1/99	11/23/99	83	1	H	1
9/1/99	11/26/99	86	1	H	1
9/1/99	12/6/99	96	1	H	0
9/1/99	12/6/99	96	1	H	1
9/1/99	12/9/99	99	1	H	1
9/1/99	12/11/99	101	1	N	0
9/1/99	12/13/99	103	1	H	0
9/1/99	12/14/99	104	1	H	1
9/1/99	12/27/99	117	1	H	0
9/1/99	1/2/00	123	1	N	1
9/1/99	1/3/00	124	1	N	0
9/1/99	1/5/00	126	1	N	1
9/1/99	1/21/00	142	1	N	0
9/1/99	1/22/00	143	1	N	1
9/1/99	1/23/00	144	1	N	0
9/1/99	2/7/00	159	1	N	0

(*continues*)

TABLE 14.3 (continued)

Date radio-collared	Final date observed	Number of days observed[a]	Fate[b]	Cause of death[c]	Migratory status[d]
9/1/99	2/9/00	161	1	N	0
9/1/99	2/23/00	175	1	N	0
9/1/99	2/27/00	179	1	N	0
9/1/99	2/27/00	179	1	N	0
9/1/99	3/6/00	187	1	N	0
9/1/99	3/15/00	196	0	A	0
9/1/99	3/15/00	196	0	A	0
9/1/99	3/15/00	196	0	A	0
9/1/99	3/15/00	196	0	A	0
9/1/99	3/15/00	196	0	A	0
9/1/99	3/15/00	196	0	A	0
9/1/99	3/15/00	196	0	A	0
9/1/99	3/15/00	196	0	A	0
9/1/99	3/15/00	196	0	A	0
9/1/99	3/15/00	196	0	A	0
9/1/99	3/15/00	196	0	A	0

[a]Date radio-collared = day 0.
[b]1, mortality; 0, censored.
[c]H, hunting mortality; N, natural mortality; A, alive.
[d]1, migratory; 0, nonmigratory.

interval. If the animals are relocated daily, it is simply the number dying divided by the number at risk. See White and Garrott (1990), Lee (1992), and Allison (1995) for excellent detailed descriptions of the hazard function.

The hazard function is useful for describing the pattern of mortality (Fig. 14.5). If the hazard function increases [$h_1(t)$, Fig. 14.5], then the probability that an animal will die in the next small time interval, given that it is alive at the beginning of the interval, is increased. If the hazard function decreases [$h_2(t)$, Fig. 14.5] then the reverse is true. Of particular interest is the situation where the hazard function is constant [$h_3(t)$, Fig. 14.5] because this indicates that use of the Mayfield Method would be appropriate. Hazard functions tell a biological story and that story needs to make biological sense.

Unfortunately, most hazard functions from wildlife studies do not produce the perfectly smooth curves seen in texts. Unlike the hazard curves from medical studies, hazard functions from wildlife studies (Fig. 14.6b) are generally multimodal (White and Garrott 1990). The modes should correspond to

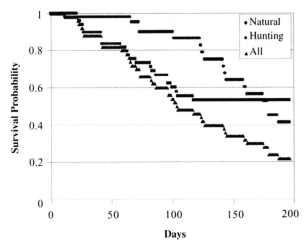

FIGURE 14.4   Results of Kaplan–Meier analyses of data in Table 14.3 demonstrating cause-specific survival. Circles reflect natural deaths, with all other fates censored. Squares reflect hunting deaths, with all other fates censored. Triangles reflect both natural and hunting deaths, with only individuals alive at the end of study censored. Data are fictitious.

events that represent periods of higher risks of mortality, such as hunting seasons. Comparing hazard functions (e.g., between males and females) should point out the different times that each group is at higher risk. Tsai et al. (1999b) suggest smoothing hazard functions to allow for easier comparisons. Tsai (1996) reviews the use of a variety of smoothing functions (e.g., running averages, kernels, spline, and generalized additive models) that can be used to make interpretation easier. Allison (1995) provides a SAS macro that produces smoothed hazard functions using a kernel smoothing method.

In addition to smoothing, hazard functions may be made easier to interpret if the length of the observation period is made coarse (e.g., from daily to weekly). For example, weekly hazard functions for males and females (Fig. 14.6c) are easier to interpret than the daily hazard functions for the same data (Fig. 14.6b). Males experienced two pulses of mortality, in week 1 and weeks 4–5. However, females experienced a constant level of mortality in every week of the study. These differences are also present in the daily hazard functions (Fig. 14.6b) and the survival curves (Fig. 14.6a), but they are less apparent.

An additional use of the hazard function is to test for the constant survival rate required in the Mayfield Method. Bart and Robson (1982) provide guidelines for a graphical method for testing for constant survival. A flat hazard function [$h_3(t)$, Fig. 14.5; females, Fig. 14.6c] indicates that survival is constant and that use of the Mayfield Method would be appropriate.

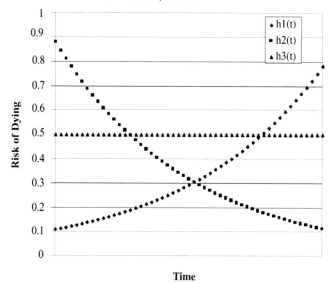

Time

FIGURE 14.5 Hypothetical hazard functions (see Lee 1992). $h_1(t)$ = increasing hazard function, $h_2(t)$ = decreasing hazard function, and $h_3(t)$ = constant hazard function.

Comparing Survival Curves

The log-rank test (see Pollock et al. 1989b) is commonly used to compare survival curves (e.g., males vs. females) when a Kaplan–Meier analysis has been done. It is simple to calculate, available in most software packages, easy to interpret, and can be used when there is staggered entry. It is the locally most powerful test statistic when the two hazard functions have a constant ratio in time (Breslow 1983), meaning that the hazard functions for the two groups compared are proportional to one another. If they are not proportional (or nearly so), the log-rank test loses power and should be avoided in favor of tests such as that developed by Breslow et al. (1984). Also, the log-rank test is sensitive to censoring. The two groups being compared must have similar censoring patterns (Pyke and Thompson 1986). The log-rank test is also more likely to detect differences in survival curves that diverge later in the study (Breslow 1983). If the proportional hazards assumption can be met and the survival curves diverge early in the study, the generalized Wilcoxon test (Breslow 1983, Lee 1992) is a more powerful choice. It is worth noting that since the log-rank test and generalized Wilcoxon test are sensitive to the pattern of censoring, grouping of data into larger intervals (e.g., lumping daily observations into weekly observations) for testing purposes should be avoided (Breslow 1983).

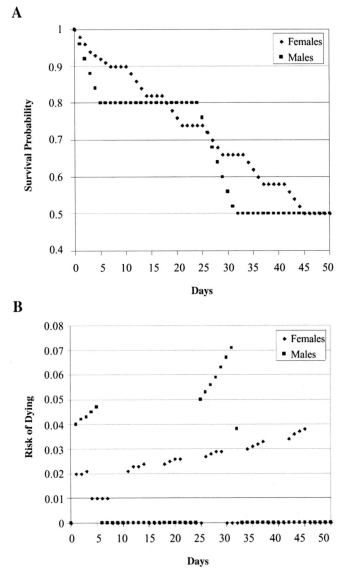

FIGURE 14.6 Survival (a), daily hazard (b), and weekly hazard (c) curves for males and females. Data are fictitious.

C

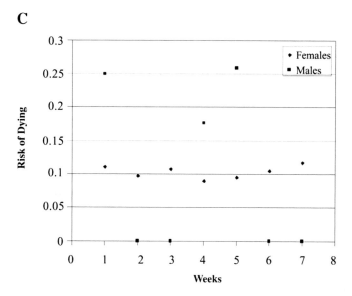

FIGURE 14.6 (*continued*)

One type of nonproportional hazard that is common in wildlife studies occurs when survival curves cross (Fig. 14.6a). The intersecting of survival curves is generally referred to as *acceleration* (Breslow et al. 1984, Riggs and Pollock 1992). In a study of black duck (*Anas rubripes*) survival (Conroy et al. 1989), the log-rank test failed to detect a difference in the survival curves between hatch-year and after-hatch-year birds (Pollock et al. 1989a) (Fig. 14.7). However, using a test developed by Breslow et al. (1984) that accounts for acceleration, they found that hatch-year birds had significantly higher survival than after-hatch-year birds.

Breslow (1983), Pyke and Thompson (1986), and White and Garrott (1990) discuss a variety of tests, including the log-rank test, that can be used to compare survival curves. However, no easy "canned" software exists for many of the tests described. In addition, the labels given to particular tests in a software package may be (unintentionally) misleading. For example, the survival analysis module in SPSS allows researchers to compare survival curves using the Breslow test. This is not the test from Breslow et al. (1984) that accounts for acceleration but rather is the generalized Wilcoxon test. We recommend that researchers who are unsure about which test to use seek advice from a qualified statistician.

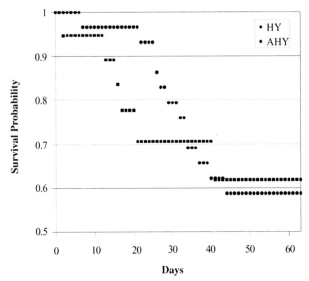

FIGURE 14.7   Survival curves for hatch-year (HY) and after-hatch-year (AHY) black ducks. Data are from Conroy et al. 1989; analysis follows Pollock et al. 1989a.

Testing Covariates

A variety of methods are available that allow survival to be related to covariates. The best known and most commonly used is the proportional hazards model originally developed by Cox (1972, 1983). The proportional hazards model allows the researcher to examine whether, for example, an animal's subsequent survival is influenced by its physical condition, which could be measured at the time of capture. Another example might be in a lead toxicity study on waterfowl in which a bird's subsequent survival would be related to the dosage of lead it received. The proportional hazards model also can be used to examine the influence of categorical variables, such as gender.

As the name implies, the model assumes that the hazards being compared are proportional. The proportional hazards model is valid as long as the hazard functions are nearly proportional and plots of the respective hazard functions for the strata being compared do not cross (Riggs and Pollock 1992). Riggs and Pollock (1992) provide a detailed example of the application of the proportional hazards model. They also introduce the concept of the risk ratio to wildlife studies. This is an underutilized concept in wildlife radiotelemetry studies that provides far more useful information than the simple publication of coefficients and P values usually associated with published results from Cox proportional hazards analyses. The risk ratio (or hazards ratio) can be used to compare the effects of different levels (or states) of a covariate on the risk of death (Riggs and Pollock 1992). It can be used to address such questions as, "How much more likely are males to die than females?" or "How much more likely is an animal to die if it experiences a 25% decrease in body weight?".

The data in Table 14.3, in addition to being classified by cause of death, are categorized by migratory status. The survival curves for the migratory and nonmigratory animals are plotted separately in Fig. 14.8. The two curves are for the most part proportional, and a log-rank test is appropriate to determine if a difference exists. The curves are significantly different ($\chi^2 = 43.06$, d.f. $= 1, P > 0.001$). However, if the curves are examined using the Cox proportional hazards model (Riggs and Pollock 1992) in addition to a significant test for differences, we also obtain a parameter estimate of 2.509 and a risk ratio of 12.299 [$2.509 = \ln(12.299)$]. This tells us that the risk of a migratory animal dying in this example is more than 12 times that of a nonmigratory animal! We contend that of all of the statistics that can be reported when testing covariates, the risk ratio is the most important and should always be reported.

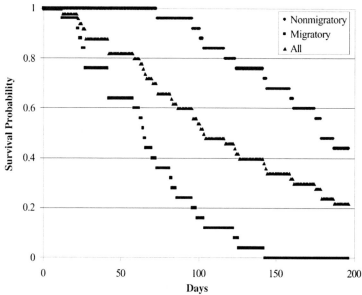

FIGURE 14.8   Results of Kaplan–Meier analyses of data in Table 14.3 demonstrating the impact of migratory status on survival. Circles reflect the survival of nonmigratory animals, squares reflect the survival of migratory animals, and triangles reflect the survival of all animals pooled. Data are fictitious.

SAMPLE SIZE

As with most statistical techniques, required sample sizes can be determined for Kaplan–Meier analysis if the researcher has information on the level of variability expected and the degree of precision required (Kalbfleisch and Prentice 1980, George 1983, Cox and Oakes 1984). Sometimes this information may be available from an earlier, comparable study. In many cases this information is not available and the required sample size becomes "as many radio-collars as you can afford to purchase and monitor."

We know of no reference with tabulated sample sizes for desired levels of precision for a particular survival estimate. However, using the Kaplan–Meier variance formula presented by Pollock et al. (1989b) it is possible to determine that for $\hat{S} = 0.50$ and a 95% confidence interval of $\pm 20\%$, a sample size of 50 is required. Note that this means that there must be 50 individuals at risk at the particular time that $\hat{S} = 0.50$ is obtained. If this is the estimate on the last day of the study, there must be 50 radio-collared animals at risk on that day. For $\hat{S} = 0.75$ and a 95% confidence interval of $\pm 20\%$, the required sample size is 25 individuals.

George (1983) provides formulas for calculating required sample sizes for comparing survival curves. If, for example, $\alpha = 0.10$ and power $= 0.80$, and if the ratio of the hazards for the two groups is 1.5, we would require 75 radio-collared individuals in each group at the onset of our study. However, if the ratio of the hazards was 2.0, $\alpha = 0.10$, and power $= 0.80$, we would need only 25 individuals in each group.

Based on the above values and our experience, we believe that a beginning sample size of less than 25 (per treatment) rarely provides adequate results, and in most cases at least 50 animals per treatment is required. If a sufficiently large sample cannot be obtained, we recommend use of the Mayfield Method over Kaplan–Meier.

WHAT MUST WE DO?

Researchers, aided by consulting statisticians and watched over (and bullied if necessary) by peer reviewers, need to

(1) Indicate that they have examined and met the required assumptions,
(2) Provide a clear statement of the observation period used and why it is appropriate,
(3) Provide a clear statement of the t_0 used and why it was selected,
(4) Pay attention to the information in the hazard function,

(5) Indicate that, when appropriate, they have examined their data for proportional hazards,
(6) Report estimates with confidence intervals, and
(7) Provide information on risk ratios in addition to (or instead of) reports of hypothesis tests.

FUTURE DIRECTIONS

Even though we need to return to the more basic steps in the analysis of survival data, there are also topics that must be addressed to conduct more sophisticated, realistic analyses. As the field of spatial analysis grows, we need to integrate survival and spatial analyses. As our ability to assess actual and potential habitat quality improves, we must model the impact of an animal's movements through, and use of, a mosaic of habitat types on its survival. We need to develop stage-structured survival models. These models will in many cases mimic time-dependent models, allowing us to examine the impact on survival of an animal moving into a new age class or moving to a lower or higher quality territory. We need to develop user-friendly software for smoothing hazard functions and modeling the hazard function with covariates. We need to better understand how to model long-term studies that require pooling results among years.

SUMMARY

Kaplan–Meier and Mayfield Method analysis continue to be viable methods for evaluating survival data from radiotelemetry studies. However, greater availability of software and the lack of a need to assume a constant survival rate have resulted in an increased use of Kaplan–Meier. The assumptions underlying the proper application of Kaplan–Meier and the Mayfield Method need to be recognized, examined, and, if at all possible, validated prior to employing either method. Certain assumptions (use of random samples, experimental units are independent, observation periods are independent, radios do not impact survival) will best be met by the careful consideration of the research design. Other assumptions (working radios are always located, use of random censoring) may be relaxed by employing recently reported modifications (e.g, Bunck et al. 1995, Pollock et al. 1995, Tsai et al. 1999a) of the traditional formulas. Despite the advances in technology, radio-collars undoubtedly impact behavior and survival for some animals (see Chapter 3). Therefore, researchers need to identify, prior to the onset of their study if possible, an acclimation period for which an animal must survive to be included in the analysis. Staggered entry of animals into a radiotelemetry study is probably the norm, requiring the researcher to specify the appropriate point to begin measuring survival.

A wealth of software exists to assist researchers in analyzing their telemetry data. The major commercially available statistical packages tend to be easy to use, but they lack flexibility and may unintentionally mislead unfamiliar "point and click" users into incorrect analyses. More specialized programs, such as MARK, tend to be difficult to use but provide an almost limitless analytical flexibility for the experienced user.

Survival analysis of telemetry data should always include an examination of the hazard function. The hazard function provides data on the biology of the animal being investigated and can be used to test the assumption of constant survival required by the Mayfield Method. Hazard functions from wildlife studies tend to be multimodal, making their interpretation difficult. A variety of smoothing functions exist to generate smoothed hazard functions; however, few easy-to-use software exist to conduct such analyses.

The log-rank test is the most commonly used test for comparing between survival curves. However, ease of use and availability in software packages has led to an overreliance on the log-rank test. It should only be used when the hazard functions of the two groups being compared are proportional to one another. Alternative tests exist (e.g., Breslow et al. 1984) to account for deviations from proportionality, but again little software is available. The risk ratio (change in the probability of death associated with a discrete change in the level of a covariate) can be easily derived from the Cox proportional hazards model, which tests the impact of continuous and categorical covariates on survival. The risk ratio provides more meaningful biological information than does the result of hypothesis tests and should always be reported.

The required sample size for any survival analysis will be a function of a number of factors (e.g., amount of censoring, number of deaths, number of animals that fail to acclimate to the radio-collars, desired precision and power), but most studies will require at least 50 radio-collared animals per treatment.

ACKNOWLEDGMENTS

We thank the editors of this volume, J. Millspaugh and J. Marzluff, and G. White for their detailed review of an earlier draft of this manuscript. Portions of the research cited herein were supported by grants from the Michigan Agricultural Experiment Station and Federal Aid in Wildlife Restoration Project W-127-R (administered by the Michigan Department of Natural Resources, Wildlife Bureau) to SRW. Finally, SRW wishes to thank Jane Thompson, the "queen of formulas," for assistance in preparing this manuscript.

PART VII

Concluding Remarks

CHAPTER 15

Radio-Tracking and Animal Populations: Past Trends and Future Needs

JOSHUA J. MILLSPAUGH AND JOHN M. MARZLUFF

Past Trends
Future Needs
 Addressing Problematic Assumptions
 Coping with Technology
 Additional Analytical Needs and Approaches
 Integrating Space Use and Demographics: The Key to
 Future Studies

The preceding chapters illustrated how radiotelemetry has been and will continue to be a dominant and essential tool in wildlife research. Steady technological advancements including subminiaturization of telemetry components, smaller and more accurate global positioning systems (GPS), and increasingly available geographic information systems (GIS) will continue to influence the way researchers collect and analyze wildlife radio-tracking data. These developments have greatly increased the diversity of animals that can be studied and boosted the number and accuracy of location estimates, within a framework of contemporary conservation issues (e.g., habitat loss and fragmentation; Collins and Barrett 1997). We contend that during the past 30 years, radiotelemetry has been one of the most versatile and important tools in wildlife research. As technology continues to advance, we suspect the reliance on radiotelemetry will increase. In most cases, radiotelemetry is the best available technology to determine the movement patterns, resource selection, and population demographics of elusive and secretive animals. In this chapter, we briefly review past trends in wildlife radio-tracking studies. We refer readers to Chapter 1 for a perspective on how radiotelemetry has influenced

Radio Tracking and Animal Populations
Copyright © 2001 by Academic Press. All rights of reproduction in any form reserved.

wildlife research. Then we will discuss several important future analytical needs and design issues for wildlife radio-tracking studies.

PAST TRENDS

We conducted a literature review using the Wildlife Worldwide (WW) database to assess the specific topics and number of studies using radiotelemetry in wildlife research. Databases within the WW search engine include Wildlife Review Abstracts (formerly Wildlife Review), Swiss Wildlife Information Service, Wildlife Database, BIODOC, Waterfowl and Wetlands database, and three files from The World Conservation Union (Chrisman and Brekke 1996). Our findings indicated that wildlife studies using radiotelemetry are primarily investigations of animal movements (i.e., simple movements and home range), resource selection, and demographics. These three research areas account for 87–94% of the published wildlife radiotelemetry studies during 1975–1996 (Fig. 15.1). Also notable is the five-fold increase in publications from 1975 to 1996 that address radiotelemetry issues or studies that have used radiotelemetry (Fig. 15.1). This rise was primarily due to increased

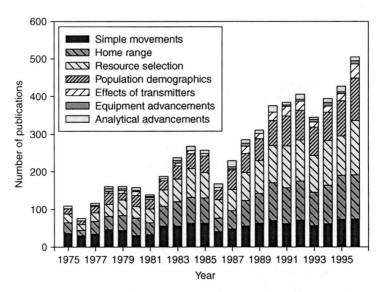

FIGURE 15.1   Type and number of radiotelemetry studies reported in the literature during 1975–1996. Annual counts of publications were obtained from a search of the *Wildlife Worldwide* database that yielded 4734 articles. Review of the search engine was limited through 1996 to ensure all databases had been updated.

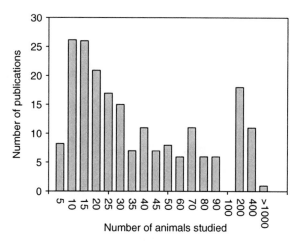

FIGURE 15.2   Number of animals used in studies involving radiotelemetry. Counts are from 205 articles published in *The Journal of Wildlife Management* during 1975–1998.

numbers of studies examining home range, resource selection, and population demographics. Analyses of simple animal movements were reported at moderate but constant rates throughout this period (Fig. 15.1).

To assess the prevalence and characteristics of wildlife radiotelemetry studies at a finer scale, we also reviewed papers published over the past 23 years in *The Journal of Wildlife Management* (*JWM*). Given this review focused on only one journal, it is intended to be an illustrative rather than an exhaustive evaluation. Our survey indicated that wildlife radiotelemetry studies published in *JWM* typically involve the marking and monitoring of 10 to 30 individual animals (Fig. 15.2) for 1–3 years (Fig. 15.3). Only 15% of these studies involved more than 100 animals and only about half (49%) lasted longer than 2 years. Given potential variation in ranging behavior, resource selection, and demographics between seasons, within and among years, and even among individuals, we need to exercise caution when generalizing from such studies. Research that determines how well a small sample of individuals characterizes a larger population is lacking in the radiotelemetry literature (but see Chapter 11). Studies involving reptiles and amphibians represent another important gap in the radiotelemetry literature; only 0.5% of the published studies in *JWM* investigated reptile or amphibian ecology (Fig. 15.4). This is not unexpected given that transmitters small enough for many reptiles and amphibians have only recently been available.

Experimental design of most telemetry studies is rudimentary, as studies are typically correlative and descriptive. We rarely uncover causal relationships

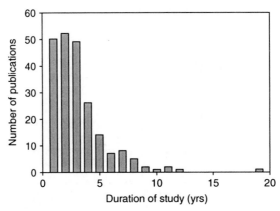

FIGURE 15.3 Duration of studies involving the use of radiotelemetry and published in *The Journal of Wildlife Management*. Counts are from 198 articles published during 1975–1998.

with radiotelemetry studies because only 11% of studies involved manipulative treatments (Fig. 15.5). A description of previously unknown activities is valuable, but creative manipulations allow researchers to test specific hypotheses, which allows for greater inference. Therefore in many research areas (e.g., resource selection), we advocate the addition of creative manipulations to standard study designs.

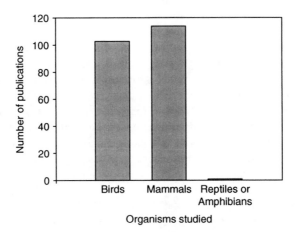

FIGURE 15.4 Classes of vertebrates studied during 1975–1998 using radiotelemetry. Counts are from a survey of 218 articles published in *The Journal of Wildlife Management*.

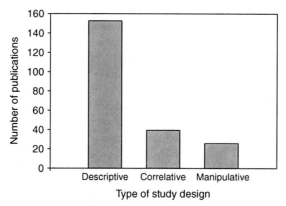

FIGURE 15.5 Types of research designs used in studies involving radio-tagged animals. Counts are from 218 articles published in *The Journal of Wildlife Management* during 1975–1998.

FUTURE NEEDS

The patterns observed in our literature search and the important considerations discussed by contributors to this book provide insight regarding the scope and requirements of future wildlife radio-tracking studies.

ADDRESSING PROBLEMATIC ASSUMPTIONS

Inherent in any radio-tracking study is a set of underlying assumptions. Some assumptions are dismissed as being irrelevant (e.g., effects of transmitters; Chapter 3), whereas others drastically influence study design (e.g., independence between locations; Chapter 5). Every researcher using radiotelemetry must determine which assumptions influence the validity of their study most and strive to meet them. Many times the importance of assumptions will be related to the objectives and question of interest. Below is a discussion of three separate assumptions that have hindered and affected the success of radio-tracking studies.

The most critical assumption in any radiotelemetry study is that the transmitter has no appreciable effect on the study animal (see Chapter 3). That is, we assume that a marked animal behaves, functions, and survives in a manner similar to that of an unmarked animal (White and Garrott 1990). Despite this critical assumption, the impact of radio-marking animals has often been subjectively evaluated (see Chapter 3). Generally, researchers assume that large animals are not affected by radio-marking whereas smaller animals,

particularly birds, are impacted more. Early researchers suggested that transmitters on birds weighing 5% or less of the total body mass would have little
effect on the bird's ecology (Cochran 1980). In fact, U.S. banding regulations
stipulate that transmitters weigh less than 3% of a bird's weight. Despite this
recommendation, there is no strong rationale for the 5% (Caccamise and
Hedin 1985) or 3% rule. Furthermore, these weight rules do not take into
consideration how the transmitter is attached to the animal. Transmitters
attached with a harness weighing 5% of a bird's body weight may effect a
bird differently than an implanted transmitter of the same weight. A harness
alone may substantially influence flight energetics (Gessaman and Nagy 1988,
Gessaman et al. 1991). Also, smaller birds can proportionately carry more
mass (Caccamise and Hedin 1985).

We strongly recommend that researchers determine the *magnitude* of the
effects of transmitter attachment techniques on their study animal. Preferably,
this would involve control and treatment animals (e.g., Schulz et al. 1998).
Prior to widespread use of any transmitter attachment technique, the efficacy
of the technique should be evaluated in a wide range of situations and
conditions. The ramifications of violating this assumption are immense. Imagine using survival estimates to model a population from a study that used a
transmitter attachment technique that decreased survival. Management actions
based on population projections may be inappropriate due to the inaccuracy
of the survival data used in the model. We further recommend that researchers
consider the physiological and pathological responses of animals to transmitter attachment techniques. Physiological responses to a transmitter may provide an innovative way to monitor an animal's response. New, noninvasive
techniques to assess stress (Wasser et al. 2000) coupled with behavioral
observation might be very valuable in this respect. Remember, just because a
tagged animal survives and produces young does not mean the transmitter had
no effect!

A second assumption that has distracted investigators using radiotelemetry
over the past 15 years is that of location "independence," both temporally
(Swihart and Slade 1985a,b, 1997) and spatially (Millspaugh et al. 1998b). *We
believe that too much emphasis has been placed on obtaining temporally and
spatially independent location estimates.* Obtaining a representative sample of
tagged animals is more important than ensuring that successive observations
are statistically independent (McNay et al. 1994, Otis and White 1999; see also
Chapter 5). Successive observations on the same animal should always be
considered dependent, and much can be learned about animal space use by
exploring the pattern of dependency rather than lamenting over the independence assumption. Individual animals are the experimental unit in telemetry
studies and as we focus on them our concern with independence among
locations vanishes. We agree with Kenward (Chapter 1), Kernohan et al.

(Chapter 5), and Otis and White (1999) that our attention should focus on the collection of more data from each individual study animal. This does not mean that researchers should ignore experimental design altogether. It does mean that researchers must carefully consider how best to collect data to address their study objectives. Assumptions will be violated; strive to meet those assumptions that most affect the validity of the study.

A third assumption that has adverse effects on the utility of radio-tracking studies involves the use of "resource availability" in resource selection studies. Nearly all of the analytical techniques used to assess resource selection rely on the assumption that we can accurately define resource use *and* resource availability. This assumption is problematic. Even with today's technology, we are limited in our ability to determine resource availability (see Chapter 12). The use of discrete choice modeling (see Chapter 9) narrows the scale at which we test for selection by specifying a spatial and temporal boundary around what is considered "available." Despite such advancements, we still do not understand the temporal and spatial scales at which most animals base choices (Wiens 1989a).

Given the problems and constraints of defining resource availability, why should researchers continue to incorporate resource availability in their assessment of resource selection? *We argue that the need to define resource availability ultimately weakens the utility of management recommendations based on resource selection assessments.* An alternative approach is to focus on analytical techniques that emphasize those resources *used* by animals (e.g., North and Reynolds 1996; see also Chapter 12). However, even studies focusing only on resource use must relate these choices to some fitness measure. Knowing what resource animals actually use *and* how use affects fitness will allow for the development of effective management prescriptions.

COPING WITH TECHNOLOGY

Technological advancements have had a tremendous influence on wildlife radio-tracking studies (see Chapter 4). Not only has technology increased the number of species that can be studied, but it also has greatly enhanced our ability to remotely monitor what animals are doing. Location estimates are becoming increasingly more accurate (e.g., GPS), reducing the problem of misclassification of movements and resource use by animals. However, increasing ease of monitoring also brings with it increasing challenges of data analysis.

One such challenge is the management and analysis of huge volumes of data. In most ecological studies, researchers have scant amounts of data, which are easily managed. Ecologists are continuously concerned with sample size constraints. In contrast, GPS technologies may produce more data than researchers are used to handling, which presents interesting data management issues and concerns. More importantly, traditional analyses may not be appropriate to handle so many data. Very high statistical power resulting from increased sample sizes potentially leads to rejection of statistical null hypotheses based on nonmeaningful biological differences. Thus, despite technological advancements, the necessary analytical techniques to cope with these data have not kept pace. Consequently, we are ill-equipped to take full advantage of the data generated.

Consider a case where a researcher is interested in quantifying resource selection for a group of animals equipped with GPS collars. Further assume that each collar is capable of collecting 10,000 individual locations per year. An analytical technique such as logistic regression may not be appropriate to assess resource selection because high power would likely result in statistically significant but biologically insignificant results. The solution to this dilemma will be in knowing when a statistically significant result is biologically important, which is dependent on the biology of the population under study and the objectives of the study. New statistical techniques to evaluate the statistical significance of results will also be needed. For example, information-theoretical approaches (Burnham and Anderson 1998) may be helpful if researchers explicitly state *a priori* models to be tested and then rank those models. This example further illustrates that the most appropriate sampling unit is the radio-marked animal and not the individual observation point (Otis and White 1999; see also Chapter 8).

Documentation of resource use has been increasingly possible with the advent of remote sensing technologies and advances in radiotelemetry equipment (e.g., GPS; see Chapters 8 and 10). However, Marzluff et al. (Chapter 12) and Cooper and Millspaugh (Chapter 9) claim that increasing reliance on these technologies can distract us from the essential task of actually observing the behavior of our radio-marked animals. Without knowing why an animal is found in a given habitat or how often it performs behavior x in each habitat, we cannot make specific statements about why particular habitats are used selectively (see Chapters 12 and Chapter 9). The importance of knowing why animals are using specific resources is essential for successful conservation actions.

We have also become increasingly dependent on technology in defining resource availability (see Chapter 8), possibly violating key assumptions in the process. For example, one would expect that resource availability would change in both quality and quantity both seasonally and annually. Despite these expectations, many researchers using GIS technology designate only a single point in time as their definition of availability, whereas their data on resource use spans many seasons and years. A single temporal definition of availability should only be used if one can assume that the landscape is constant over the course of the study. Such validation can only be done by spending time in the field and collecting data necessary to determine if resource availability varies among seasons or years.

There is an even more insidious abuse of technology when quantifying resource availability. Some researchers use buffers centered on a radiotelemetry location to define resource availability. Such buffers are easily computed in any GIS. However, intuitively, what is available to the animal is likely not some circle centered on the habitat patch actually chosen but rather a shape that encompasses where the animal was located last and all of the patches that it could have theoretically chosen along the way (Cooper and Millspaugh 1999; see also Chapter 9). Such inappropriate definitions of availability could drastically bias the results of resource selection analyses. Discrete choice analysis (McFadden 1981, Ben-Akiva and Lerman 1985) offers one solution to this dilemma. With discrete choice models, the definition of availability may differ for each use event, thus allowing differences to be both spatial and temporal (see Chapter 9). By nesting multiple discrete choice models, it is possible to analyze resource selection for many behaviors in a single model (see Chapter 9). Discrete choice analyses are well established and frequently used in the econometrics literature, and further applications of discrete choice analysis to the wildlife field should be considered. Discrete choice analyses have been so useful to the field of economics that Daniel McFadden of the University of California at Berkeley won a Nobel Prize in economics in 2000 for his pioneering work on these models.

Equipment that allows more precise and continuous monitoring involves many study design trade-offs. An important trade-off exists between the number of locations per animal and the number of animals that can be marked and successfully monitored (Otis and White 1999 Chapter 2). This trade-off should be explicitly considered in any radiotelemetry study. Few people have an appreciation for the effect of sample size on telemetry analyses (Seaman et al. 1999; see also Chapter 11). Differences reported among populations could be due to differences in sample size alone. We recommend that sample sizes, analysis options (e.g., bandwidth in home range analyses), and the exact analytical technique used be reported in any radio-tracking study (Seaman et al. 1999). This information is essential to determine the validity of subsequent comparisons. Similarly, we suggest that researchers explicitly identify the computer software used for their analyses along with program options; this too ensures valid comparisons can be made (Lawson and Rodgers 1997; see Appendix A).

Additional Analytical Needs and Approaches

In the "Animal Movements" section of this book (Part IV), contributors reviewed historical techniques (e.g., home range; see Chapter 5) and described animal movement modeling using advanced descriptive (e.g., fractal dimensions; see Chapter 6) and visualization (see Chapter 7) approaches, general movement models (e.g., random walk models), and biological models (Moorcroft et al. 1999; see also Chapter 5). We contend that many traditional analyses (e.g., home range) may become obsolete as we now have the ability to examine more detailed aspects of animal ecology (e.g., modeling of the movement process). We recommend that the use of radio-tracking in animal studies move beyond simple, general descriptions of animal movements (e.g., home range) to the use of model-driven approaches to examine the key factors determining why and how an animal uses space (see Chapter 5). We challenge researchers to integrate the outlined approaches into a more rigid study design to compare specific biological hypotheses and models regarding animal space use and movements, while considering important factors such as sampling error (see Chapter 7).

One approach to help unify and summarize animal movements and resource selection analyses is an old concept: the utilization distribution (UD; see Chapters 5 and 12). The UD is a powerful analytical tool that describes an individual's or group of individuals' relative use of space. It portrays the amount of use throughout a species' range as a function of the density of points at each location throughout the range (White and Garrott 1990:146; see also Chapter 5). We believe the UD would enhance future studies of animal movements patterns, interspecific and intraspecific interactions, and resource selection because it focuses on the probability of use throughout the entire range of an individual animal or an entire population. Thus, incorporating the UD explicitly considers use patterns in a probabilistic framework. Important assumptions, including appropriate sample sizes (see Chapter 5), must be considered to effectively compute a UD. Time-series analyses should also be incorporated into the analysis of space use and resource selection studies.

Throughout this book, critical aspects of experimental design and data analysis were discussed. The decision of how to collect and analyze telemetry data is not easy. We encourage users of radiotelemetry to carefully consider their objectives and species under investigation. In many ways, the most appropriate design or analytical procedure may be found by using a decision-tree process, excluding those techniques least appropriate and utilizing those that apply most in your particular situation.

Radiotelemetry will undoubtedly continue to be a useful and important tool to document population demographics of wild animals. White and Shenk (see Chapter 13) described how radio-marked animals help reduce bias in population size estimates, and we suspect that continued uses will develop as technology advances. Survival analysis will continue to play a central role in radiotelemetry studies (see Chapter 14). We agree with Winterstein et al.

(Chapter 14) that the development of stage-structured survival models would be a significant improvement by allowing researchers to examine the effect of animal choices on survival. Assuredly, advancements in radio-tracking technology will improve our understanding of population demographics for a diversity of species.

INTEGRATING SPACE USE AND DEMOGRAPHICS: THE KEY TO FUTURE STUDIES

Authors in this book raised many important issues for future wildlife radiotelemetry studies. Of paramount importance is the need to relate animal movement, resource selection, and behavior information to population demographics (Boyce and McDonald 1999; see also Chapters 2 and 14). Currently, most radiotelemetry studies consider either animal movement and resource choices or survival; very few studies do both. Such a link would help determine the underlying mechanisms affecting animal fitness within different landscapes. We contend that population demographics is the key response variable that should be monitored to determine the success of an individual based on the choices it makes. Clearly, radiotelemetry could play a role in fulfilling this important step needed in wildlife conservation (see Chapter 2), although meeting this challenge will not be easy (see Chapter 2). To do so will require the integration of several tools (e.g., DNA analyses, manipulative studies, and an understanding of species life history). Additionally, important assumptions will need to be met, studies will need to be of longer duration, more animals will need to be monitored, and we will need to employ manipulative study designs. Understanding demographic processes would allow us great insight into how differences in resource abundance affect a species' growth rates. This insight would allow us to understand how individual decisions and behaviors shape population responses. Armed with that understanding, we will be able to better conserve and manage wildlife resources.

ACKNOWLEDGMENTS

We thank Gary Brundige, Andy Cooper, Mark Fuller, Bob Gitzen, Brian Kernohan, and Steve Knick for their thought-provoking questions. Bob Gitzen, Brian Kernohan, and Brian Washburn provided helpful suggestions about this manuscript. We are grateful to Mike Hubbard, Chad Rittenhouse, John Schulz and Steve Sheriff, for recent discussions on radio-tracking assumptions and analyses.

APPENDICES

A Catalog of Software to Analyze Radiotelemetry Data

MICHAEL A. LARSON

The purpose of this appendix is to help readers select and obtain the most appropriate software for collecting and analyzing radiotelemetry data. Although I conducted a thorough search for radiotelemetry software, the appendix is not exhaustive in its listing of programs. Certainly, there are useful programs I could not find, and I selectively excluded programs whose developers could not be contacted or no longer support them.

The appendix generally follows the organization of the major sections in the book. There are brief reviews of software related to preliminary analyses (e.g., data proofing), animal movements, resource selection, demographics, and general statistics. Each review, except the one for general statistics, is followed by a table summarizing and comparing characteristics of specific software. Some programs have entries in several tables because they perform diverse analyses. However, to avoid repeating information about their availability, all programs are listed alphabetically in the next to last section (Availability of Software) along with addresses for downloading the soft-

Radio Tracking and Animal Populations
Copyright © 2001 by Academic Press. All rights of reproduction in any form reserved.

ware from the Internet or for contacting the developer. Last, there is a list of vendors and distributors of radiotelemetry equipment (e.g., radiotransmitters, antennas, receivers). Their products and services are not reviewed in this appendix.

New software is developed continuously, so I recommend that biologists consult the *Wildlife Society Bulletin* and other journals describing software advancements (e.g., *Bulletin of the Ecological Society of America*, *The Journal of Wildlife Management*) for reports of new software and reviews of existing software related to the analysis of radiotelemetry data. Furthermore, several organizations maintain Internet sites that help disseminate software for radiotelemetry analyses. For example, The Wildlife Society has supported the exchange of software-related information for more than a decade (Samuel 1988) and continues to facilitate electronic distribution of published software through the Bird Monitor Bulletin Board System and the Internet (http://www.im.nbs.gov/tws/cse.html; Smith 1996). The GIS, Remote Sensing, and Telemetry Working Group of The Wildlife Society also has a useful web site (http://fwie.fw.vt.edu/tws-gis/index.htm).

PRELIMINARY ANALYSES

Prior to analyzing radiotelemetry data, researchers must collect, format, and edit their data. Using a triangulation program to enter bearings and save animal location data electronically on a laptop computer in the field can save valuable time. Software for formatting and editing radiotelemetry data can also save time, but the more important purpose of all preliminary analyses (e.g., triangulation, outlier removal, autocorrelation) is to help ensure the validity of subsequent analyses. Despite the importance of preliminary analyses, few have compared software designed for these purposes (Table 1).

In one comparison, Biggins et al. (1997; see abstract at http://www.npwrc.usgs.gov/resource/tools/telemtry/azimuth.htm) examined the accuracy of programs LOCATE and TRITEL when using azimuths from three stations. They reported that the estimated telemetry locations from LOCATE, which used all three azimuths to compute a confidence ellipse for a location, were closer to the true transmitter locations determined by a global positioning system (GPS). However, TRITEL, which selected only the two azimuths that resulted in the smallest of three possible error quadrangles to estimate a telemetry location, was more likely to yield an error polygon containing the GPS location. Therefore, the estimated location of an animal can differ among programs, and the difference may affect subsequent analyses.

ANIMAL MOVEMENTS

SIMPLE MOVEMENTS

Several programs are available to compute angles, distances, and rates of animal movement between successive telemetry locations, and a few provide animated graphics of the movement (Table 2). Some programs also compute home range overlap or site fidelity indices. At their most complex, these programs analyze the dynamic interactions among marked individuals (i.e., the relationship between the location and movements of two animals). The sophistication of "simple movements" analyses varies widely, as does the ability of programs to conduct them.

HOME RANGE

In contrast to other radiotelemetry software, the comparison of home range programs has received some attention. Larkin and Halkin (1994) were the first to rigorously compare multiple home range programs. They reviewed many characteristics of CALHOME, DC80, DIXON, HOMER, HOME RANGE, KERNELHR, McPAAL, RANGES IV, and several programs that are no longer commonly used. The capacity to subset data (e.g., by time, animal age, or animal sex) within the program makes it easier to analyze differences between groups of animals, but only CALHOME, RANGES IV, and HOMER provided this capability (Larkin and Halkin 1994). Also, RANGES IV was the only program that could combine, plot, and compare home ranges from more than one animal (Larkin and Halkin 1994).

More recently, Lawson and Rodgers (1997) compared minimum convex polygon (MCP), harmonic mean, and adaptive kernel estimates from CALHOME, HOME RANGE, RANGES IV, RANGES V, and TRACKER. They found large differences in home range estimates among programs when using the same dataset. Although Lawson and Rodgers (1997:728) stated that "RANGES V may be the most suitable program for analyses of large data sets obtained from automated tracking systems," their most striking conclusion was that valid comparisons of home range size can only be made when the same program, user-defined options, and sample size are used.

As discussed in Chapter 5, the most popular estimators for home range analyses are kernel-based. Larkin and Halkin (1994) identified that choice of smoothing parameter for kernel-based home ranges was an important difference among programs. The significance of selecting the smoothing parameter

TABLE 1 Comparison of Software with Features for the Preliminary Analysis of Radiotelemetry Data

Name	OS[a]	Cost[b]	Data formatting and error checking	Triangulation	Autocorrelation	Comments
WILDTRAK	Mac	C	x	x	x	See functions in Table 2.
ArcView						
Animal Movements	Win	F	x	x	x	See functions in Tables 2 and 3.
Tracking Analyst	Win	C	x			Can use real-time GPS data.
TRACKER	Win	C	x	x	x	See functions in Table 2.
BIOTAS	Win	F/C[c]			x	A GIS that is compatible with LOAS triangulations.
GTM	Win	F		x		See functions in Tables 2 and 4.
LOAS	Win	F/C[d]	x	x		Formats locations for BIOTAS.
SAS	Win	C		x		General statistics application.
ELSA	Win	C	x			Links to Service Argos system.
Excel	Win	C	x			Spreadsheet for plotting locations.
LOTE	Win	C	x			Converts among geographical projections.
Lotus 1-2-3	Win	C	x			Spreadsheet for plotting locations.
QuattroPro	Win	C	x			Spreadsheet for plotting locations.
TELVIS	Win	F	x			Generates SAS-ready data; integrated with UTOOLS.
RANGES V	Dos	C	x		x	See functions in Tables 2 and 3.
SPADS[e]	Dos	F	x	x		See functions in Table 2.
MAP	Dos	F	x			Converts XYLOG or TRIANG to AutoCAD[f]
TELSTAR	Dos	F	x			
HOME RANGE	Dos	F			x	See functions in Table 2.

400

Name	OS[a]	Cost[b]	Data formatting and error checking	Triangulation	Autocorrelation	Comments
TRIANG	Fortran	F		x		Allows station-specific error estimates.
TRITEL	Basic	F		x		For checking errors in locations.
BIOCHECK	Fortran	F	x			
BIOPLOT	BasicA	F	x			Interactive plot for checking errors.
FIELDS	Fortran	F	x			Prepares data for BIOCHECK, BIOPLOT, and HOMER.

[a]OS = Operating system; Mac = Macintosh; and Win = Microsoft Windows; Basic, Basic A, and Fortran indicate that the source code is available to be compiled for the desired OS.
[b]C, Commercial; F, free.
[c]BIOTAS is free as a temporary prebeta version. A password must be purchased once version 1 is complete.
[d]LOAS is free as a temporary beta version. A password must be purchased once version 1 is complete.
[e]SPADS may not be accurate "unless the data were taken on or very close to the equator" (Larkin and Halkin 1994:275).
[f]AutoCad is a product of Autodesk, Inc., 2320 Marinship Way, Sausalito, CA 94965.

TABLE 2   Comparison of Software with Features for Analyzing Animal Movements Using Radiotelemetry Data

Name	OS[a]	Cost[b]	Simple movements — Travel path analysis[c]	Animal interactions[d]	Home range estimators — Kernel[e]	Harmonic mean	Bivariate normal	Fourier	MCP[f]	Cluster	Grid cell	Digitized polygon[g]	Comments
ANTELOPE	Mac	F					x	x	x				
HomeRange	Mac	F					x		x				Contains modules adapted from ANTELOPE.
THE KERNEL	Mac	F			x								
WILDTRAK	Mac	C	x	x					x		x		See functions in Table 1.
ArcView													
Animal Movements	Win	F	x	x	F	x	x		x				See functions in Tables 1 and 3.
Spatial Analyst	Win	C		x									Not designed specifically for wildlife.
Tracking Analyst	Win	C	x										Can use real-time GPS data.
TRACKER	Win	C[h]		x	A	x	x		x				See functions in Table 1.
BIOTAS	Win	F/C			F/A	x	x		x				GIS program compatible with LOAS locations.
Home Ranger	Win	F			F/A								Includes a smoothed bootstrap option.
GTM	Win	F				x	x		x				See functions in Tables 1 and 4.
HR195	Win					x	x		x				HOME RANGE for use in Windows.
SAS	Win	C	x	x			x		x				General statistics application.
HOMERUN	Win	F					x		x				Calculates the restricted MCP.
IDRISI	Win	C									x	x	
ELSA	Win	C	x								x		Links to Service Argos system.
TELVIS	Win	F	x								x		Generates SAS-ready data; integrated with UTOOLS.
RANGES V	Dos	C	x	x	F/A	x	x		x	x	x		See functions in Tables 1 and 3.
CALHOME	Dos	F			A	x	x		x	x			
GRID	Dos	F			F								Large grid size.
KERNELHR	Dos	F			F/A								Flexibility and control over smoothing.

Name	Simple movements		Travel path analysis[c]	Animal interactions[d]	Home range estimators								Comments
	OS[a]	Cost[b]			Kernel[e]	Harmonic mean	Bivariate normal	Fourier	MCP[f]	Cluster	Grid cell	Digitized polygon[g]	
McPAAL	Dos	F				x	x	x	x				Includes the Koeppl et al. (1975) F statistic.
DC80	Dos	F				x	x	x	x				Includes the Koeppl et al. (1975) F statistic.
DIXON[i]	Dos	F				x							Dixon and Chapman (1980) method.
SPADS[i]	Dos	F					x		x				See functions in Table 1.
BLOSSOM	Dos	F		x									Site fidelity test.
TELSTAR	Dos	F	x										Requires Paradox[j]; see function in Table 3.
UTOOLS	Dos	F	x										See function in Table 1.
HOME RANGE	Dos	F				x	x		x				Requires Mathematica.
Ulysses	all	F			F/A				x		x		
HRI	Fortran	F				x	x		x		x		HOME RANGE for use in ARC/INFO.[k]
HOMER	Fortran	F					x		x			x	
WORLD	Fortran	F	x									x	Plots locations on user-defined maps.

[a] OS, Operating system; Mac, Macintosh; Win, Microsoft Windows; and Fortran and "all" indicate that the source code is available to be compiled for the desired OS.

[b] C, Commercial; F, free.

[c] This category includes animated graphics and analyses of angles, distances, and rates of animal movement between successive telemetry locations.

[d] This category includes analyses of site fidelity and static and dynamic interactions among marked individuals.

[e] Indicates whether programs calculate (F)ixed kernels, (A)daptive kernels, or both (F/A), when known.

[f] MCP, Minimum convex polygon.

[g] This estimator was described by Ostro et al. (1999).

[h] BIOTAS is free as a temporary prebeta version. A password must be purchased when version 1 is complete.

[i] Larkin and Halkin (1994:275) found that SPADS "did not give correct areas unless the data were taken on or very close to the equator."

[j] Paradox is a product of Corel Corporation, Ottawa, Ontario, Canada.

[k] ARC/INFO is a product of ESRI, Redlands, California, USA.

was supported by the simulation work of Worton (1995) and Seaman et al. (1999). In a comparison of kernel home range programs, Seaman et al. (1998) reviewed selected features of CALHOME, GRID, KERNEL, KERNELHR, RANGES V, and TRACKER. CALHOME had the lowest maximum number of locations allowed (500); the others allowed 3000. Although GRID was the only program that did not use adaptive kernels, its maximum grid size of 360 × 360 pixels was much larger than the grid sizes in the other programs. Only CALHOME and TRACKER did not use fixed kernels.

RESOURCE SELECTION

The comparison of resource selection software, unlike home range software, has received virtually no attention. The most likely reason is that the algorithms used for a particular resource selection analysis are limited. For example, there are several available algorithms for kernel home range estimation, but there is only one way to conduct a χ^2 test as described by Neu et al. (1974) and Byers et al. (1984). However, the ability of software to analyze resource selection varies widely. Some programs listed in Table 3 may compute resource availability within a polygon (e.g., home range outline) or count the number of telemetry locations within certain vegetation types. On the other hand, some programs compare resource use to availability within a user-defined area to compute a measure of resource selection. Because of the lack of formal reviews, this appendix was intended more as a compilation of available software than as an intensive investigation of program capabilities.

DEMOGRAPHICS

Radiotelemetry data lend themselves to several types of survival analysis. Several general statistics applications have built-in functions for Mayfield, Kaplan–Meier, and proportional hazards analyses. Many others are capable of these analyses, but they often require advanced knowledge of both the software and the method of analysis. Alternatively, there are programs designed specifically for survival analysis. Some provide a single estimator (e.g., STAGKAM), whereas others allow multiple types of survival analyses (e.g., MARK; Table 4). CONTRAST compares survival and recapture rates, but some estimation programs perform these tests themselves. Finally, a few programs estimate animal abundance from capture or survey samples of marked and unmarked individuals (e.g., NOREMARK; Table 4).

TABLE 3 Comparison of Software with Features for Analyzing Resource Selection Using Radiotelemetry Data

Name	OS[a]	Cost[b]	Resource use	X^2	Johnson (1980)[c]	Logistic regression	Compositional analysis	Discrete choice	Comments
							Resource selection analyses		
Mac Comp	Mac	F					x		Habitat use analysis.
BIOTAS	Win	F/C[d]		x	x		x		GIS compatible with LOAS locations.
SPSS	Win	C		x		x		x	General statistics application.
SAS	Win	C		x		x		x	General statistics application.
SYSTAT	Win	C		x		x			General statistics application.
Arc View									
Animal Movements	Win	F		x			x		See functions in Tables 1 and 2.
SUBSAMPL/HABUSE	Win	F	x						Determine 'use' with error polygon, not a single point.
RSW	Win/Dos	F		x	x		x		Includes several measures of habitat selection.
PREFER	Win/Dos	F			x				Johnson's (1980) analysis.
SST	Dos	C				x			
UTOOLS	Dos	F	x					x	Requires Paradox[e]; see function in Table 1.
RANGES V	Dos	C	x						See functions in Tables 1 and 2.

[a] OS, Operating system; Mac, Macintosh; Win, Microsoft Windows.
[b] C, Commercial; F, free.
[c] Programs in this category use the methods of Johnson (1980).
[d] BIOTAS is free as a temporary prebeta version. A password must be purchased when version 1 is complete.
[e] Paradox is a product of Corel Corporation, Ottawa, Ontario, Canada.

TABLE 4 Comparison of Software with Features for Analyzing Demographics Using Radiotelemetry Data

Name	OS[a]	Cost[b]	Survival analysis methods					Estimating animal abundance					Comments
			Mark-recapture	Mayfield	Kaplan-Meier	Proportional hazards	Comparing survival and recapture rates	Lincoln-Petersen	JHE[c]	IEJHE[d]	Minta-Mangel	Bowden	
MARK	Win	F	x	x	x		x	x					For single-group open populations.
POPAN-4	Win/Sun	F	x					x					For multigroup (cohort) open populations.
POPAN-5	Win/Sun	F	x					x					For multigroup (cohort) open populations.
NOREMARK	Dos	F										x	
TMSURVIV	Win/Fortran	F	x					x	x	x	x		Transient capture-recapture models (Pradel et al. 1997).
SAS	Win	C		x	x	x	x						General statistics application.
SPSS	Win	C			x	x	x						General statistics application.
SYSTAT	Win	C			x	x	x						General statistics application.
PROPHET	Win	C			x	x	x						General statistics application.
GTM	Win	F			x	x							
SURPH	Win/Unix	F											Allows individual covariates.
MSSURVIV	Fortran	F	x										Multistate open capture-recapture models.
MICROMORT	Dos	F		x			x						Compares by cause of death (Heisey and Fuller 1985).
MAYFIELD	Dos	F		x									Uses methods of Bart and Robson (1982).
STAGKAM	Dos	F			x								
SURVIV	Fortran	F	x				x						Uses methods of White (1983).
RDSURVIV	Fortran	F	x										Robust design capture-recapture models (Kendall et al. 1997).
CONTRAST	Fortran	F					x						

[a] OS, Operating system; Win, Microsoft Windows; Sun, Sun Microsystems; Fortran indicates that the source code is available to be compiled for the desired OS.
[b] C, Commercial; F, free.
[c] JHE, Joint hypergeometric maximum likelihood estimator.
[d] IEJHE, JHE extended to incorporate immigration and emigration.

GENERAL STATISTICS

Flexible and powerful statistical analysis programs may be useful for analyzing radiotelemetry data (Table 5). SAS appears in Tables 1–4 because the source codes necessary to make the software perform specific radiotelemetry analyses are widely available. PROPHET, SPSS, SST, and SYSTAT appear in Tables 3 and 4 because some radiotelemetry analyses are sufficiently general in nature that specialized software is not necessary. Other general statistics programs could perform equally well. Interested readers may find BMDP, MATLAB, and NCSS to be useful alternatives to more specialized programs as well. Information about their availability is given in the next section. The following Internet site contains links to many general statistical applications: http://nb.vse.cz/kstp/win/jirkauvo/sysel/software/pakety/stata.htm.

AVAILABILITY OF SOFTWARE

The following organizations host Internet sites containing many links to radiotelemetry software:

Illinois Natural History Survey
http://detritus.inhs.uiuc.edu

The Wildlife Society's GIS, Remote Sensing, and Telemetry Working Group
http://fwie.fw.vt.edu/tws-gis/wwwsrce.htm

The Wildlife Society Bulletin
http://www.im.nbs.gov/tws/cse.html

TABLE 5 Comparison of Selected General Statistical Software with Potential for Analyzing Radiotelemetry Data

Name	OS[a]	Cost[b]
BMDP	Dos/UNIX	C
NCSS	Windows	C
PROPHET	Windows	C
SAS	Windows	C
S-PLUS	Windows	C
SPSS	Windows	C

[a]OS, Operating system.
[b]C, Commercial.

Below is an alphabetical list of software mentioned in this appendix. The list contains information about program availability and appropriate literature citations only. The capabilities of specific programs can be found in Tables 1–4.

Animal Movements (see **ArcView**).

ANTELOPE
Jack Bradbury and Sandra Vehrencamp, University of California, San Diego, CA, USA.
Documentation and downloading:
http://www-biology.ucsd.edu/research/vehrenbury/
programs.html#antelope

ArcView
Environmental Systems Research Institute, Inc., Redlands, CA 92373–8100 USA.
Description and information about ordering:
http://www.esri.com/software/arcview/index.html

Animal Movements extension for ArcView
Philip Hooge, U.S. Geological Survey, Alaska Biological Science Center, Gustavus, AK 99826 USA.
Requires ArcView. The Spatial Analyst extension is not required, but many basic functions are limited without it.
Documentation and downloading:
http://www.absc.usgs.gov/glba/gistools/

Spatial Analyst extension for ArcView
Environmental Systems Research Institute, Inc., Redlands, CA 92373–8100 USA.
Requires ArcView.
Description and information about ordering:
http://www.esri.com/software/arcview/extensions/spatext.html

Tracking Analyst extension for ArcView
Environmental Systems Research Institute, Inc., Redlands, CA 92373–8100 USA.
Requires ArcView.
Description and information about ordering:
http://www.esri.com/software/arcview/extensions/
trackingext.html

BIOCHECK
User's guide is in White and Garrott (1990:295–297).
Downloading: http://www.cnr.colostate.edu/~gwhite/software.html
(see "Radiotelemetry Programs")

BIOPLOT
User's guide is in White and Garrott (1990:299).
Downloading: http://www.cnr.colostate.edu/~gwhite/software.html
(see "Radiotelemetry Programs")

BIOTAS
Ecological Software Solutions.
Description and availability information:
http://www.ecostats.com/software/biotas/biotas.htm

BLOSSOM
Paul W. Mielke, Jr., U.S. Geological Survey, Midcontinent Ecological
Science Center, Fort Collins, CO 80525 USA.
Documentation and downloading:
http://www.mesc.usgs.gov/blossom/blossom.html

BMDP
BMDP Statistical Software, Inc., Los Angeles, CA 90025 USA.
Description and information about ordering:
http://business.software-directory.com/software-2.cdprod1/002/
377.BMDP.Statistical.Software.shtml

CALHOME
Kie et al. (1996).
Documentation and downloading:
http://www.im.nbs.gov/tws/contr.html

CONTRAST
Hines and Sauer (1989, Sauer and Hines 1989).
Documentation and downloading:
http://www.mbr.nbs.gov/software.html

DC80
Available from John Carey, address unknown.
Source: White and Garrott (1990:9,173).

DIXON
Description and downloading:
   http://detritus.inhs.uiuc.edu/wes/home_range.html

ELSA
NACLS, Largo, MD 20774 USA.
Software description: http://www.cls.fr/html/argos/general/elsa_en.html
Vendor information: http://www.nacls.com

Excel
Microsoft Corporation, Redmond, WA 98052–6399 USA.
Description and information about ordering:
   http://www.microsoft.com/office/excel/default.htm

FIELDS
User's guide is in White and Garrott (1990:293).
Downloading: http://www.cnr.colostate.edu/~gwhite/software.html
   (see "Radiotelemetry Programs")

GRID
Available from Beat Naef-Daenzer, Swiss Ornithological Institute, CH-6204 Sempach, Switzerland. E-mail: naefb@orninst.ch

GTM
Available on CD-ROM from Joel Sartwell, Missouri Department of Conservation, Columbia, MO 65201 USA.
   E-mail: sartwj@mail.conservation.state.mo.us
Upgrades and information:
   http://www.conservation.state.mo.us/mnrc/gtm.html

HOMER
User's guide is in White and Garrott (1990:301-305).
Downloading: http://www.cnr.colostate.edu/~gwhite/software.html
   (see "Radiotelemetry Programs")

HOME RANGE
User's manual is by Ackerman et al. (1990).
Citation abstract and software downloading:
   http://www.uidaho.edu/cfwr/fishwild/homerange.html

HomeRange
Robert Huber and Jack Bradbury, addresses unknown.
Description and downloading:
   http://detritus.inhs.uiuc.edu/wes/home_range.html

Home Ranger
Fred Hovey, Ursus Software, Revelstoke, B.C., Canada V0E 3K0.
Description and downloading:
 http://detritus.inhs.uiuc.edu/wes/home_range.html

HOMERUN
Weinstein et al. (1997).
Abstract:
 http://www.npwrc.usgs.gov/resource/tools/telemtry/homerun.htm
Available from Mike Weinstein, Department of Wildlife and Fisheries,
P.O. Box 9690, Mississippi State University, Mississippi State, MS
39759 USA.

HRI and HRI95 (Home Range Interface)
Anderson and Dixon (1997).
Abstract:
 http://www.npwrc.usgs.gov/resource/tools/telemtry/range.htm

IDRISI
Eastman (1995).
Description and information about ordering:
 http://www.clarklabs.org/01home/01home.htm

THE KERNEL
Jack Bradbury and Sandra Vehrencamp, University of California, San
Diego, CA, USA.
Documentation and downloading:
 http://www-biology.ucsd.edu/research/vehrenbury/
 programs.html#kernel

KERNELHR
Seaman et al. (1998).
Description and downloading:
 http://detritus.inhs.uiuc.edu/wes/home_range.html

LOAS
Ecological Software Solutions.
Description and availability information:
 http://www.ecostats.com/software/loas/loas.htm

LOTE
Ecological Software Solutions.

Documentation and availability information:
http://www.ecostats.com/software/lote/lote.htm

Lotus 1-2-3
Lotus Development Corporation, Cambridge, MA 02142 USA.
Documentation and information about ordering:
http://www.lotus.com/home.nsf/tabs/lotus123

Mac Comp
John Carroll, California University of Pennsylvania, California, PA
15419 USA.
Description and downloading: http://detritus.inhs.uiuc.edu/wes/
habitat.html

MAP
Tanner et al. (1992).
Citation text and software downloading:
http://www.im.nbs.gov/tws/contr.html

MARK
Gary White, Colorado State University, Fort Collins, CO 80523 USA.
Documentation and downloading:
http://www.cnr.colostate.edu/~gwhite/mark/mark.htm

MATLAB
MathWorks, Natick, MA 01760–2098 USA.
Description and information about ordering:
http://www.mathworks.com

MAYFIELD
James Hines, U.S. Geological Survey, Patuxent Wildlife Research Center,
Laurel, MD 20708–4017 USA. E-mail: jim_hines@usgs.gov
Documentation and downloading:
http://www.mbr.nbs.gov/software.html

McPAAL
Stüwe and Blohowiak (1985).
Downloading: http://detritus.inhs.uiuc.edu/wes/home_range.html

MICROMORT
Available from Todd Fuller, Department of Forestry and Wildlife,
University of Massachusetts, Amherst, MA 01003 USA.

E-mail: tkfuller@forwild.umass.edu

MSSURVIV
User's manual is by Hines (1994).
Documentation and downloading:
http://www.mbr.nbs.gov/software.html

NCSS
NCSS Statistical Software, Kaysville, UT 84037 USA.
Description and information about ordering:
http://www.ncss.com

NOREMARK
Gary White (1996), Colorado State University, Fort Collins, CO 80523 USA.
Documentation and downloading:
http://www.cnr.colostate.edu/~gwhite/software.html

POPAN-4
Neil Arnason and Len Baniuk (University of Manitoba, Winnipeg, Manitoba R3T 2N2 Canada) and Carl Schwarz (Simon Fraser University).
Documentation and downloading: http://www.cs.umanitoba.ca/~popan/

POPAN-5
A.N. Arnason and C. J. Schwarz, see POPAN-4 above.
Documentation and downloading: http://www.cs.umanitoba.ca/~popan/

PREFER
Douglas Johnson, U.S. Geological Survey, Northern Prairie Wildlife Research Center, Jamestown, ND, USA.
E-mail: douglas_h_johnson@usgs.gov
Description and downloading:
http://www.npwrc.usgs.gov/resource/tools/software/prefer/prefer.htm

PROPHET
MarketMiner Inc., Charlottesville, VA 22911 USA.
Description and information about ordering:
http://www.prophet.abtech.com/default.htm

QuattroPro
Corel Corporation, Ottawa, Ontario K1Z 8R7 Canada.

Available only as a component of WordPerfect Suite.
Information about ordering: http://www.corel.com

RANGES V
Kenward and Hodder (1996).
Description and information about ordering: http://www.nmw.ac.uk/ite/
ranges.html

RDSURVIV
User's manual is by James Hines, U.S. Geological Survey, Patuxent
Wildlife Research Center, Laurel, MD 20708–4017 USA.
E-mail: jim_hines@usgs.gov
Documentation and downloading:
http://www.mbr.nbs.gov/software.html

RSW (Resource Selection for Windows)
Fred Leban. E-mail: fleban@saintmail.net
Downloading: http://members.xoom.com/fred_leban/reselect.html

S-PLUS
MathSoft Inc., Seattle, WA 98109 USA.
Description and information about ordering:
http://www.splus.mathsoft.com

SAS
SAS Institute Inc., Cary, NC 27513 USA.
Description and information about ordering:
http://www.sas.com/software/sas_system/
Programs were given by White and Garrott (1990:55–56, 64–68,
130–131, 197, 236–239, 250, 343–359).
Downloading SAS programs given by White and Garrott (1990):
http://www.cnr.colostate.edu/~gwhite/software.html
(see "Radiotelemetry Programs")

SPADS
Juneja et al. (1991).
Citation text and software downloading:
http://www.im.nbs.gov/tws/contr.html#Contr

Spatial Analyst (see ArcView).

SPSS
SPSS Inc., Chicago, IL 60606 USA.

Description and information about ordering:
http://www.spss.com/solutions/products/#analy

SST (Statistical Software Tools)
J. A. Dubin and R. D. Rivers.

STAGKAM
Available from Steve Sheriff, Missouri Department of Conservation,
Columbia, MO 65201 USA.
E-mail: sheris@mail.conservation.state.mo.us

SYSTAT
SPSS Inc., Chicago, IL 60606 USA.
Description and information about ordering:
http://www.spssscience.com/systat/

SURPH
John R. Skalski, School of Fisheries, Columbia Basin Research, University
of Washington, 1325 Fourth Ave., Suite 1820, Seattle, WA, USA.
Documentation and downloading:
http://www.cqs.washington.edu/surph/index.html

SURVIV
Gary White (1983), Colorado State University, Fort Collins, CO 80523
USA.
User's manual is in White and Garrott (1990:307–341).
Downloading: http://www.cnr.colostate.edu/~gwhite/software.html

TELSTAR
Alan Ager, U.S. Forest Service, Pendleton, OR 97801 USA.
Description and downloading:
http://www.fs.fed.us/r6/uma/ager/telstar.htm

TELVIS
Alan Ager, U.S. Forest Service, Pendleton, OR 97801 USA.
TELVIS is the Windows version of TELSTAR.
Description and downloading:
http://www.fs.fed.us/r6/uma/ager/telvis.htm

TMSURVIV
User's manual is by James Hines, U.S. Geological Survey, Patuxent
Wildlife Research Center, Laurel, MD 20708–4017 USA.
E-mail: jim_hines@usgs.gov

Documentation and downloading:
    http://www.mbr.nbs.gov/software.html

TRACKER

Camponotus AB and Radio Location Systems AB, Stockholm, Sweden.
Description and information about ordering:
    http://ethology.zool.su.se/tracker/

Tracking Analyst (see **ArcView**).

TRIANG

White and Garrott (1984).
Downloading: http://www.cnr.colostate.edu/~gwhite/software.html
    (see "Radiotelemetry Programs")

TRITEL

Available from Dean Biggins, U.S. Geological Survey, Midcontinent
    Ecological Science Center, Fort Collins, CO 80525–3400 USA.
Summary of Biggins et al. (1997):
    http://www.npwrc.usgs.gov/resource/tools/telemtry/azimuth.htm

Ulysses

Paolo Cavallinii (Università di Firenze; E-mail: cavallini@unisi.it), Sara
    Colombini, and Francesco Romani.
Requires Mathematica.
Description and downloading:
    http://www.di.unipi.it/~romani/volpi.html

UTOOLS

Robert McGaughey (U.S. Forest Service, Pacific Northwest Research
    Station, Seattle, WA 98195–2100 USA) and Alan Ager (U.S. Forest
    Service, Pendleton, OR 97801 USA).
Description, including text of McGaughey and Ager (1996), and
    downloading: http://forsys.cfr.washington.edu/utools_uview.html

WILDTRAK

Ian Todd, University of Hertfordshire, Hatfield, U.K.
Reviewed by Gorman (1993).
Description and information about ordering:
    http://www.geocities.com/RainForest/3722/

WORLD

Available from Philip Voxland, Social Sciences Research Facilities,
    University of Minnesota, Minneapolis, MN 55455 USA.

EQUIPMENT VENDORS AND DISTRIBUTORS

The source of much of the following information was the Directory of Biotelemetry Equipment Manufacturers (http://www.biotelem.org/manufact.htm).

Advanced Telemetry Systems, Inc.
470 First Avenue North, PO Box 398, Isanti, MN 55040, USA.
Phone: 612–444–9267; Fax: 612–444–9384
E-mail: atstrack@compuserve.com
http://www.atstrack.com

AF Antronics, Inc.
RR 1, Box 82, White Heath, IL 61884 USA.
Phone: 217–687–2786
(Receiving antennas only)

Andreas Wagener Telemetrieanlagen
Herwarthstr. 22, D–50672 Koeln, Germany.
Phone: int 221–514966; Fax: int 221–514966
E-mail: Andreas.Wagener@t-online.de

Austec Electronics, Ltd.
1006 11025–82 Avenue, Edmonton, Alberta, Canada T6G 0T1.
Phone: 403–432–1878; Fax: 403–489–3697

AVM Instrument Company, Ltd.
2356 Research Drive, Livermore, CA 94550 USA.
Phone: 1–925–449–2286; Fax: 1–925–449–3980
http://www.avminstrument.com
E-mail: bckermeen@avminstrument.com

Ayama Radio Tracking
133 Bajos, Camíi Ral, Mataróó 08301, Barcelona, Spain.
Phone: (93) 7905862; Fax: (93) 7964932
E-mail: ayamast@lix.intercom.es
http://www.ayama.com

B & R Ingenieorgesellschaft mbH
Johann-Schill-Str.22, 77806 March-Buchheim, Germany.
Phone: 7665–3885; Fax: 761–123794

Biotrack Ltd.
52 Furzebrook Road, Wareham, Dorset, BH20 5AX, UK.
Phone: 44 (0)1929 552992; Fax: 44 (0)1929 554948
http://www.biotrack.co.uk; E-mail: info@biotrack.co.uk

CEIS TM
Division LCD, Immeuble Centreda 2–2eme etage,
    4 avenue Didier Daurat-BP 48, 31702 Blagnac Cedex, France.
Phone: 33–5 61 16 32 30; Fax: 33–5 61 17 32 31
(Satellite transmitters only)

Custom Electronics of Urbana, Inc.
2009 Silver Court West, Urbana, IL 61801 USA.
Phone: 1–217–344–3460; Fax: 1–217–344–3460
http://members.aol.com/~customel/; E-mail: CUSTOMEL@aol.com

Custom Telemetry Co.
1050 Industrial Drive, Watkinsville, GA 30677 USA.
Phone: 1–706–769–4024; Fax: 1–706–769–4026

Detlef Burchard, Dipl.-Ing.
Box 14426, Riverside Drive No. 45, Nairobi, Kenya.
Phone: 442371; Fax: 580454 (include D. Burchard and 442371 in fax)

GFT – Gesellschaft füür Telemetriesysteme mbH
Eichenweg 26, D–25358 Horst, Germany.
Phone: 49–041–263–8798; Fax: 49–041–263–8794
E-mail: rls.gftmbh@T-Online.de

Hi-Tech Services
9 Devon Place, Camillus, NY 13031 USA.
Phone: 315–487–2484
(Transmitters only)

Holohil Systems, Ltd.
112 John Cavanagh Road, Carp, Ontario K0A 1L0, Canada.
Phone: 1–13–839–0676; Fax: 1–613–839–0675
E-mail: holohil@logisys.com
http://www.holohil.com
(Transmitters only)

Instituto de Pesquisas Espaciais

Avenue dos Astronautas 1758, CP 515, 12200 San Jose dos Campos-SP, Brazil.
Phone: 55–123 22 99 77; Fax: 55–123 21 87 43
(Satellite transmitters only)

L.L. Electronics

PO Box 420/#2 Pearl Drive, Mahomet, IL 61853 USA.
Phone: 215–586–5327 or 800–553–5328

Lotek Engineering Inc.

115 Pony Drive, Newmarket, Ontario, L3Y 7B5, Canada.
Phone: 905–836–6680; Fax: 905–836–6455
E-mail: telemetry@lotek.com
http://www.lotek.com

Mariner Radar Ltd.

Dridleway Campsheath, Lowestoft, Suffolk, NR 32 5DN, UK.
Phone: 44–1–502–567–195; Fax: 44–1–502–508–762
(Transmitters for satellite system)

Microwave Telemetry Inc.

10280 Old Columbia Road, Suite 260, Columbia, MD 21046 USA.
Phone: 1–410–290–8672; Fax: 1–410–290–8847
E-mail: microwt@aol.com
(Specializes in miniature Argos satellite transmitters)

North Star Science and Technology

1450 South Rolling Road, Room 4.036, Baltimore, MD 21227 USA.
Phone: 410–961–6692; Fax: 603–462–5144
E-mail: blakehenke@msn.com
http://www.northstarst.com
(Satellite transmitters only)

Seimac, Ltd.

271 Brownlow Avenue, Dartmouth, Nova Scotia, B3B 1W6 Canada.
Phone: 902–468–3007; Fax: 902–468–3009; E-mail: phill@seimac.com
http://www.seimac.com
(Satellite transmitters only)

Service Argos, Inc.

Toulouse, France.

Phone: 33–5 61 39 47 00; Fax: 33–5 61 75 10 14
E-mail: useroffice@cls.fr
   Or
1801 McCormick Drive, Suite 10, Landover, MD 20785 USA.
Phone: 301–925–4411; Fax: 301–925–8995
E-mail: useroffice@argosinc.com
   Or
Seattle, WA, USA.
Phone: 425–672–4699; Fax: 425–672–8926
E-mail: seattle@argosinc.com
Information about services and certified transmitter manufacturers:
   http://www.argosinc.com
Description of wildlife applications:
   http://www.cls.fr /html/argos/wildlife/wildlife_en.html
(Provides a satellite-based system for geopositioning.)

Sirtrack Limited
Private Bag 1403, Goddard Lane, Havelock North, New Zealand.
Phone: 64–6–877–7736; Fax: 64–6–877–5422
E-mail: wardd@landcare.cri.nz
http://sirtrack.landcare.cri.nz/sirtrack/sirtrack.html

Synergetics International
c/o Harsh International, 600 Oak Avenue, Eaton, CO 80615 USA.
Phone: 970–353–0800; Fax: 970–353–0884
(Satellite transmitters only)

Telemetry Solutions
1130 Burnett Avenue, Suite J, Concord, CA 94520 USA.
Phone: 925–798–2373; Fax: 925–798–2375
http://www.track-it.com; E-mail: qkermeen@ix.netcom.com

Telemetry Systems, Inc.
PO Box 187, Mequon, WI 53092 USA.
Phone: 414–241–8335

Televilt International AB
Box 53, S-711 22 Lindesberg, Sweden.
Phone: 46–581–17195; Fax: 46–581–17196
http://www.televilt.se; E-mail: per-arne.lemnell@televilt.se

Telonics
932 East Impala Avenue, Mesa, AZ 85204–66990 USA.

Phone: 1–602–892–4444; Fax: 1–602–892–9139
E-mail: darrel@telonics.com
http://www.telonics.com

Titley Electronics Pty Ltd.
PO Box 19, Ballina NSW 2478, Australia.
Phone: (02)66–811017; International Phone & Fax: 61–2–66866617
http://www.titley.com.au; E-mail: titley@nor.com.au

Toyocom USA, Inc.
617 East Golf Road, Suite 112, Arlington Heights, IL 60005 USA.
Phone: 847–593–8786; Fax: 847–593–5678
E-mail: info@toyocom.com
(Satellite transmitters only)

Wildlife Computers
1650 Northeast 85th Street # 226, Redmond, WA 98052 USA.
Phone: 425–881–3048; Fax: 425–881–3405
E-mail: Wildlife_Computers@compuserve.com
http://www.wildlifecomputers.com
(Satellite-linked timed data recorders and software to format Argos data)

Wildlife Materials Inc.
1031 Autumn Ridge Road, Carbondale, IL 62901 USA.
Phone: 800–842–4537; Fax: 618–457–3340
E-mail: wmi@wildlifematerials.com
http://www.wildlifematerials.com
(Specializes in miniaturization of transmitters)

Wildlink
2924 98th Avenue North, Brooklyn Park, MN 55444 USA.
Phone: 612–424–8340 or 800–421–8340
(Large mammal data acquisition and recapture)

Ziboni Ornitecnica, s.r.l.
Costa Volpino (Bergamo), Italy.
Phone: 035–970434; Fax: 035–972488

LITERATURE CITED

Abrams, P. A. 1994. Should prey overestimate the risk of predation? The American Naturalist 144:317–328.

Ackerman, B. B. 1988. Visibility bias of mule deer aerial census procedures in southeast Idaho. Ph.D. Dissertation. University of Idaho, Moscow, Idaho, USA.

Ackerman, B. B., F. A. Leban, M. D. Samuel, and E. O. Garton. 1990. User's manual for Program HOME RANGE. Second edition. Technical Report 15. Forestry, Wildlife, and Range Experiment Station. University of Idaho, Moscow, Idaho, USA.

Adams, L., and S. D. Davis. 1967. The internal anatomy of home range. Journal of Mammalogy 48:529–536.

Aebischer, N. J., and P. A. Robertson. 1994. Testing for resource use and selection by marine birds: a comment. Journal of Field Ornithology 65:210–213.

Aebischer, N. J., P. A. Robertson, and R. E. Kenward. 1993. Compositional analysis of habitat use from animal radio-tracking data. Ecology 74:1313–1325.

Aitchison, J. 1986. The statistical analysis of compositional data. Chapman & Hall, London, England.

Aldridge, H. D. J. N., and R. M. Brigham. 1988. Load carrying and maneuverability in an insectivorous bat: a test of the 5% "rule" of radio-telemetry. Journal of Mammalogy 69:379–382.

Alho, J. M. 1990. Logistic regression in capture-recapture models. Biometrics 46:623–635.

Alldredge, J. R., and J. T. Ratti. 1986. Comparison of some statistical techniques for analysis of resource selection. Journal of Wildlife Management 50:157–165.

Alldredge, J. R., and J. T. Ratti. 1992. Further comparison of some statistical techniques for analysis of resource selection. Journal of Wildlife Management 56:1–9.

Alldredge, J. R, D. L. Thomas, and L. L. McDonald. 1998. Survey and comparison of methods for study of resource selection. Journal of Agricultural, Biological, and Environmental Statistics 3:237–253.

Allison, P. D. 1995. Survival analysis using the SAS system: a practical guide. SAS Institute, Cary, North Carolina, USA.

Alpizar-Jara, R., E. N. Brooks, K. H. Pollock, D. F. Steffen, J. C. Pack, and G. W. Norman. In press. An eastern wild turkey population dynamics model for the Virginias. Journal of Wildlife Management.

Althoff, D. P., G. L. Storm, T. W. Collins, and V. B. Kuechle. 1989. Remote sensing system for monitoring animal activity, temperature, and light. Pages 116–124 in C. J. Amlaner, Jr., editor. Biotelemetry X. Proceedings of the Tenth International Symposium on Biotelemetry. University of Arkansas Press, Fayetteville, Arkansas, USA.

Amemiya, T. 1978. On a two-step estimation of a multivariate logit model. Journal of Econometrics 8:13–21.

Amlaner, C. J., Jr., and D. W. Macdonald, editors. 1980. A handbook on biotelemetry and radio tracking. Pergamon Press, Oxford, England.

Amlaner, C. J., Jr., and D. W. Macdonald. 1980. A practical guide to radio tracking. Pages 143–159 in C. J. Amlaner, Jr. and D. W. Macdonald, editors. A handbook on biotelemetry and radio tracking. Pergamon Press, Oxford, England.

Anderka, F. W. 1980. Modulators for miniature tracking transmitters. Pages 181–184 in C. J. Amlaner, Jr. and D. W. Macdonald, editors. A handbook on biotelemetry and radio tracking. Pergamon Press, Oxford, England.

Andersen, D. E. 1994. Longevity of solar-powered radio transmitters on buteonine hawks in eastern Colorado. Journal of Field Ornithology 65:122–132.

Andersen, D. E., and O. J. Rongstad. 1989. Home-range estimates of red-tailed hawks based on random and systematic relocations. Journal of Wildlife Management 53:802–807.

Anderson, C. R., Jr., and F. G. Lindzey. 1996. Moose sightability model developed from helicopter surveys. Wildlife Society Bulletin 24:247–259.

Anderson, D. J. 1982. The home range: a new nonparametric estimation technique. Ecology 63:103–112.

Anderson, S. R., and K. R. Dixon. 1997. Integrating GIS and home range analysis. Presentation at the Forum on Wildlife Telemetry, Snowmass Village, Colorado, USA.

Anderson-Sprecher, R. 1994. Robust estimates of wildlife location using telemetry data. Biometrics 50:406–416.

Anderson-Sprecher, R., and J. Ledolter. 1991. State-space analysis of wildlife telemetry data. Journal of the American Statistical Association 86:596–602.

Andries, A. M., H. Gulinch, and M. Herremans. 1994. Spatial modeling of the barn owl, *Tyto alba*, using landscape characteristics derived from SPOT data. Ecography 17:278–287.

Appleby, S. 1996. Multifractal characterization of the distribution pattern of the human population. Geographical Analysis 28:147–160.

Armstrong, J. D., M. Lucas, J. French, L. Vera, and I. G. Priede. 1988. A combined radio and acoustic transmitter for fixing direction and range of freshwater fish (RAFIX). Journal of Fish Biology 33:879–884.

Arnason, A. N., C. J. Schwarz, and J. M. Gerrard. 1991. Estimating closed population size and number of marked animals from sighting data. Journal of Wildlife Management 55:716–730.

Arthur, S. M., and C. C. Schwartz. 1999. Effects of sample size on accuracy and precision of brown bear home range models. Ursus 11:139–148.

Arthur, S. M., B. F. J. Manly, L. L. McDonald, and G. W. Garner. 1996. Assessing habitat selection when availability changes. Ecology 77:215–227.

Augustin, N. H., M. A. Mugglestone, and S. T. Buckland. 1996. An autologistic model for the spatial distribution of wildlife. Journal of Applied Ecology 33:339–347.

Baker, R. R. 1978. The evolutionary ecology of animal migration. Hodder & Stoughton, London, United Kingdom.

Bakken, G. S., P. S. Reynolds, K. P. Kenow, C. E. Korschgen, and A. F. Boysen. 1996. Thermoregulatory effects of radiotelemetry transmitters on mallard ducklings. Journal of Wildlife Management 60:669–678.

Ball, I. J., D. S. Gilmer, L. M. Cowardin, and J. H. Riechmann. 1975. Survival of wood duck and mallard broods in north-central Minnesota. Journal of Wildlife Management 39:776–780.

Ball, N. J., and C. J. Amlaner, Jr. 1980. Changing heart rates of herring gulls when approached by humans. Pages 589–594 in C. J. Amlaner, Jr. and D. W. Macdonald, editors. A handbook on biotelemetry and radio tracking. Pergamon Press, Oxford, England.

Ballard, W. B., D. J. Reed, S. G. Fancy, and P. R. Krausman. 1995. Accuracy, precision, and performance of satellite telemetry for monitoring wolf movements. Pages 461–467 in L. N. Carbyn, S. H. Fritts, and D. R. Seip, editors. Ecology and conservation of wolves in a changing world. Canadian Circumpolar Institute, University of Alberta, Edmonton, Canada.

Bart, J., and D. S. Robson. 1982. Estimating survivorship when subjects are visited periodically. Ecology 63:1078–1090.

Bartmann, R. M., G. C. White, L. H. Carpenter, and R. A. Garrott. 1987. Aerial mark-recapture estimates of confined mule deer in pinyon-juniper woodland. Journal of Wildlife Management 51:41–46.

Bascompte, J., and C. Vila. 1997. Fractals and search paths in mammals. Landscape Ecology 12:213–221.

Batschelet, E. 1981. Circular statistics in biology. Academic Press, New York, New York, USA.

Bauman, P. J. 1998. The Wind Cave National Park elk herd: home ranges, seasonal movements, and alternative control methods. M.S. Thesis. South Dakota State University, Brookings, South Dakota, USA.

Beaty, D., and S. Tomkiewicz. 1990. Polarization: the effect on range performance. Telonics Quarterly 3:2–3.

Beck, T. D. I. 1977. Sage grouse flock characteristics and habitat selection in winter. Journal of Wildlife Management 41:18–26.

Bekoff, M., and L. D. Mech. 1984. Simulation analyses of space use: home range estimates, variability, and sample size. Behavior Research Methods, Instruments, and Computers 16:32–37.

Ben-Akiva, M., and S. R. Lerman. 1985. Discrete choice analysis: theory and application to travel demand. MIT Press, Cambridge, Massachusetts, USA.

Bennett, K. D., J. R. Biggs, and P. R. Fresquez. 2001. Determination of locational error associated with global positioning system radio collars in relation to vegetation and topography. West North American Naturalist.

Bergmann, P. J., L. D. Flake, and W. L. Tucker. 1994. Influence of brood rearing on female mallard survival and effects of harness-type transmitters. Journal of Field Ornithology 65:151–159.

Berteaux, D., F. Masseboeuf, J. M. Bonzom, J. M. Bergeron, D. W. Thomas, and H. Lapierre. 1996. Effect of carrying a radiocollar on expenditure of energy by meadow voles. Journal of Mammalogy 77:359–363.

Beyer, D. E., Jr., and J. B. Haufler. 1994. Diurnal versus 24–hour sampling of habitat use. Journal of Wildlife Management 58:178–180.

Biggins, D. E., M. R. Matchett, and J. L. Godbey. 1997. A comparison of strategies for processing azimuth data from multi-station radio-tracking systems. Presentation at the Forum on Wildlife Telemetry, Snowmass Village, Colorado, USA.

Biggs, J. R., K. D. Bennett, and P. R. Fresquez. 2001. Relationship between home range characteristics and the probability of obtaining successful global positioning system (GPS) collar positions for elk in New Mexico. West North American Naturalist.

Biondini, M. E., P. W. Mielke, and E. F. Redente. 1988. Permutation techniques based on Euclidean analysis spaces: a new and powerful statistical method for ecological research. Coenoses 3:155–174.

Block, B. A., H. Dewer, T. Williams, E. D. Prince, C. Farwell, and D. Fudge. 1998. Archival tagging of Atlantic bluefin tuna (*Thunnus thynnus thynnus*). Marine Technology Society Journal 32:37–47.

Blouin, F., J. F. Giroux, J. Ferron, G. Gauthier, and C. J. Doucet. 1999. The use of satellite telemetry to track greater snow geese. Journal of Field Ornithology 70:187–199.

Boag, D. A. 1972. Effect of radio packages on behavior of captive red grouse. Journal of Wildlife Management 36:511–518.

Boag, D. A., A. Watson, and R. Parr. 1973. Radio-marking versus back-tabbing red grouse. Journal of Wildlife Management 37:410–412.

Boal, C. W., and R. W. Mannan. 1998. Nest-site selection by Cooper's hawks in an urban environment. Journal of Wildlife Management 62:864–871.

Bodie, W. L., E. O. Garton, E. R. Taylor, and M. McCoy. 1995. A sightability model for bighorn sheep in canyon habitats. Journal of Wildlife Management 59:832–840.

Boone, R. B., and M. L. Hunter, Jr. 1996. Using diffusion models to simulate the effects of land use on grizzly bear dispersal in the Rocky Mountains. Landscape Ecology 11:51–64.

Boulanger, J. G., and G. C. White. 1990. A comparison of home-range estimators using Monte Carlo simulation. Journal of Wildlife Management 54:310–315.

Bowden, D. C. 1993. A simple technique for estimating population size. Department of Statistics, Colorado State University, Fort Collins, Colorado, USA.

Bowden, D. C., and R. C. Kufeld. 1995. Generalized mark-sight population size estimation applied to Colorado moose. Journal of Wildlife Management 59:840–851.

Bowen, W. D., D. J. Boness, and S. J. Iverson. 1999. Diving behaviour of lactating harbour seals and their pups during maternal foraging trips. Canadian Journal of Zoology 77:978–988.

Bowman, A. W., and A. Azzalini. 1997. Applied smoothing techniques for data analysis: the kernel approach with S-PLUS illustrations. Oxford University Press, New York, New York, USA.

Bowman, J. L., C. O. Kochanny, S. Demarais, and B. D. Leopold. 2000. Evaluation of a GPS collar for white-tailed deer. Wildlife Society Bulletin 28:141–145.

Box, G. E. P., and G. C. Tiao. 1992. Bayesian inference in statistical analysis. John Wiley & Sons, New York, New York, USA.

Boyce, M. S., and L. L. McDonald. 1999. Relating populations to habitats using resource selection functions. Trends in Ecology and Evolution 14:268–272.

Bradbury, J., R. M. Gibson, C. E. McCarthy, and S. L. Vehrencamp. 1989. Dispersion of displaying male sage grouse. II. The role of female dispersion. Behavioral Ecology and Sociobiology 24:15–24.

Brander, R. B., and W. W. Cochran. 1971. Radio-location telemetry. Pages 95–105 in R. J. Giles, editor. Wildlife management techniques manual. Third edition. The Wildlife Society, Washington, D.C., USA.

Brennan, L. A., W. M. Block, and R. J. Gutierrez. 1986. The use of multivariate statistics for developing habitat suitability index models. Pages 177–182 in J. Verner, M. L. Morrison, and C. J. Ralph, editors. Wildlife 2000: modeling habitat relationships of terrestrial vertebrates. University of Wisconsin Press, Madison, Wisconsin, USA.

Breslow, N. E. 1983. Comparison of survival curves. Pages 381–406 in M. E. Buyse, M. J. Staquet, and R. J. Sylvester, editors. Cancer clinical trials. Oxford University Press, New York, New York, USA.

Breslow, N. E., L. Edler, and J. Berger. 1984. A two-sample censored-data test for acceleration. Biometrics 40:1049–1062.

Britten, M. W., P. L. Kennedy, and S. Ambrose. 1999. Performance and accuracy evaluation of small satellite transmitters. Journal of Wildlife Management 63:1349–1358.

Bro, E., J. Clobert, and F. Reitz. 1999. Effects of radiotransmitters on survival and reproductive success of gray partridge. Journal of Wildlife Management 63:1044–1051.

Brodeur, S., R. Decarie, D. M. Bird, and M. Fuller. 1996. Complete migration cycle of golden eagles breeding in northern Quebec. Condor 98:293–299.

Brothers, N., R. Gales, A. Hedd, and G. Robertson. 1998. Foraging movements of the shy albatross Diomedea cauta breeding in Australia; implications for interactions with longline fisheries. Ibis 140:446–457.

Brown, C. G. 1992. Movement and migration patterns of mule deer in southeastern Idaho. Journal of Wildlife Management 56:246–253.

Brownie, C., D. R. Anderson, K. P. Burnham, and D. S. Robson. 1985. Statistical inference from band recovery data – a handbook. Second edition. Resource Publication 156. U.S. Fish and Wildlife Service, Washington, D.C., USA.

Bryant, L. D., J. W. Thomas, and M. M. Rowland. 1993. Techniques to build New Zealand woven-wire fence. U. S. Forest Service General Technical Report PNW-GTR-313.

Buckland, S. T., D. R. Anderson, K. P. Burnham, and J. L. Laake. 1993. Distance sampling: estimating abundance of biological populations. Chapman & Hall, London, England.

Buechner, H. K., F. C. Craighead, Jr., J. J. Craighead, and C. E. Côté. 1971. Satellites for research on free-roaming animals. BioScience 21:1201–1205.

Buehler, D. A., J. D. Fraser, M. R. Fuller, L. S. McAllister, and J. K. D. Seegar. 1995. Captive and field-tested radio transmitter attachments for bald eagles. Journal of Field Ornithology 66:173–180.

Bunck, C. M., and K. H. Pollock. 1993. Estimating survival of radiotagged birds. Pages 51–63 in J. C. Lebreton and P. M. North, editors. Marked individuals in the study of bird populations. Birkhauser-Verlag, Basel, Switzerland.

Bunck, C. M., C. Chen, and K. H. Pollock. 1995. Robustness of survival estimates from radio-telemetry studies with uncertain relocation of individuals. Journal of Wildlife Management 59:790–794.

Burchardt, D. 1989. Direction finding in wildlife research by Doppler effect. Pages 169–177 in C. J. Amlaner, Jr., editor. Biotelemetry X. Proceedings of the Tenth International Symposium on Biotelemetry. University of Arkansas Press, Fayetteville, Arkansas, USA.

Burger, L. W., Jr., M. R. Ryan, D. P. Jones, and A. P. Wywialowski. 1991. Radio transmitters bias estimation of movements and survival. Journal of Wildlife Management 55:693–697.

Burkey, T. V. 1989. Extinction in nature reserves: the effect of fragmentation and the importance of migration between reserve fragments. Oikos 55:75–81.

Burnham, K. P., and D. R. Anderson. 1992. Data-based selection of an appropriate biological model: the key to modern data analysis. Pages 16–30 in D. R. McCullough and R. H. Barrett, editors. Wildlife 2001: populations. Elsevier Science Publishers, New York, New York, USA.

Burnham, K. P., and D. R. Anderson. 1998. Model selection and inference: a practical information-theoretic approach. Springer-Verlag, New York, New York, USA.

Burnham, K. P., and W. S. Overton. 1978. Estimation of the size of a closed population when capture probabilities vary among animals. Biometrika 65:625–633.

Burnham, K. P., and W. S. Overton. 1979. Robust estimation of population size when capture probabilities vary among animals. Ecology 60:927–936.

Burnham, K. P., D. R. Anderson, and J. L. Laake. 1980. Estimation of density from line transect sampling of populations. Wildlife Monograph 72:1–202.

Burt, W. H. 1943. Territoriality and home range concepts as applied to mammals. Journal of Mammalogy 24:346–352.

Butler, P. J. 1980. The use of radio telemetry in the studies of diving and flying of birds. Pages 569–577 in C. J. Amlaner, Jr. and D. W. Macdonald, editors. A handbook on biotelemetry and radio tracking. Pergamon Press, Oxford, England.

Byers, C. R., R. K. Steinhorst, and P. R. Krausman. 1984. Clarification of a technique for analysis of utilization-availability data. Journal of Wildlife Management 48:1050–1053.

Caccamise, D. F., and R. S. Hedin. 1985. An aerodynamic basis for selecting transmitter loads in birds. Wilson Bulletin 97:306–318.

Capen, D. E., editor. 1981. The use of multivariate statistics in studies of wildlife habitat. General Technical Report RM-87. USDA Forest Service Rocky Mountain Forest and Range Experiment Station, Fort Collins, Colorado, USA.

Carrel, W. K., R. R. Ockenfels, J. A. Wennerlund, and J. C. Devos. 1997. Topographic mapping, LORAN-C, and GPS accuracy for aerial telemetry locations. Journal of Wildlife Management 61:1406–1412.

Carroll, J. P. 1990. Winter and spring survival of radio-tagged gray partridge in North Dakota. Journal of Wildlife Management 54:657–662.

Caughley, G. 1974. Bias in aerial survey. Journal of Wildlife Management 38:921–933.

Caughley, G. 1977. Analysis of vertebrate populations. John Wiley & Sons, New York, New York, USA.

Cederlund, G., and P. A. Lemnell. 1980. A simplified technique for mobile radio tracking. Pages 319–322 in C. J. Amlaner, Jr. and D. W. Macdonald, editors. A handbook on biotelemetry and radio tracking. Pergamon Press, Oxford, England.

Chao, A. 1988. Estimating animal abundance with capture frequency data. Journal of Wildlife Management 52:295–300.

Chapin, T. G., D. J. Harrison, and D. M. Phillips. 1997. Seasonal habitat selection by marten in an untrapped forest preserve. Journal of Wildlife Management 61:707–717.

Chapman, D. G. 1951. Some properties of the hypergeometric distribution with applications to zoological sample censuses. University of California Publication in Statistics 1:131–160.

Cheeseman, C. L., and R. B. Mitson, editors. 1982. Telemetric studies of vertebrates. Symposia of the Zoological Society of London, No. 49. Academic Press, London, United Kingdom.

Cherry, S. 1996. A comparison of confidence interval methods for habitat use-availability studies. Journal of Wildlife Management 60:653–658.

Cherry, S. 1998. Statistical tests in publications of The Wildlife Society. Wildlife Society Bulletin 26:947–953.

Chrisman, J. K., and E. Brekke. 1996. Comparing coverage in 2 indexes: Wildlife Review and Zoological Record. Wildlife Society Bulletin 24:149–152.

Chu, D. S., B. A. Hoover, M. R. Fuller, and P. H. Geissler. 1988. Telemetry location error in a forested habitat. Pages 188–194 in C. J. Amlaner Jr., editor. Biotelemetry X. Proceedings of the Tenth International Symposium on Biotelemetry. University of Arkansas Press, Fayetteville, Arkansas, USA.

Clark, J. D., J. E. Dunn, and K. G. Smith. 1993. A multivariate model of female black bear habitat use for a geographic information system. Journal of Wildlife Management 57:519–526.

Clark, W. R., R. A. Schmitz, and T. R. Bogenschutz. 1999. Site selection and nest success of ring-necked pheasants as a function of location in Iowa landscapes. Journal of Wildlife Management 63:976–989.

Claussen, D. L., M. S. Finkler, and M. M. Smith. 1997. Thread trailing of turtles: methods for evaluating spatial movements and pathway structure. Canadian Journal of Zoology 76:387–389.

Clute, R. K., and J. J. Ozoga. 1983. Icing of transmitter collars on white-tailed deer fawns. Wildlife Society Bulletin 11:70–71.

Cochran, W. G. 1977. Sampling techniques. Third edition. John Wiley & Sons, New York, New York, USA.

Cochran, W. W. 1980. Wildlife telemetry. Pages 507–520 in S. D. Schemnitz, editor. Wildlife management techniques manual. Fourth edition. The Wildlife Society, Washington, D.C., USA.

Cochran, W. W., and R. D. Lord. 1963. A radio-tracking system for wild animals. Journal of Wildlife Management 27:9–24.

Cochran, W. W., D. W. Warner, J. R. Tester, and V. B. Keuchle. 1965. Automatic radio-tracking system for monitoring animal movements. BioScience 15:98–100.

Cogan, R. D., and D. R. Diefenbach. 1998. Effect of undercounting and model selection on a sightability-adjustment estimator for elk. Journal of Wildlife Management 62:269–279.

Cohn, J. P. 1999. Tracking wildlife. BioScience 49:12–17.

Cole, L. C. 1949. The measurement of interspecific association. Ecology 30:411–424.

Collins, R. J., and G. W. Barrett. 1997. Effects of habitat fragmentation on meadow vole (Microtus pennsylvanicus) population dynamics in experiment landscape patches. Landscape Ecology 12:63–76.

Connelly, J. W., H. W. Browers, and R. J. Gates. 1988. Seasonal movements of sage grouse in southeastern Idaho. Journal of Wildlife Management 52:116–122.

Conner, L. M., and B. D. Leopold. 1998. A multivariate habitat model for female bobcats: A GIS approach. Proceedings of the Annual Conference of the Southeastern Association of Fisheries and Wildlife Agencies 52:232–243.

Conover, W. J. 1999. Practical non-parametric statistics. Third edition. John Wiley & Sons, New York, New York, USA.

Conroy, M. J. 1993. Testing hypotheses about the relationship of habitat to animal survivorship. Pages 331–342 in J. D. Lebreton and P. M. North, editors. The use of marked individuals in the study of bird population dynamics. Birkhauser-Verlag, Basel, Switzerland.

Conroy, M. J., G. R. Costanzo, and D. B. Stotts. 1989. Winter survival of female American black ducks on the Atlantic coast. Journal of Wildlife Management 53:99–109.

Conroy, M. J., Y. Cohen, F. C. James, Y. G. Matsinos, and B. A. Maurer. 1995. Parameter estimation, reliability, and model improvement for spatially explicit models of animal populations. Ecological Applications 5:17–19.

Cook, R. S., M. White, D. O. Trainer, and W. C. Glazener. 1967. Radio-telemetry for fawn mortality studies. Wildlife Disease Association Bulletin 3:160–165.

Cooper, A. B., and J. J. Millspaugh. 1999. The application of discrete choice models to wildlife resource selection studies. Ecology 80:566–575.

Copeland, J. P. 1996. Biology of the wolverine in Central Idaho. M.S. Thesis. University of Idaho, Moscow, Idaho, USA.

Côté, S. D., M. Festa-Bianchet, and F. Fournier. 1998. Life-history effects of chemical immobilization and radiocollars on mountain goats. Journal of Wildlife Management 62:745–752.

Cottam, D. F. 1988. Accuracy and efficiency associated with radio tracking deer. Pages 195–204 in C. J. Amlaner Jr., editor. Biotelemetry X. Proceedings of the Tenth International Symposium on Biotelemetry. University of Arkansas Press, Fayetteville, Arkansas, USA.

Cotter, R. C., and C. J. Gratto. 1995. Effects of nest and brood visits and radio transmitters on rock ptarmigan. Journal of Wildlife Management 59:93–98.

Cox, D. R. 1972. Regression models and life tables. Journal of the Royal Statistical Society, Series B 34:187–220.

Cox, D. R. 1983. A remark on censoring and surrogate response variables. Journal of the Royal Statistical Society Bulletin 45:391–393.

Cox, D. R., and D. Oakes. 1984. Analysis of survival data. Chapman & Hall, New York, New York, USA.

Craighead, F. C., Jr., and J. J. Craighead. 1970. Grizzly bear prehibernation and denning activities as determined by radio-tracking. Wildlife Monograph 32:1–35.

Craighead, J. J., F. C. Craighead, Jr., J. R. Varney, and C. E. Cote. 1971. Satellite monitoring of black bear. BioScience 21:1206–1212.

Cresswell, E. 1960. Ranging behaviour studies with Romney Marsh and Cheviot Sheep in New Zealand. Animal Behaviour 8:32–38.

Crist, T. O., D. S. Guertin, J. A. Wiens, and B. T. Milne. 1992. Animal movement in heterogeneous landscapes: an experiment with *Elodes* beetles in shortgrass prairie. Functional Ecology 6:536–544.

Croll, D. A., J. K. Jansen, M. E. Goebel, P. L. Boveng, and J. L. Bengston. 1996. Foraging behavior and reproductive success in chinstrap penguins: the effects of transmitter attachment. Journal of Field Ornithology 67:1–9.

Cudak, M. C., G. W. Swenson, Jr., and W. W. Cochran. 1991. Airborne measurements of incidental radio noise from cities. Radio Science 26:773–781.

Cupal, J. J., and R. W. Weeks. 1989. Digital encoding techniques for the telemetering of biological data. Pages 39–50 *in* C. J. Amlaner, Jr., editor. Biotelemetry X. Proceedings of the Tenth International Symposium on Biotelemetry. University of Arkansas Press, Fayetteville, Arkansas, USA.

Cypher, B. L. 1997. Effects of radiocollars on San Joaquin kit foxes. Journal of Wildlife Management 61:1412–1423.

Daganzo, C. 1980. Optimal sampling strategies for statistical models with discrete dependent variables. Transportation Science 14:324–345.

Danielson, B. J. 1992. Habitat selection, interspecific interactions and landscape composition. Evolutionary Ecology 6:399–411.

Danielson, B. J., and R. K. Swihart. 1987. Home-range dynamics and activity patterns of *Microtus ochrogaster* and *Synaptomys cooperi* in syntopy. Journal of Mammalogy 68:160–165.

Darroch, J. N. 1958. The multiple recapture census: I. Estimation of a closed population. Biometrika 45:343–359.

Daubenmire, R. F., and J. B. Daubenmire. 1968. Forest vegetation of eastern Washington and northern Idaho. Agricultural Experiment Station Bulletin 60. Washington State University, Pullman, Washington, USA.

Davis, B. J., D. L. Miller, R. M. Kaminski, M. P. Vrtiska, and D. M. Richardson. 1999. Evaluation of a radio transmitter for wood duck ducklings. Journal of Field Ornithology 70:107–113.

Davis, J. R., A. F. Von Recum, D. D. Smith, and D. C. Guynn, Jr. 1984. Implantable telemetry in beaver. Wildlife Society Bulletin 12:322–324.

Deat, A., C. Mauget, R. Mauget, D. Maurel, and A. Sempere. 1980. The automatic, continuous and fixed radio tracking system of the Chizé Forest: theoretical and practical analysis. Pages 439–451 in C. J. Amlaner, Jr. and D. W. Macdonald, editors. A handbook on biotelemetry and radio tracking. Pergamon Press, Oxford, England.

Delgiudice, G. D., K. E. Kunkel, L. D. Mech, and U. S. Seal. 1990. Minimizing capture-related stress on white-tailed deer with a capture collar. Journal of Wildlife Management 54:299–303.

Delong, R. L., B. S. Stewart, and R. D. Hill. 1992. Documenting migrations of northern elephant seals using day length. Marine Mammal Science 8:156–159.

Derleth, E. L., and G. F. Sepik. 1990. Summer – fall survival of American woodcock. Journal of Wildlife Management 54:97–106.

De Solla, S. R., R. Bonduriansky, and R. J. Brooks. 1999. Eliminating autocorrelation reduces biological relevance of home range estimates. Journal of Animal Ecology 68:221–234.

Dewsbury, D. A., editor. 1985. Studying animal behavior. University of Chicago Press, Chicago, Illinois, USA.

Diehl, R. H., and R. P. Larkin. 1998. Wingbeat frequency of two Catharus thrushes during nocturnal migration. Auk 115:591–601.

Diggle, P. J. 1990. Time series: a biostatistical introduction. Oxford Statistical Science Series, Volume 5. Clarendon Press, Oxford, United Kingdom.

Dixon, K. R., and J. A. Chapman. 1980. Harmonic mean measure of animal activity areas. Ecology 61:1040–1044.

Dixon, R. D. 1995. Ecology of the white-headed woodpecker in the Central Oregon Cascades. M.S. Thesis. University of Idaho, Moscow, Idaho, USA.

Doncaster, C. P. 1990. Non-parametric estimates of interaction from radio-tracking data. Journal of Theoretical Biology 143:431–443.

Doncaster, C. P., and D. W. Macdonald. 1991. Drifting territoriality in the red fox (Vulpes vulpes). Journal of Animal Ecology 60:423–439.

Dueser, R. D., and H. H. Shugart, Jr. 1979. Niche pattern in a forest-floor small-mammal fauna. Ecology 60:108–118.

Dumke, R. T., and C. M. Pils. 1973. Mortality of radio-tagged pheasants on the Waterloo wildlife area. Department of Natural Resources, Wisconsin, Technical Bulletin 72.

Dunn, J. E. 1979. A complete test for dynamic territorial interaction. Pages 159–169 in F. M. Long, editor. Proceedings of the Second International Conference on Wildlife Biotelemetry. International Conference on Wildlife Biotelemetry, University of Wyoming, Laramie, Wyoming, USA.

Dunn, J. E., and I. L. Brisbin, Jr. 1982. Characterizations of the multivariate Ornstien-Uhlenbeck diffusion process in the context of home range analysis. Technical Report 16. Statistics Laboratory, University of Arkansas, Fayetteville, Arkansas, USA.

Dunn, J. E., and P. S. Gipson. 1977. Analysis of radio telemetry data in studies of home range. Biometrics 33:85–101.

Dunn, P. O., and C. E. Braun. 1986. Summer habitat use by adult female and juvenile sage grouse. Journal of Wildlife Management 50:228–235.

Dunning, J. B., D. J. Stewart, B. J. Danielson, B. R. Noon, T. L. Root, R. H. Lamberson, and E. E. Stevens. 1995. Spatially explicit population models: current forms and future uses. Ecological Applications 5:3–11.

Dussault, C., R. Courtois, J.-P. Ouellet, and J. Huot. 1999. Evaluation of GPS telemetry collar performance for habitat studies in the boreal forest. Wildlife Society Bulletin 27:965–972.

Dwyer, T. J. 1972. An adjustable radio-package for ducks. Bird-Banding 43:282–284.

Dzus, E. H., and R. G. Clark. 1996. Effects of harness-style and abdominally implanted transmitters on survival and return rates of mallards. Journal of Field Ornithology 67:549–557.

Eagle, T. C., J. Choromanski-Norris, and V. B. Kuechle. 1984. Implanting radio transmitters in mink and Franklin's ground squirrels. Wildlife Society Bulletin 12:180–184.

Eastman, J. R. 1995. IDRISI. Version 1.0. The IDRISI Project. Clark Labs for Cartographic Technology and Geographic Analysis, Clark University, Worcester, Massachusetts, USA.

Eberhardt, L. E., R. G. Anthony, and W. H. Rickard. 1989. Survival of juvenile Canada geese during the rearing period. Journal of Wildlife Management 53:372–377.

Eberhardt, L. L. 1990. Using radio-telemetry for mark-recapture studies with edge effects. Journal of Applied Ecology 27:259–271.

Edenius, L. 1997. Field test of a GPS location system for moose (*Alces alces*) under Scandinavian boreal conditions. Wildlife Biology 3:39–43.

Edge, W. D., C. L. Marcum, S. L. Olson, and J. F. Lehmkuhl. 1986. Nonmigratory cow elk herd range as management units. Journal of Wildlife Management 50:660–663.

Edge, W. D., C. L. Marcum, and S. L. Olson-Edge. 1987. Summer habitat selection by elk in western Montana: a multivariate approach. Journal of Wildlife Management 51:844–851.

Ehrlich, P. R., and R. W. Holm. 1963. The process of evolution. McGraw-Hill, New York, New York, USA.

Eiler, J. H. 1995. A remote satellite-linked tracking system for studying Pacific salmon with radio telemetry. Transactions of the American Fisheries Society 124:184–193.

Eliassen, E. 1960. A method for measuring the heart rate and stroke/pulse pressures of birds in normal flight. Årbok Universitet Bergen, Matematisk Naturvitenskapelig 12:1–22.

Ellis, R. J. 1964. Tracking raccoons by radio. Journal of Wildlife Management 28:363–368.

Ely, C. R., D. C. Douglas, A. C. Fowler, C. A. Babcock, D. V. Derksen, and J. Y. Takekawa. 1997. Migration behavior of tundra swans from the Yukon-Kuskokwim Delta, Alaska. Wilson Bulletin 109:679–692.

Erickson, W. P., T. L. McDonald, and R. Skinner. 1998. Habitat selection using GIS data: a case study. Journal of Agricultural, Biological and Environmental Statistics 3:296–310.

Erikstad, K. E. 1979. Effects of radio packages on reproductive success of willow grouse. Journal of Wildlife Management 43:170–175.

Esler, D., D. M. Mulcahy, and R. L. Jarvis. 2000. Testing assumptions for unbiased estimation of survival of radiomarked harlequin ducks. Journal of Wildlife Management 64:591–598.

ESRI. 1997. Environmental Systems Research Institute, Inc., Redlands, California, USA.

Etzenhouser, M. J., M. Owens, D. E. Spalinger, and S. B. Murden. 1998. Foraging behavior of browsing ruminants in a heterogeneous landscape. Landscape Ecology 13:55–64.

Fagerstone, K. A., and B. E. Johns. 1987. Transponders as permanent identification markers for domestic ferrets, black-footed ferrets, and other wildlife. Journal of Wildlife Management 51:294–297.

Fancy, S. G., L. F. Pank, D. C. Douglas, C. H. Curby, G. W. Garner, S. C. Amstrup, and W. L. Regelin. 1988. Satellite telemetry: a new tool for wildlife research and management. Resource Publication 172. U. S. Fish and Wildlife Service, Washington, D.C., USA.

Ferguson, S. H., M. K. Taylor, E. W. Born, and F. Messier. 1998. Fractals, sea-ice landscape and spatial patterns of polar bears. Journal of Biogeography 25:1081–1092.

Findholt, S. L., B. K. Johnson, L. D. Bryant, and J. W. Thomas. 1996. Corrections for position bias of a LORAN-C radio-telemetry system using DGPS. Northwest Science 70:273–280.

Findholt, S. L., B. K. Johnson, L. L. Mcdonald, J. W. Kern, A. Ager, R. J. Stussy, and L. D. Bryant. In press. Estimating habitat use from error associated with radiotelemetry positions. USDA Forest Service Research Paper, PNW-RP-xxx.

Fischer, R. A., K. P. Reese, and J. W. Connelly. 1996. Influence of vegetal moisture content and nest fate on timing of sage grouse migration. Condor 98:868–872.

Flint, P. L., K. H. Pollock, D. Thomas, and J. S. Sedinger. 1995. Estimating prefledging survival: allowing for brood mixing and dependence among brood mates. Journal of Wildlife Management 59:448–455.

Focardi, S., P. Marcellini, and P. Montanaro. 1996. Do ungulates exhibit a food density threshold? A field study of optimal foraging and movement patterns. Journal of Animal Ecology 65:606–620.

Foran, D. R., S. C. Minta, and K. S. Heinemeyer. 1997. DNA-based analysis of hair to identify species and individuals for population research and monitoring. Wildlife Society Bulletin 25:840–847.

Forsman, E. D., E. C. Meslow, and H. M. Wight. 1984. Distribution and biology of the spotted owl in Oregon. Wildlife Monograph 87:1–64.

Foster, C. C., E. D. Forsman, E. C. Meslow, G. S. Miller, J. A. Reid, F. F. Wagner, A. B. Carey, and J. B. Lint. 1992. Survival and reproduction of radio-marked adult spotted owls. Journal of Wildlife Management 56:91–95.

Fretwell, S. D., and H. L. Lucas. 1969. On territorial behavior and other factors influencing habitat distribution in birds. I. Theoretical development. Acta Biotheoretica 19:16–36.

Friedman, M. 1937. The use of ranks to avoid the assumption of normality implicit in the analysis of variance. Journal of the American Statistical Association 32:675–701.

Fritts, S. H., and L. D. Mech. 1981. Dynamics, movements and feeding ecology of a newly protected wolf population in northwestern Minnesota. Wildlife Monograph 80:1–79.

Frontier, S. 1987. Applications of fractal theory to ecology. Pages 335–378 in P. Legendre and L. Legendre, editors. Developments in numerical ecology. Springer, New York, New York, USA.

Fry, M. D., B. W. Wilson, N. D. Ottum, J. T. Yamamoto, R. W. Stein, J. N. Seiber, M. M. McChesney, and E. Richardson. 1999. Radiotelemetry and GIS computer modeling as tools for analysis of exposure to organophosphate pesticides in red-tailed hawks. Pages 67–84 in L. Brewer and K. Fagerstone, editors. Radiotelemetry applications for wildlife toxicology field studies. SETAC, Pensacola, Florida, USA.

Fuller, M. R., G. B. Groom, and A. R. Jones. 1994. The land cover map of Great Britain: an automated classification of Landsat Thematic Mapper data. Photogrammetric Engineering and Remote Sensing 60:553–562.

Fuller, M. R., W. S. Seegar, and P. W. Howey. 1995. The use of satellite systems for the study of bird migration. Israel Journal of Zoology 41:243–252.

Furlow, R. C., M. Haderlie, and R. Van den Berge. 1981. Estimating a bighorn sheep population by mark-recapture. Desert Bighorn Council Transactions 1981:31–33.

Furnival, G. M. 1971. All possible regressions with less computation. Technometrics 13:403–408.

Gammonley, J. H., and J. R. Kelley, Jr. 1994. Effects of back-mounted radio packages on breeding wood ducks. Journal of Field Ornithology 65:530–533.

Gardner, S. N., and M. Mangel. 1996. Mark-resight population estimation with imperfect observations. Ecology 77:880–884.

Garner, G. W., S. C. Amstrup, D. C. Douglas, and C. L. Gardner. 1988. Performance and utility of satellite telemetry during field studies of free-ranging polar bears in Alaska. Pages 66–76 in C. J. Amlaner, Jr., editor. Biotelemetry X. Proceedings of the Tenth International Symposium on Biotelemetry. University of Arkansas Press, Fayetteville, Arkansas, USA.

Garrettson, P. R., F. C. Rohwer, and E. B. Moser. 2000. Effects of backpack and implanted radiotransmitters on captive blue-winged teal. Journal of Wildlife Management 64:216–222.

Garrott, R. A., R. M. Bartmann, and G. C. White. 1985. Comparison of radio-transmitter packages relative to deer fawn mortality. Journal of Wildlife Management 49:758–759.

Garrott, R. A., G. C. White, R. M. Bartmann, and D. L. Weybright. 1986. Reflected signal bias in biotelemetry triangulation systems. Journal of Wildlife Management 50:747–752.

Garrott, R. A., G. C. White, R. M. Bartmann, L. H. Carpenter, and A. W. Alldredge. 1987. Movements of female mule deer in northwest Colorado. Journal of Wildlife Management 51:634–643.

Garshelis, D. L. 1992. Mark-recapture density estimation for animals with large home ranges. Pages 1098–1111 in D. R. McCullough and R. H. Barrett, editors. Wildlife 2001: populations. Elsevier Science Publishers, London, England.

Garshelis, D. L., and D. B. Siniff. 1983. Evaluation of radio-transmitter attachments for sea otters. Wildlife Society Bulletin 11:378–383.

Gautestad, A. O., and I. Mysterud. 1993. Physical and biological mechanisms in animal movement processes. Journal of Applied Ecology 30:523–535.

Gautestad, A. O., and I. Mysterud. 1994. Fractal analysis of population ranges: methodological problems and challenges. Oikos 69:154–157.

Gautestad, A. O., and I. Mysterud. 1995. The home range ghost. Oikos 74:195–204.

Gavin, T. A. 1991. Why ask "why": the importance of evolutionary biology in wildlife science. Journal of Wildlife Management 55:760–766.

George, S. L. 1983. The required size and length of a phase III clinical trial. Pages 287–310 in M. E. Buyse, M. J. Staquet, and R. J. Sylvester, editors. Cancer clinical trials. Oxford University Press, New York, New York, USA.

Gerard, P. D., D. R. Smith, and G. Weerakkody. 1998. Limits of retrospective power analysis. Journal of Wildlife Management 62:801–807.

Gese, E. M., D. E. Anderson, and O. J. Rongstad. 1990. Determining home range size of resident coyotes from point and sequential locations. Journal of Wildlife Management 54:501–506.

Gessaman, J. A. 1980. An evaluation of heart rate as an indirect measure of daily energy metabolism of the American kestrel. Comparative Biochemistry and Physiology 65:273–289.

Gessaman, J. A., and K. A. Nagy. 1988. Transmitter loads affect the flight speed and metabolism of homing pigeons. The Condor 90:662–668.

Gessaman, J. A., G. W. Workman, and M. R. Fuller. 1991. Flight performance, energetics, and water turnover of tippler pigeons with a harness and dorsal load. The Condor 93:546–554.

Giesen, K. M., T. J. Schoenberg, and C. E. Braun. 1982. Methods for trapping sage grouse in Colorado. Wildlife Society Bulletin 10:224–231.

Giroux, J. F., D. V. Bell, S. Percival, and R. W. Summers. 1990. Tail-mounted radio transmitters for waterfowl. Journal of Field Ornithology 61:303–309.

Goldstein, M. I., T E. Lacher, Jr., B. Woodbridge, M.J. Bechard, S. B. Cnavelli, M. E. Zaccagnini, G. P. Cobb, R. Tribolet, and M. J. Hooper. 1998. Monocrotophos-induced mass mortality of Swainson's hawks in Argentina, 1995–1996. Ecotoxicology 8:201–214.

Goodchild, M., and S. Gopal, editors. 1992. Accuracy of spatial databases. Taylor and Francis, Bristol, Pennsylvania, USA.

Goodyear, J. D. 1993. A sonic/radio tag for monitoring dive depths and underwater movements of whales. Journal of Wildlife Management 57:503–513.

Gorman, M. 1993. Wildtrak. A suite of non-parametric home range analyses for the Macintosh computer. Animal Behaviour 45:1253.

Gormley, A. 1996. Movements, habitat use and survival of ruffed grouse (Bonasa umbellus) in northern Michigan. M.S. Thesis. Michigan State University, East Lansing, Michigan, USA.

Goss-Custard, J. D. 1996. The Oystercatcher: from individuals to populations. Oxford University Press, Oxford, England.

Green, R. A., and G. D. Bear. 1990. Seasonal cycles and daily activity patterns of Rocky Mountain elk. Journal of Wildlife Management 54:272–279.

Greenwood, R. J., and A. B. Sargeant. 1973. Influence of radio packs on captive mallards and blue-winged teal. Journal of Wildlife Management 37:3–9.

Guyn, K. L., and R. G. Clark. 1999. Decoy trap bias and effects of markers on reproduction of northern pintails. Journal of Field Ornithology 70:504–513.

Guynn, D. C., Jr., J. R. Davis, and A. F. Von Recum. 1987. Pathological potential of intraperitoneal transmitter implants in beavers. Journal of Wildlife Management 51:605–606.

Hagen, C. A. 1999. Sage grouse habitat use and seasonal movements in a naturally fragmented landscape, northwestern Colorado. M.S. Thesis. University of Manitoba, Winnipeg, Manitoba, Canada.

Haney, C. J. 1994. Testing for resource use and selection by marine birds: a reply to Aebischer and Robertson. Journal of Field Ornithology 65:214–220.

Haney, C. J., and A. R. Solow. 1992. Testing for resource use and selection by marine birds. Journal of Field Ornithology 63:43–52.

Hansen, J. A., S. L. Garman, B. Marks, and D. L. Urban. 1993. An approach for managing vertebrate diversity across multiple-use landscapes. Ecological Applications 3:481–496.

Hanski, I., and M. E. Gilpin, editors. 1997. Metapopulation biology: ecology, genetics and evolution. Academic Press, San Diego, California, USA.

Hansson, L., and P. Angelstam. 1991. Landscape ecology as a theoretical basis for nature conservation. Landscape Ecology 5:191–201.

Hansteen, T. L., H. P. Andreassen, and R. A. Ims. 1997. Effects of spatiotemporal scale on autocorrelation and home range estimators. Journal of Wildlife Management 61:280–290.

Harris, R. B., S. G. Fancy, D. C. Douglas, G. W. Garner, S. C. Amstrup, T. R. McCabe, and L. F. Pank. 1990. Tracking wildlife by satellite: current systems and performance. Fish and Wildlife Technical Report 30. U.S. Fish and Wildlife Service, Washington, D.C., USA.

Harris, S. 1980. Home ranges and patterns of distribution of foxes (*Vulpes vulpes*) in an urban area, as revealed by radio tracking. Pages 685–690 *in* C. J. Amlaner, Jr. and D. W. Macdonald, editors. A handbook on biotelemetry and radio tracking. Pergamon Press, Oxford, England.

Harris, S., W. J. Cresswell, P. G. Forde, W. J. Trewhella, T. Woolard, and S. Wray. 1990. Home-range analysis using radio-tracking data – a review of problems and techniques particularly as applied to the study of mammals. Mammal Review 20:97–123.

Hartigan, J. A. 1987. Estimation of a convex density contour in two dimensions. Journal of the American Statistical Association 82:267–270.

Harvey, M. J., and R. W. Barbour. 1965. Home range of *Michrotus ochrogaster* as determined by a modified minimum area method. Journal of Mammalogy 46:398–402.

Hastie, T. J., and D. Pregibon. 1992. Generalized linear models. Pages 195–247 *in* J. M. Chambers and T. J. Hastie, editors. Statistical models in S. Cole Advanced Books and Software, Pacific Grove, California, USA.

Hastings, H. M., and G. Sugihara. 1993. Fractals: a user's guide for the natural sciences. Oxford University Press, Oxford, England.

Hayne, D. W. 1949. Calculation of size of home range. Journal of Mammalogy 30:1–18.

Heezen, K. L., and J. R. Tester. 1967. Evaluation of radio-tracking by triangulation with special reference to deer movements. Journal of Wildlife Management 31:124–141.

Hein, E. W., and W. F. Andelt. 1995. Estimating coyote density from mark-resight surveys. Journal of Wildlife Management 59:164–169.

Heisey, D. M. 1985. Analyzing selection experiments with log-linear models. Ecology 66:1744–1748.

Heisey, D. M., and T. K. Fuller. 1985. Evaluation of survival and cause-specific mortality rates using telemetry data. Journal of Wildlife Management 49:668–674.

Hensler, G. L. 1985. Estimation and comparison of functions of daily nest survival probabilities using the Mayfield Method. Pages 289–301 in P. M. North and B. J. T. Morgan, editors. Statistics in ornithology. Springer-Verlag, New York, New York, USA.

Hensler, G. L., and J. D. Nichols. 1981. The Mayfield method of estimating nesting success: a model, estimators and simulation results. Wilson Bulletin 93:42–53.

Hentschel, H. G. E., and I. Procaccia. 1983. The infinite number of generalized dimensions of fractals and strange attractors. Physica 8D:435–444.

Herbst, L. 1991. Pathological and reproductive effects of intraperitoneal telemetry devices on female armadillos. Journal of Wildlife Management 55:628–631.

Herzog, P. W. 1979. Effects of radio-marking on behavior, movements, and survival of spruce grouse. Journal of Wildlife Management 43:316–323.

Hessler, E. J. R., J. R. Tester, D. B. Siniff, and M. M. Nelson. 1970. A biotelemetry study of survival of pen-reared pheasants released in selected habitats. Journal of Wildlife Management 34:267–274.

Hickey, M. B. C. 1992. Effect of radiotransmitters on the attack success of hoary bats, Lasiurus cinereus. Journal of Mammalogy 73:344–346.

Higuchi, H., K. Ozaki, G. Fujita, J. Minton, M. Ueta, M. Soma, and N. Mita. 1996. Satellite tracking of white-naped crane migration and the importance of the Korean demilitarized zone. Conservation Biology 10:806–812.

Hilborn, R., and M. Mangel. 1997. The ecological detective. Princeton University Press, Princeton, New Jersey, USA.

Hill, L. A., and L. G. Talent. 1990. Effects of capture, handling, banding, and radio-marking on breeding least terns and snowy plovers. Journal of Field Ornithology 61:310–319.

Hill, R. D. 1994. Theory of geolocation by light levels. Pages 227–236 in B. J. Le Boeuf, editor. Elephant seals. University of California Press, Berkeley, California, USA.

Hinch, S. G., R. E. Diewert, T. J. Lissimore, A. M. J. Prince, M. C. Healey, and M. A. Henderson. 1996. Use of electromyogram telemetry to assess difficult passage areas for river-migrating adult sockeye salmon. Transactions of the American Fisheries Society 125:253–260.

Hines, J. E. 1994. MSSURVIV user's manual. National Biological Service, Patuxent Wildlife Research Center, Laurel, Maryland, USA.

Hines, J. E., and J. R. Sauer. 1989. Program CONTRAST: a general program for the analysis of several survival or recovery rate estimates. Fish and Wildlife Technical Report 24. U.S. Fish and Wildlife Service, Washington, D.C., USA.

Hines, J. E., and F. C. Zwickel. 1985. Influence of radio packages on young blue grouse. Journal of Wildlife Management 49:1050–1054.

Hiraldo, F., J. A. Donazar, and J. J. Negro. 1994. Effects of tail-mounted radio-tags on adult lesser kestrels. Journal of Field Ornithology 65:466–471.

Hobbs, N. T., and T. A. Hanley. 1990. Habitat evaluation: do use/availability data reflect carrying capacity? Journal of Wildlife Management 54:515–522.

Hodder, K. H., R. E. Kenward, S. S. Walls, and R. T. Clarke. 1998. Estimating core ranges: a comparison of techniques using the common buzzard (*Buteo buteo*). Journal of Raptor Research 32:82–89.

Hoetting, J. A., M. Leecaster, and D. Bowden. 2000. An improved model for spatially correlated binary responses. Journal of Agricultural, Biological and Environmental Statistics 5:102–114.

Holter, N. J. 1961. New method for heart studies. Science 134:1214–1220.

Hooge, P. N. 1991. The effects of radio weight and harnesses on time budgets and movements of acorn woodpeckers. Journal of Field Ornithology 62:230–238.

Hooge, P. N., and B. Eichenlaub. 1997. Animal movement extension to Acview: version 1.1. Alaska Biological Science Center, U.S. Geological Survey, Anchorage, Alaska, USA.

Horton, G. I., and M. K. Causey. 1984. Brood abandonment by radio-tagged American woodcock hens. Journal of Wildlife Management 48:606–607.

Hoskinson, R. L. 1976. The effect of different pilots on aerial telemetry error. Journal of Wildlife Management 40:137–139.

Hosmer, D. W., and S. Lemeshow. 1989. Applied logistic regression. John Wiley & Sons, New York, New York, USA.

Houston, R. A., and R. J. Greenwood. 1993. Effects of radio transmitters on nesting captive mallards. Journal of Wildlife Management 57:703–709.

Howey, P. W. 2000. Innovations from Microwave Telemetry, Inc. Argos Newsletter 56:15.

Howey, P. W., W. S. Seegar, M. R. Fuller, and K. Titus. 1989. A coded tracking telemetry system. Pages 103–107 *in* C. J. Amlaner, Jr., editor. Biotelemetry X. Proceedings of the Tenth International Symposium on Biotelemetry. University of Arkansas Press, Fayetteville, Arkansas, USA.

Hubbard, M. W., L. C. Tsao, E. E. Klaas, M. Kaiser, and D. H. Jackson. 1998. Evaluation of transmitter attachment techniques on growth of wild turkey poults. Journal of Wildlife Management 62:1574–1578.

Hudson, D. J. 1971. Interval estimation from the likelihood function. Journal of the Royal Statistical Society, Series B 33:256–262.

Huggins, R. M. 1989. On the statistical analysis of capture-recapture experiments. Biometrika 76:133–140.

Huggins, R. M. 1991. Some practical aspects of a conditional likelihood approach to capture experiments. Biometrics 47:725–732.

Hünerbein, K., W. Wiltschko, R. Müller, and D. Klauer. 1997. A GPS based flight recorder for homing pigeons. Pages 561–570 *in* Proceedings of GNSS 97. Deutsche Gesellschaft für Ortung und Navigation, Bonn, Germany.

Hunt, W. G. 1998. Raptor floaters at Moffat's equilibrium. Oikos 82:191–197.

Hupp, J. W., and C. E. Braun. 1989. Topographic distribution of sage grouse foraging in winter. Journal of Wildlife Management 53:823–829.

Hupp, J. W., and J. T. Ratti. 1983. A test of radio telemetry triangulation accuracy in heterogeneous environments. Pages 31–46 in D. G. Pincock, editor. Proceedings of the Fourth Wildlife Biotelemetry Conference. Applied Microelectronics Institute and Technical University of Nova Scotia, Halifax, Nova Scotia, Canada.

Hurlbert, S. H. 1984. Pseudoreplication and the design of ecological field experiments. Ecological Monographs 54:187–211.

Ims, R. A. 1988. Spatial clumping of sexually receptive females induces space sharing among male voles. Nature 335:541–543.

Irwin, L. L., and J. M. Peek. 1983. Elk habitat use relative to forest succession in Idaho. Journal of Wildlife Management 47:664–672.

Ivlev, V. S. 1961. Experimental ecology of the feeding of fishes. Yale University Press, New Haven, Connecticut, USA.

Jaremovic, R. V., and D. B. Croft. 1987. Comparison of techniques to determine eastern gray kangaroo home range. Journal of Wildlife Management 51:921–930.

Jelinski, D. E. 1991. On the use of chi-square analyses in studies of resource utilization. Canadian Journal of Forest Research 21:58–65.

Jennrich, R. I., and F. B. Turner. 1969. Measurements of non-circular home range. Journal of Theoretical Biology 22:227–237.

Jewell, P. A. 1966. The concept of home range in mammals. Symposium of the Zoological Society of London 18:85–109.

Johannesen, E., H. P. Andreassen, and H. Steen. 1997. Effect of radiocollars on survival of root voles. Journal of Mammalogy 78:638–642.

Johnson, A. R., B. T. Milne, and J. A. Wiens. 1992. Diffusion in fractal landscapes: simulations and experimental studies of tenebrionid beetle movements. Ecology 73:1968–1983.

Johnson, B. K., A. A. Ager, S. L. Findholt, M. J. Wisdom, D. B. Marx, J. W. Kern, and L. D. Bryant. 1998. Mitigating spatial differences in observation rate of automated telemetry systems. Journal of Wildlife Management 62:958–967.

Johnson, B. K., J. W. Kern, M. J. Wisdom, S. L. Findholt, and J. G. Kie. 2000. Resource selection and spatial separation of mule deer and elk in spring. Journal of Wildlife Management 64:685–697.

Johnson, D. E. 1998. Applied multivariate methods for data analysts. Duxbury Press, New York, New York, USA

Johnson, D. H. 1979. Estimating nest success: the Mayfield method and an alternative. Auk 96:651–661.

Johnson, D. H. 1980. The comparison of usage and availability measurements for evaluating resource preference. Ecology 61:65–71.

Johnson, D. H. 1994. Population analysis. Pages 419–444 in T. A. Bookhout, editor. Research and management techniques for wildlife and habitats. Fifth edition. The Wildlife Society, Bethesda, Maryland, USA.

Johnson, D. H. 1999. The insignificance of statistical significance testing. Journal of Wildlife Management 63:763–772.

Johnson, R. N., and A. H. Berner. 1980. Effects of radio transmitters on released cock pheasants. Journal of Wildlife Management 44:686–689.

Jones, M. C., J. S. Marron, and S. J. Sheather. 1996a. A brief survey of bandwidth selection for density estimation. Journal of the American Statistical Association 91:401–407.

Jones, M. C., J. S. Marron, and S. J. Sheather. 1996b. Progress in data-based bandwidth selection for kernel density estimation. Computational Statistics 11:337–381.

Jorgensen, E. E., S. Demarais, S. M. Sell, and S. P. Lerich. 1998. Modeling habitat suitability for small mammals in Chihuahuan Desert foothills of New Mexico. Journal of Wildlife Management 62:989–996.

Jouventin, P., and H. Weimerskirch. 1990. Satellite tracking of wandering albatrosses. Nature 343:746–748.

Juneja, N., G. Singh, and T. L. Bashore. 1991. Simplified plotting analysis and data storage (SPADS) for telemetry. Wildlife Society Bulletin 19:226–227.

Kalas, J. A., L. Lofaldli, and P. Fiske. 1989. Effects of radio packages on great snipe during breeding. Journal of Wildlife Management 53:1155–1158.

Kalbfleisch, J. D., and L. Prentice. 1980. The statistical analysis of failure data. John Wiley & Sons, New York, New York, USA.

Kalcounis, M. C., and R. M. Brigham. 1998. Secondary use of aspen cavities by tree-roosting big brown bats. Journal of Wildlife Management 62:603–611.

Kaplan, E. L., and P. Meier. 1958. Nonparametric estimation from incomplete observations. Journal of the American Statistical Association 53:457–481.

Kareiva, P. M., and N. Shigesada. 1983. Analyzing insect movement as a correlated random walk. Oecologia 56:234–238.

Katnik, D. D., D. J. Harrison, and T. P. Hodgman. 1994. Spatial relations in a harvested population of marten in Maine. Journal of Wildlife Management 58:600–607.

Keating, K.A. 1995. Mitigating elevation-induced errors in satellite telemetry locations. Journal of Wildlife Management 59:801–808.

Keating, K. A., W. G. Brewster, and C. H. Key. 1991. Satellite telemetry: performance of animal-tracking systems. Journal of Wildlife Management 55:160–171.

Keister, G. P., Jr., C. E. Trainer, and M. J. Willis. 1988. A self-adjusting collar for young ungulates. Wildlife Society Bulletin 16:321–323.

Kendall, W. L., J. D. Nichols, and J. E. Hines. 1997. Estimating temporary emigration and breeding proportions using capture-recapture data with Pollock's robust design. Ecology 78:563–578.

Kenkel, N. C. 1993. Modeling Markovian dependence in populations of *Aralia nudicaulis*. Ecology 74:1700–1706.

Kenkel, N. C., and D. J. Walker. 1996. Fractals in the biological sciences. Coenoses 11:77–100.

Kenward, R. E. 1977. Predation on released pheasants (*Fasianus colchicus*) by goshawks (*Accipiter gentilis*) in central Sweden. Swedish Game Research 10:79–112.

Kenward, R. E. 1982. Techniques for monitoring the behaviour of grey squirrels by radio. Pages 175–196 in C. L. Cheeseman and R. B. Mitson, editors. Telemetric studies of vertebrates. Academic Press, London, England.

Kenward, R. E. 1985. Ranging behaviour and population dynamics in grey squirrels. Pages 319–330 in R. M. Sibly and R. H. Smith, editors. Telemetric studies of vertebrates. Academic Press, London, England.

Kenward, R. E. 1987. Wildlife radio tagging: equipment, field techniques and data analysis. Academic Press, London, England.

Kenward, R. E. 1990. RANGES IV. Software for analysing animal location data. Institute of Terrestrial Ecology, Wareham, England.

Kenward, R. E. 1992. Quantity versus quality: programmed collection and analysis of radio-tracking data. Pages 231–246 in I. G. Priede and S. M. Swift, editors. Wildlife telemetry: remote monitoring and tracking of animals. Ellis Horwood, West Sussex, United Kingdom.

Kenward, R. E. 1993. Modelling raptor populations: to ring or to radio tag? Pages 157–167 in J. D. Lebreton and P. M. North, editors. The use of marked individuals in the study of bird population dynamics: models, methods and software. Birkhauser-Verlag, Basel, Switzerland.

Kenward, R. E. 2001. A manual for wildlife radio tagging. Academic Press, London, England.

Kenward, R. E., and K. H. Hodder. 1996. RANGES V: An analysis system for biological location data. Institute of Terrestrial Ecology, Wareham, United Kingdom.

Kenward, R. E., V. Marcström, and M. Karlbom. 1981a. Goshawk winter ecology in Swedish pheasant habitats. Journal of Wildlife Management 45:397–408.

Kenward, R. E., M. Marquiss, and I. Newton. 1981b. What happens to goshawks trained for falconry. Journal of Wildlife Management 45:802–806.

Kenward, R. E., V. Marcström, and M. Karlbom. 1993. Post-nestling behaviour in goshawks, Accipiter gentilis: I. The causes of dispersal. Animal Behaviour 46:365–70.

Kenward, R. E., V. Marcström, and M. Karlbom. 1999. Demographic estimates from radio-tagging: models of age-specific survival and breeding in the goshawk. Journal of Animal Ecology 68:1020–1033.

Kenward, R. E., S. S. Walls, K. H. Hodder, M. Pahkala, S. N. Freeman, and V. R. Simpson. 2000. The prevalence of non-breeders in raptor populations: evidence from radio-tagging and survey data. Oikos 91:271–279.

Kenward, R. E., R. H. Pfeffer, M. A. Al-Bowardi, N. C. Fox, K. E. Riddle, Y. A. Bragin, A. S. Levin, S. S. Walls, and K. H. Hodder. In press a. New techniques for demographic studies of falcons. Journal of Field Ornithology.

Kenward, R. E., R. T. Clarke, K. H. Hodder, and S. S. Walls. In press b. Distance and density estimators of home range in raptors and squirrels: I. Defining multi-nuclear cores by nearest-neighbor clustering. Ecology.

Kenward, R. E., S. S. Walls, and K. H. Hodder. In press c. Life path analysis: scaling indicates priming effects of social and habitat factors on dispersal distances. Journal of Animal Ecology.

Kenward, R. E., N. J. Aebischer, P. A. Robertson, M. R. Fuller, R. J. Rose, and S. S. Walls. In review. Distance and density estimators of home-range: assessing habitat dependence from area and composition.

Kernohan, B. J. 1994. Winter/spring population characteristics of white-tailed deer in an agricultural/wetland complex in Northeastern South Dakota. M.S. Thesis. South Dakota State University, Brookings, South Dakota, USA.

Kernohan, B. J., J. A. Jenks, and D. E. Naugle. 1994. Movement patterns of white-tailed deer at Sand Lake National Wildlife Refuge, South Dakota. Prairie Naturalist 26:293–300.

Kernohan, B. J., J. J. Millspaugh, J. A. Jenks, and D. E. Naugle. 1998. Use of home range estimators in a GIS environment to identify habitat use patterns. Journal of Environmental Management 53:83–89.

Kie, J. G., J. A. Baldwin, and C. J. Evans. 1996. CALHOME: a program for estimating animal home ranges. Wildlife Society Bulletin 24:342–344.

Kjellen, N., M. Hake, and T. Alerstam. 1997. Strategies of two ospreys *Pandion haliaetus* migrating between Sweden and tropical Africa as revealed by satellite tracking. Journal of Avian Biology 28:15–23.

Klaassen, M., P. H. Becker, and M. Wagener. 1992. Transmitter loads do not affect the daily energy expenditure of nesting common terns. Journal of Field Ornithology 63:181–185.

Klett, A. T., H. F. Duebbert, C. A. Fannes, and K. F. Higgins. 1986. Techniques for studying nest success of ducks in upland habitats in the prairie pothole region. Resource Publication 158. U.S. Fish and Wildlife Service, Washington, D.C., USA.

Klugman, S. S., and M. R. Fuller. 1990. Effects of implanted transmitters on captive Florida sandhill cranes. Wildlife Society Bulletin 18:394–399.

Knick, S. T., and D. L. Dyer. 1997. Distribution of black-tailed jackrabbit habitat determined by GIS in southwestern Idaho. Journal of Wildlife Management 61:75–85.

Knick, S. T., and J. T. Rotenberry. 1998. Limitations to mapping habitat use areas in changing landscapes using Mahalanobis distance statistic. Journal of Agricultural, Biological and Environmental Statistics 3:311–322.

Knight, R. R. 1970. The Sun River elk herd. Wildlife Monograph 23:1–66.

Knopf, F. L., J. A. Sedgewick, and D. B. Inkley. 1990. Regional correspondence among shrubsteppe habitats. Condor 92:42–53.

Koeppl, J. W., N. A. Slade, and R. S. Hoffmann. 1975. A bivariate home range model with possible application to ethological data analysis. Journal of Mammalogy 56:81–90.

Kohn, B. E., and J. J. Mooty. 1971. Summer habitat of white-tailed deer in north-central Minnesota. Journal of Wildlife Management 35:476–487.

Kolz, A. L., J. W. Lentfer, and H. G. Fallek. 1980. Satellite radio tracking of polar bears instrumented in Alaska. Pages 743–752 in C. J. Amlaner, Jr. and D. W. Macdonald, editors. A handbook on biotelemetry and radio tracking. Pergamon Press, Oxford, England.

Korschgen, C. E., S. J. Maxson, and V. B. Kuechle. 1984. Evaluation of implanted radio transmitters in ducks. Journal of Wildlife Management 48:982–987.

Korschgen, C. E., K. P. Kenow, J. E. Austin, C. O. Kochanny, W. L. Green, C. H. Simmons, and M. Janda. 1995. An automated telemetry system for studies of migrating diving ducks. Pages 179–184 in C. Cristalli, C. J. Amlaner, Jr., and M. R. Neuman, editors. Biotelemetry XIII. Proceedings of the Thirteenth International Symposium on Biotelemetry. Williamsburg, Virginia, USA.

Korschgen, C. E., K. P. Kenow, W. L. Green, D. H. Johnson, M. D. Samuel, and L. Sileo. 1996a. Survival of radiomarked canvasback ducklings in northwestern Minnesota. Journal of Wildlife Management 60:120–132.

Korschgen, C. E., K. P. Kenow, A. Gendron-Fitzpatrick, W. L. Green, and F. J. Dein. 1996b. Implanting intra-abdominal radiotransmitters with external whip antennas in ducks. Journal of Wildlife Management 60:132–137.

Krauss, G. D., H. B. Graves, and S. M. Zervanos. 1987. Survival of wild and game-farm cock pheasants released in Pennsylvania. Journal of Wildlife Management 51:555–559.

Kufeld, R. C., D. C. Bowden, and J. M. Siperek, Jr. 1987. Evaluation of a telemetry system for measuring habitat usage in mountainous terrain. Northwest Science 61:249–256.

Lacroix, G. L., and P. McCurdy. 1996. Migratory behaviour of post-smolt Atlantic salmon during initial stages of seaward migration. Journal of Fish Biology 49:1086–1101.

Lamberson, R. H., B. R. Noon, C. Voss, and K. S. McKelvey. 1994. Reserve design for a territorial species: the effects of patch size and spacing on the viability of the northern spotted owl. Conservation Biology 8:185–195.

Lance, A. N., and A. Watson. 1977. Further tests of radio-marking on red grouse. Journal of Wildlife Management 41:579–582.

Lance, A. N., and A. Watson. 1980. A comment on the use of radio tracking in ecological research. Pages 355–359 in C. J. Amlaner, Jr. and D. W. Macdonald, editors. A handbook on biotelemetry and radio tracking. Pergamon Press, Oxford, England.

Land, E. D., D. R. Garman, and G. A. Holt. 1998. Monitoring female Florida panthers via cellular telephone. Wildlife Society Bulletin 26:29–31.

Lariviere, S., and F. Messier. 1998. The influence of close-range radio-tracking on the behavior of free-ranging striped skunks, Mephitis mephitis. Canadian Field-Naturalist 112:657–660.

Larkin, R. P., and D. Halkin. 1994. A review of software packages for estimating animal home ranges. Wildlife Society Bulletin 22:274–287.

Larkin, R. P., A. Raim, and R. H. Diehl. 1996. Performance of a non-rotating direction-finder for automatic radio tracking. Journal of Field Ornithology 67:59–71.

Laundré, J. W., and B. L. Keller. 1981. Home-range use by coyotes in Idaho. Animal Behavior 29:449–461.

Laundré, J. W., and B. L. Keller. 1984. Home-range size of coyotes: a critical review. Journal of Wildlife Management 48:127–139.

Lawler, A. 2000. Scientists gain access to sharper GPS signal. Science 288:783.

Lawson, E. J. G., and A. R. Rodgers. 1997. Differences in home-range size computed in commonly used software programs. Wildlife Society Bulletin 25:721–729.

Leban, F. A. 1999. Performance of five resource selection methods under different sampling designs: a case study with elk radio-telemetry data. M.S. Thesis. University of Idaho, Moscow, Idaho, USA.

Lebreton, J. D., K. P. Burnham, J. Clobert, and D. R. Anderson. 1992. Modeling survival and testing biological hypotheses using marked animals: a unified approach with case studies. Ecological Monographs 62:67–118.

Leclerc, J. 1991. Optimal foraging strategy of the sheet-web spider *Lepthyphantes flavipes* under perturbation. Ecology 72:1267–1272.

Le Cren, E. D. 1965. A note on the history of mark-recapture population estimates. Journal of Animal Ecology 34:77–91.

Lee, E. T. 1980. Statistical methods for survival data analysis. Lifetime Learning Publications, Belmont, California, USA.

Lee, E. T. 1992. Statistical methods for survival data analysis. John Wiley & Sons, New York, New York, USA.

Lee, J. E., G. C. White, R. A. Garrott, R. M. Bartmann, and A. W. Alldredge. 1985. Assessing accuracy of a radiotelemetry system for estimating animal locations. Journal of Wildlife Management 49:658–663.

Lehmkuhl, J. F., and M. G. Raphael. 1993. Habitat pattern around northern spotted owl locations on the Olympic Peninsula, Washington. Journal of Wildlife Management 57:302–315.

Lemnell, P. A. 1980. An automatic telemetry system for tracking and physiology. Pages 453–456 *in* C. J. Amlaner, Jr. and D. W. Macdonald, editors. A handbook on biotelemetry and radio tracking. Pergamon Press, Oxford, England.

Lemnell, P. A., G. Johnsson, H. Helmersson, O. Holmstrand, and L. Norling. 1983. An automatic radio-telemetry system for position determination and data acquisition. Pages 76–93 *in* D. G. Pincock, editor. Proceedings of the Fourth Wildlife Biotelemetry Conference. Applied Microelectronics Institute and Technical University of Nova Scotia, Halifax, Nova Scotia, Canada.

Le Munyan, C. D., W. White, E. Nybert, and J. J. Christian. 1959. Design of a miniature radio transmitter for use in animal studies. Journal of Wildlife Management 23:107–110.

Lenth, R. V. 1981. On finding the source of a signal. Technometrics 23:149–154.

Leptich, D. J., D. G. Beck, and D. E. Beaver. 1994. Aircraft-based LORAN-C and GPS accuracy for wildlife research on inland study sites. Wildlife Society Bulletin 22:561–565.

Leslie, D. M., Jr., and C. L. Douglas. 1979. Desert bighorn sheep of the River Mountains, Nevada. Wildlife Monographs 66:1–56.

Leslie, D. M., Jr., and C. L. Douglas. 1986. Modeling demographics of bighorn sheep: current abilities and missing links. North American Wildlife and Natural Resources Conference Transactions 51:62–73.

Levin, S. A. 1992. The problem of pattern and scale in ecology. Ecology 61:65–71.

Levins, R. 1969. Some demographic and genetic consequences of environmental heterogeneity for biological control. Bulletin of the Entomological Society of America 15:237–240.

Lewis, J. C., and T. L. Haithcoat. 1986. TELEMPC: personal computer package for analyzing radio-telemetry data. Geographic Resources Center, University of Missouri, Columbia, Missouri, USA.

Lewis, M., and J. Murray. 1993. Modeling territoriality and wolf deer interactions. Nature 366:738–740.

Li, H., and J. F. Reynolds. 1994. A simulation experiment to quantify spatial heterogeneity in categorical maps. Ecology 75:2446–2455.

Lincoln, F. C. 1930. Calculating waterfowl abundance on the basis of banding returns. U.S. Department of Agriculture Circular 118:1–4.

Loader, C. R. 1999. Bandwidth selection: classical or plug-in? Annals of Statistics 27:415–438.

Loehle, C. 1990. Home range: a fractal approach. Landscape Ecology 5:39–52.

Loehle, C., and B. L. Li. 1996. Statistical properties of ecological and geologic fractals. Ecological Modelling 85:271–284.

Lotimer, J. S. 1980. A versatile coded wildlife transmitter. Pages 185–191 in C. J. Amlaner, Jr. and D. W. Macdonald, editors. A handbook on biotelemetry and radio tracking. Pergamon Press, Oxford, England.

Lubin, Y., S. Ellner, and M. Kotzman. 1993. Web relocation and habitat selection in a desert widow spider. Ecology 74:1915–1928.

Lyon, L. J., and A. L. Ward. 1982. Elk and land management. Pages 443–447 in J. W. Thomas and D. E. Toweill, editors. Elk of North America. Stackpole Books, Harrisburg, Pennsylvania, USA.

Macdonald, D. W. 1979. Helpers in fox society. Nature 282:69–71.

Macdonald, D. W., and C. J. Amlaner, Jr. 1980. A practical guide to radio tracking. Pages 143–159 in C. J. Amlaner, Jr. and D. W. Macdonald, editors. A handbook on biotelemetry and radio tracking. Pergamon Press, Oxford, England.

Macdonald, D. W., F. G. Ball, and N. G. Hough. 1980. The evaluation of home range size and configuration using radio tracking data. Pages 405–424 in C. J. Amlander, Jr. and D. W. Macdonald, editors. A handbook on biotelemetry and radio tracking. Pergamon Press, Oxford, England.

Mace, R. D., S. C. Minta, T. L. Manley, and K. E. Aune. 1994. Estimating grizzly bear population size using camera sightings. Wildlife Society Bulletin 22:74–83.

MacInness, C. D., and R. K. Misra. 1972. Predation on Canada goose nests at McConnell River, Northwest Territories. Journal of Wildlife Management 36:414–422.

Madsen, E. R. 1995. A finite population bootstrap method for interval estimates of animal populations. M.S. Thesis. Colorado State University, Fort Collins, Colorado, USA.

Mandelbrot, B. B. 1983. The fractal geometry of nature. W. H. Freeman, San Francisco, California, USA.

Manly, B. F. J. 1974. A model for certain types of selection experiments. Biometrics 30:281–294.

Manly, B. F. J. 1992. The design of research studies. Cambridge University Press, Cambridge, United Kingdom.

Manly, B. F. J. 1997. Randomization, bootstrap and other Monte Carlo methods in biology. Second edition. Chapman & Hall, London, England.

Manly, B. F. J., L. L. McDonald, and D. L. Thomas. 1993. Resource selection by animals: statistical design and analysis for field studies. Chapman & Hall, London, England.

Mannering, F. 1998. Modeling driver decision-making: a review of methodological alternatives. In W. Barfield and T. Dingus, editors. Human factors in intelligent transportation systems. Lawrence Erlbaum, Hillsdale, New Jersey, USA.

Manski, C. 1977. The structure of random utility models. Theory and Decision 8:229–254.

Marcstrom, V., R. E. Kenward, and M. Karlbom. 1989. Survival of ring-necked pheasants with backpacks, necklaces, and leg bands. Journal of Wildlife Management 53:808–810.

Marcum, C. L., and D. O. Loftsgaarden. 1980. A non-mapping technique for studying habitat preferences. Journal of Wildlife Management 44:963–968.

Mardia, K. V. 1972. Statistics of directional data. Academic Press, New York, New York, USA.

Marks, J. S., and V. S. Marks. 1987. Influence of radio collars on survival of sharp-tailed grouse. Journal of Wildlife Management 51:468–471.

Marquiss, M., and I. Newton. 1982. Habitat preference in male and female sparrowhawks (Accipiter nisus). Ibis 124:324–328.

Marshall, W. H., G. W. Gullion, and R. G. Schwab. 1962. Early summer activities of porcupines as determined by radio-positioning techniques. Journal of Wildlife Management 26:75–79.

Marzluff, J. M., M. S. Vekasy, and C. Coody. 1994. Comparative accuracy of aerial and ground telemetry locations of foraging raptors. Condor 96:447–454.

Marzluff, J. M., B. A. Kimsey, L. S. Schueck, M. E. McFadzen, M. S. Vekasy, and J. C. Bednarz. 1997a. The influence of habitat, prey abundance, sex, and breeding success on the ranging habits of Prairie Falcons. Condor 99:567–584.

Marzluff, J. M., S. T. Knick, M. S. Vekasy, L .S. Schueck, and T. J. Zarriello. 1997b. Spatial use patterns and habitat selection of Golden Eagles in southwestern Idaho. Auk 114:673–687.

Marzluff, J. M., M. S. Vekasy, M. N. Kochert, and K. Steenhof. 1997c. Productivity of golden eagles wearing backpack radiotransmitters. Journal of Raptor Research 31:223–227.

Mate, B. R., S. L. Nieukirk, and S. D. Kraus. 1997. Satellite-monitored movements of the northern right whale. Journal of Wildlife Management 61:1393–1405.

Mauser, D. M., and R. L. Jarvis. 1991. Attaching radio transmitters to 1–day-old mallard ducklings. Journal of Wildlife Management 55:488–491.

Mauser, D. M., R. L. Jarvis, and D. L. Gilmer. 1994. Survival of radio-marked mallard ducklings in northeastern California. Journal of Wildlife Management 58:82–87.

Mayfield, H. 1961. Nesting success calculated from exposure. Wilson Bulletin 73:255–261.

Mayfield, H. 1975. Suggestions for calculating nest success. Wilson Bulletin 87:456–466.

McClean, S. A., M. A. Rumble, R. M. King, and W. L. Baker. 1998. Evaluation of resource selection methods with different definitions of availability. Journal of Wildlife Management 62:793–801.

McCullagh, P., and J. A. Nelder. 1989. Generalized linear models. Second edition. Monographs on statistics and applied probability No. 37. Chapman & Hall, London, England.

McCulloch, C. E., and M. L. Cain. 1989. Analyzing discrete movement data as a correlated random walk. Ecology 70:383–388.

McFadden, D. 1974. Conditional logit analysis of qualitative choice behavior. Pages 105–142 in P. Zarembka, editor. Frontiers in econometrics. Academic Press, New York, New York, USA.

McFadden, D. 1978. Modeling the choice of residential location. Pages 79–96 in A. Karlquist, editor. Spatial interaction theory and planning models. North-Holland, Amsterdam, Holland.

McFadden, D. 1981. Econometric models of probabilistic choice. Pages 198–272 in C. Manski and D. McFadden, editors. Structural analysis of discrete data with econometric applications. MIT Press, Cambridge, Massachusetts, USA.

McGaughey, R. J., and A. A. Ager. 1996. UTOOLS and UVIEW: analysis and visualization software. Presentation at the Sixth Biennial Forest Service Remote Sensing Conference, Denver, Colorado, USA.

McKee, G., M. R. Ryan, and L. M. Mechlin. 1998. Predicting greater prairie-chicken nest success from vegetation and landscape characteristics. Journal of Wildlife Management 62:314–321.

McNay, R. S., J. A Morgan, and F. L. Bunnell. 1994. Characterizing independence of observations in movements of Columbian black-tailed deer. Journal of Wildlife Management 58:422–429.

Mech, L. D. 1967. Telemetry as a technique in the study of predation. Journal of Wildlife Management 31:492–496.

Mech, L. D. 1983. Handbook of animal radio-tracking. University of Minnesota Press, Minneapolis, Minnesota, USA.

Mech, L. D, and E. M. Gese. 1992. Field testing the Wildlink capture collar on wolves. Wildlife Society Bulletin 20:221–223.

Mech, L. D., and M. Korb. 1977. An unusually long pursuit of a deer by a wolf. Journal of Mammalogy 59:860–861.

Mech, L. D., V. B. Keuchle, D. W. Warner, and J. R. Tester. 1965. A collar for attaching radio transmitters on rabbits, hares and raccoons. Journal of Wildlife Management 29:898–902.

Mech, L. D., R. C. Chapman, W. W. Cochran, L. Simmons, and U. S. Seal. 1984. A radio-triggered anesthetic-dart collar for recapturing large mammals. Wildlife Society Bulletin 12:69–74.

Mech, L. D., K. E. Kunkel, R. C. Chapman, and T. J. Kreeger. 1990. Field testing of commercially manufactured capture collars on white-tailed deer. Journal of Wildlife Management 54:297–299.

Melvin, S. M., R. C. Drewien, S. A. Temple, and E. G. Bizeau. 1983. Leg-band attachment of radio transmitters for large birds. Wildlife Society Bulletin 11:282–285.

Merrill, S. B., L. G. Adams, M. E. Nelson, and L. D. Mech. 1998. Testing releasable GPS radiocollars on wolves and white-tailed deer. Wildlife Society Bulletin 26:830–835.

Meyburg, B. U., and E. G. Lobkov. 1994. Satellite tracking of a juvenile Stellar's sea eagle Haliaeetus pelagicus. Ibis 136:105–106.

Meyburg, B. U., and C. Meyburg. 1998. The study of raptor migration using satellite telemetry: some goals, achievements and limitations. Biotelemetry 14:415–420.

Mielke, P. W., Jr. 1986. Non-metric statistical analyses: some metric alternatives. Journal of Statistical Planning and Inference 13:377–387.

Mielke, P. W., Jr., and K. J. Berry. 1982. An extended class of permutation techniques for matched pairs. Communications in Statistics: Theory and Methods 11:1197–1207.

Mielke, P. W., Jr., K. J. Berry, and E. S. Johnson. 1976. Multi-response permutation procedures for a priori classifications. Communications in Statistics: Theory and Methods 5:1409–1424.

Mikesic, D. G., and L. C. Drickamer. 1992. Effects of radiotransmitters and fluorescent powders on activity of wild house mice (Mus musculus). Journal of Mammalogy 73:663–667.

Miller, H. W., and D. H. Johnson. 1978. Interpreting the results of nesting studies. Journal of Wildlife Management 42:471–476.

Miller, S. D., E. F. Becker, and W. H. Ballard. 1987. Black and brown bear density estimates using modified capture-recapture techniques in Alaska. International Conference on Bear Research and Management 7:23–35.

Miller, S. G., G. C. White, R. A. Sellers, H. V. Reynolds, J. W. Schoen, K. Titus, V. G. Barnes, Jr., R. B. Smith, W. B. Ballard, and C. C. Schwartz. 1997. Brown and black bear density estimation in Alaska using radiotelemetry and replicated mark-resight techniques. Wildlife Monograph 133:1–55.

Mills, L. S., and F. F. Knowlton. 1989. Observer performance in known and blind radio-telemetry accuracy tests. Journal of Wildlife Management 53:340–342.

Millspaugh, J. J. 1995. Seasonal movements, habitat use patterns, and the effects of human disturbances on elk in Custer State Park, South Dakota. M.S. Thesis. South Dakota State University, Brookings, South Dakota, USA.

Millspaugh, J. J. 1999. Behavioral and physiological responses of elk to human disturbances in the southern Black Hills, South Dakota. Ph.D. Dissertation. University of Washington, Seattle, Washington, USA.

Millspaugh, J. J., J. R. Skalski, B. J. Kernohan, K. J. Raedeke, G. C. Brundige, and A. B. Cooper. 1998a. Some comments on spatial independence in studies of resource selection. Wildlife Society Bulletin 26:232–236.

Millspaugh, J. J., K. J. Raedeke, G. C. Brundige, and C. C. Willmott. 1998b. Summer bed sites of elk (Cervus elaphus) in the Black Hills, South Dakota: Implications for thermal cover management. American Midland Naturalist 139:133–140.

Millspaugh, J. J., G. C. Brundige, R. A. Gitzen, and K. J. Raedeke. 2000. Elk and hunter space-use sharing in the Southern Black Hills, South Dakota. Journal of Wildlife Management 64:994–1003.

Milne, B. T. 1997. Applications of fractal geometry in wildlife biology. Pages 32–69 in J. A. Bissonette, editor. Wildlife and landscape ecology: effects of pattern and scale. Springer, New York, New York, USA.

Minta, S. C. 1992. Tests of spatial and temporal interaction among animals. Ecological Applications 2:178–188.

Minta, S. C., and M. Mangel. 1989. A simple population estimate based on simulation for capture-recapture and capture-resight data. Ecology 70:1738–1751.

Mizutani, F., and P. A. Jewell. 1998. Home-range and movements of leopards (Panthera pardus) on a livestock ranch in Kenya. Journal of Zoology 244:269–286.

Mladenoff, D. J., T. A. Sickley, R. G. Haight, and A. P. Wydeven. 1995. A regional landscape analysis and prediction of favorable gray wolf habitat in the Northern Great Plains lakes region. Conservation Biology 9:279–294.

Moen, R., J. Pastor, Y. Cohen, and C. C. Schwartz. 1996. Effect of moose movement and habitat use on GPS collar performance. Journal of Wildlife Management 60:659–668.

Moen, R., J. Pastor, and Y. Cohen. 1997. Accuracy of GPS telemetry collar locations with differential Correction. Journal of Wildlife Management 61:530–539.

Mohr, C. O. 1947. Table of equivalent populations of North American small mammals. American Midland Naturalist 37:223–449.

Mohr, C. O., and W. A. Stumpf. 1966. Comparison of methods for calculating areas of animal activity. Journal of Wildlife Management 30:293–304.

Montgomery, D. C. 1984. Design and analysis of experiments. Second edition. John Wiley & Sons, New York, New York, USA.

Moorcroft, P. R., M. A. Lewis, and R. L. Crabtree. 1999. Home range analysis using a mechanistic home range model. Ecology 80:1656–1665.

Moran, P. A. P. 1950. Notes on continuous stochastic phenomena. Biometrika 37:17–23.

Morris, P. A. 1988. A study of home range and movements in the hedgehog (*Erinaceus europaeus*). Journal of Zoology 214:433–449.

Morris, R. D., M. C. Benkel, A. Biernacki, and J. M. Ross. 1981. A new transmitter package assembly for adult herring gulls. Journal of Field Ornithology 52:242–244.

Morrison, D. F. 1976. Multivariate statistical methods. McGraw-Hill, New York, New York, USA.

Moser, M. L., and S. W. Ross. 1995. Habitat use and movements of shortnose and Atlantic sturgeons in the Lower Cape Fear River, North Carolina. Transactions of the American Fisheries Society 124:225–234.

Muenchow, G. 1986. Ecological use of failure time analysis. Ecology 67:246–250.

Munger, J. C. 1984. Home-ranges of horned lizards (*Phrynosoma*): circumscribed and exclusive? Oecologia 62:351–360.

Murray, J. D. 1967. Dispersal in vertebrates. Ecology 48:975–978.

Mysterud, A., and R. A. Ims. 1998. Functional responses in habitat use: availability influences relative use in trade-off situations. Ecology 79:1435–1441.

Naef-Daenzer, B. 1993. A new transmitter for small animals and enhanced methods of home-range analysis. Journal of Wildlife Management 57:680–689.

Nams, V. O. 1989. Effects of radiotelemetry error on sample size and bias when testing habitat selection. Canadian Journal of Zoology 67:1631–1636.

Nams, V. O. 1996. The VFractal: a new estimator for fractal dimension of animal movement paths. Landscape Ecology 11:289–297.

Nams, V. O., and S. Boutin. 1991. What is wrong with error polygons? Journal of Wildlife Management 55:172–176.

Neal, A. K. 1990. Evaluation of mark-resight population estimates using simulations and field data from mountain sheep. M.S. Thesis. Colorado State University, Fort Collins, Colorado, USA.

Neal, A. K., G. C. White, R. B. Gill, D. F. Reed, and J. H. Olterman. 1993. Evaluation of mark-resight model assumptions for estimating mountain sheep numbers. Journal of Wildlife Management 57:436–450.

Nenno, E. S., and W. M. Healy. 1979. Effects of radio packages on behaviour of wild turkey hens. Journal of Wildlife Management 43:760–765.

Neter, J., W. Wasserman, and M. H. Kutner. 1990. Applied linear statistical models: regression, analysis of variance, and experimental designs. Third edition. Richard D. Irwin, Homewood, Illinois, USA.

Neu, C. W., C. R. Byers, and J. M. Peek. 1974. A technique for analysis of utilization-availability data. Journal of Wildlife Management 38:541–545.

Neudorf, D. L., and T. E. Pitcher. 1997. Radio transmitters do not affect nestling feeding rates by female hooded warblers. Journal of Field Ornithology 68:64–68.

Newcomer, J. A., and J. Szajgin. 1984. Accumulation of thematic map errors in digital overlay analysis. The American Cartographer 11:58–62.

Newman, S. H., J. Y. Takekawa, D. L. Whitworth, and E. E. Burkett. 1999. Subcutaneous anchor attachment increases retention of radio transmitters on Xantus' and marbled murrelets. Journal of Field Ornithology 70:520–535.

Nicholls, T. H., and D. W. Warner. 1972. Barred owl habitat use as determined by radiotelemetry. Journal of Wildlife Management 36:213–224.

North, M. P., and J. H. Reynolds. 1996. Microhabitat analysis using radiotelemetry locations and polytomous logistic regression. Journal of Wildlife Management 60:639–653.

Nygård, T., R. E. Kenward, and K. Einvik. 2000. Radio telemetry studies of dispersal and survival in juvenile White-tailed Sea Eagles in Norway. Proceedings of the Fifth World Conference on Birds of Prey and Owls. World Working Group on Birds of Prey, Berlin, Germany.

Obbard, M. E., B. A. Pond, and A. Perera. 1998. Preliminary evaluation of GPS collars for analysis of habitat use and activity patterns of black bears. Ursus 10:209–217.

Ogata, Y., and K. Katsura. 1991. Maximum likelihood estimates of the fractal dimension for random spatial patterns. Biometrika 78:463–474.

Opdam, P. 1991. Metapopulation theory and habitat fragmentation: a review of holarctic breeding bird studies. Landscape Ecology 5:93–106.

Ormsbee, P. C., and W. C. McComb. 1998. Selection of day roosts by female long-legged myotis in the central Oregon Cascade Range. Journal of Wildlife Management 62:596–603.

Orthmeyer, D. L., and I. J. Ball. 1990. Survival of mallard broods on Benton Lake National Wildlife Refuge in northcentral Montana. Journal of Wildlife Management 54:62–66.

Ostro, L. E. T., T. P. Young, S. C. Silver, and F. W. Koontz. 1999. A geographic information system method for estimating home range size. Journal of Wildlife Management 63:748–755.

Otis, D. L., and G. C. White. 1999. Autocorrelation of location estimates and the analysis of radiotracking data. Journal of Wildlife Management 63:1039–1044.

Otis, D. L., K. P. Burnham, G. C. White, and D. R. Anderson. 1978. Statistical inference from capture data on closed animal populations. Wildlife Monograph 62:1–135.

Otten, M. R., J. B. Haufler, S. R. Winterstein, and L. C. Bender. 1993. An aerial censusing procedure for elk in Michigan. Wildlife Society Bulletin 21:73–80.

Owen-Smith, N. 1994. Foraging responses of kudus to seasonal changes in food resources: elasticity in constraints? Ecology 75:1050–1062.

Oyler-McCance, S. J. 1999. Genetic and habitat factors underlying conservation strategies for Gunnison sage grouse. Ph.D. Dissertation. Colorado State University, Fort Collins, Colorado, USA.

Pace, R. M. III. 1988. Measurement error models for common wildlife radio-tracking systems. Minnesota Wildlife Reports 5:1–19.

Pace, R. M. III. 2000a. Radio tracking via triangulation: A simple moving window estimator for describing movement paths. Fifteenth International Symposium on Biotelemetry. Juneau, Alaska, USA.

Pace, R. M. III. 2000b. Radio tracking via triangulation: Are small sample size von Mises maximum likelihood estimates normal? Fifteenth International Symposium on Biotelemetry. Juneau, Alaska, USA.

Pace, R. M. III, and H. P. Weeks, Jr. 1990. A nonlinear weighted least-squares estimator for radiotracking via triangulation. Journal of Wildlife Management 54:304–310.

Pack, J. C., G. W. Norman, C. I. Taylor, D. E. Steffen, D. A. Swanson, K. H. Pollock, and R. Alpizar-Jara. 1999. Effects of fall hunting on wild turkey populations in Virginia and West Virginia. Journal of Wildlife Management 63:964–975.

Palomares, F. 1994. Site fidelity and effects of body mass on home-range size of Egyptian mongooses. Canadian Journal of Zoology 72:465–469.

Palsbøll, P. J., J. Allen, M. Bérube, P. J. Clapham, T. P. Feddersen, P. S. Hammond, R. R. Hudson, H. Jorgensen, S. Katona, A. H. Larsen, F. Larsen, J. Lien, D. K. Mattila, J. Sigurjonsson, R. Sears, T. Smith, R. Sponer, P. Stevick, and N. Oien. 1997. Genetic tagging of humpback whales. Nature 388:767–769.

Panico, J., and P. Sterling. 1995. Retinal neurons and vessels are not fractal but space-filling. Journal of Comparative Neurology 361:479–490.

Paquette, G. A., J. H. Devries, R. B. Emery, D. W. Howerter, B. L. Joynt, and T. P. Sankowski. 1997. Effects of transmitters on reproduction and survival of wild mallards. Journal of Wildlife Management 61:953–961.

Park, B. U., and J. S. Marron. 1990. Comparison of data-driven bandwidth selectors. Journal of the American Statistical Association 85:66–72.

Parker, N., A. Pascoe, and H. Moller. 1996. Inaccuracy of a radio-tracking system for small mammals: the effect of electromagnetic interference. Journal of Zoology 239:401–406.

Partridge, L. 1978. Habitat selection. Pages 351–376 in J. R. Krebs and N. B. Davies, editors. Behavioral ecology: an evolutionary approach. Blackwell Scientific Publications, Oxford, England.

Paton, P. W. C., C. J. Zabel, D. L. Neal, G. N. Steger, N. G. Tilghman, and B. R. Noon. 1991. Effects of radio tags on spotted owls. Journal of Wildlife Management 55:617–622.

Patric, E. F., and R. W. Serenbetz. 1971. A new approach to wildlife position finding telemetry. New York Fish and Game Journal 18:1–14.

Patterson, R. L. 1952. The sage grouse in Wyoming. Sage Books, Denver, Colorado, USA.

Patton, D. R., D. W. Beaty, and R. H. Smith. 1973. Solar panels: an energy source for radio transmitters on wildlife. Journal of Wildlife Management 37:236–238.

Pearson, K. 1901. On lines and planes of closest fit to systems of points in space. Philosophical Magazine 2:559–572.

Pearson, S. M. 1993. The spatial extent and relative influence of landscape-level factors on wintering bird populations. Landscape Ecology 8:3–18.

Pedlar, J., L. Fahrig, and H. G. Merriam. 1997. Raccoon habitat use at 2 spatial scales. Journal of Wildlife Management 61:102–112.

Pekins, P. J. 1988. Effects of poncho-mounted radios on blue grouse. Journal of Field Ornithology 59:46–50.

Pendleton, G. W., K. Titus, E. Degayner, G. J. Flatten, and R E. Lowell. 1998. Compositional analysis and GIS for study of habitat selection by goshawks in southeast Alaska. Journal of Agricultural, Biological, and Environmental Statistics 3:280–295.

Perry, M. C. 1981. Abnormal behavior of canvasbacks equipped with radio transmitters. Journal of Wildlife Management 45:786–789.

Persson, L., and L. A. Greenberg. 1990. Optimal foraging and habitat shift in perch (*Perca fluviatilis*) in a resource gradient. Ecology 71:1699–1713.

Petersen, C. G. J. 1896. The yearly immigration of young plaice into the Limfjord from the German Sea. Report of the Danish Biological Station 6:1–48.

Peterson, D. L., and V. T. Parker, editors. 1998. Ecological scale: theory and applications. Columbia University Press, New York, New York, USA.

Petraitis, P. S. 1979. Likelihood measures of niche breadth and overlap. Ecology 60:703–710.

Petraitis, P. S. 1981. Algebraic and graphical relationships among niche breadth measures. Ecology 62:545–548.

Petrie, S. A., and K. H. Rogers. 1997. Satellite tracking of white-faced whistling ducks in a semiarid region of South Africa. Journal of Wildlife Management 61:1208–1213.

Pettifor, R. A., R. W. G. Caldow, J. M. Rowcliffe, J. D. Goss-Custard, J. M. Black, K. H. Hodder, A. I. Houston, A. Lang, and J. Webb. In press. Spatially explicit, individually-based, behavioural models of the annual cycle of two migratory geese populations – model development, theoretical insights and applications. Journal of Animal Ecology.

Phillips, D. M., D. J. Harrison, and D. C. Payer. 1998. Seasonal changes in home-range area and fidelity of martens. Journal of Mammalogy 79:180–190.

Picozzi, N. 1975. Crow predation on marked nests. Journal of Wildlife Management 39:151–155.

Pielou, E. C. 1984. The interpretation of ecological data: a primer on classification and ordination. John Wiley & Sons, New York, New York, USA.

Pietz, P. J., G. L. Krapu, R. J. Greenwood, and J. T. Lokemoen. 1993. Effects of harness transmitters on behavior and reproduction of wild mallards. Journal of Wildlife Management 57:696–703.

Pietz, P. J., D. A. Brandt, G. L. Krapu, and D. A. Buhl. 1995. Modified transmitter attachment method for adult ducks. Journal of Field Ornithology 66:408–417.

Pollock, K. H., S. R. Winterstein, and M. H. Conroy. 1989a. Estimation and analysis of survival distributions for radiotagged animals. Biometrics 45:99–109.

Pollock, K. H., S. R. Winterstein, C. M. Bunck, and P. D. Curtis. 1989b. Survival analysis in telemetry studies: the staggered entry design. Journal of Wildlife Management 53:7–15.

Pollock, K. H., J. D. Nichols, J. E. Hines, and C. Brownie. 1990. Statistical inference for capture-recapture experiments. Wildlife Monograph 107:1–97.

Pollock, K. H., C. M. Bunck, S. R. Winterstein, and C. Chen. 1995. A capture-recapture survival analysis model for radio-tagged animals. Journal of Applied Statistics 22:661–672.

Porter, J. H., and J. L. Dooley. 1993. Animal dispersal patterns: a reassessment of simple mathematical models. Ecology 74:2436–2443.

Porter, W. F., and K. E. Church. 1987. Effects of environmental pattern on habitat preference analysis. Journal of Wildlife Management 51:681–685.

Powell, L. A., D. G. Krementz, J. D. Lang, and M. J. Conroy. 1998. Effects of radio transmitters on migrating wood thrushes. Journal of Field Ornithology 69:306–315.

Powell, L. A., J. D. Lang, M. J. Conroy, and D. G. Krementz. 2000a. Effects of forest management on density, survival, and population growth of wood thrushes. Journal of Wildlife Management 64:11–23.

Powell, L. A., M. J. Conroy, J. E. Hines, J. D. Nichols, and D. G. Krementz. 2000b. Simultaneous use of mark-recapture and radiotelemetry to estimate survival, movement, and capture rates. Journal of Wildlife Management 64:302–313.

Pradel, R., J. E. Hines, J. D. Lebreton, and J. D. Nichols. 1997. Capture-recapture survival models taking account of transients. Biometrics 53:60–72.

Priede, I. G. 1992. Wildlife telemetry: an introduction. Pages 3–24 in I. G. Priede and S. M. Swift, editors. Wildlife telemetry: remote monitoring and tracking of animals. Ellis Horwood, West Sussex, United Kingdom.

Priede, I. G., and J. French. 1991. Tracking of marine animals by satellite. International Journal of Remote Sensing 12:667–680.

Priede, I. G., and S. M. Swift, editors. 1992. Wildlife telemetry: remote monitoring and tracking of animals. Ellis Horwood, West Sussex, United Kingdom.

Pulliam, H. R. 1988. Sources, sinks, and population regulation. American Naturalist 132:652–661.

Pulliam, H. R., and B. J. Danielson. 1991. Sources, sinks, and habitat selection: a landscape perspective on population dynamics. American Naturalist 137:55–66.

Putaala A., J. Oksa, H. Rintamaki, and R. Hissa. 1997. Effects of hand-rearing and radiotransmitters on flight of gray partridge. Journal of Wildlife Management 61:1345–1351.

Pyke, D. A., and J. N. Thompson. 1986. Statistical analysis of survival and removal rate experiments. Ecology 67:240–245.

Pyke, G. H., and P. J. O'Connor. 1990. The accuracy of a radiotracking system for monitoring honeyeater movements. Australian Wildlife Research 17:501–509.

Quade, D. 1979. Using weighted rankings in the analysis of complete blocks with additive block effects. Journal of the American Statistical Association 74:680–683.

Ramakka, J. M. 1972. Effects of radio-tagging on breeding behavior of male woodcock. Journal of Wildlife Management 36:1309–1312.

Rasmussen, D. R. 1980. Clumping and consistency in primates' patterns of range use: definitions, sampling assessment and application. Folia Primatologica 34:111–139.

Ratti, J. T., and E. O. Garton. 1994. Research and experimental design. Pages 1–23 in T. A. Bookhout, editor. Research and management techniques for wildlife and habitats. Fifth edition. The Wildlife Society, Bethesda, Maryland, USA.

Reed, J. Z., D. J. Tollit, P. M. Thompson, and W. Amos. 1997. Molecular scatology: the use of molecular genetic analysis to assign species, sex and individual identity to seal faeces. Molecular Ecology 6:225–234.

Reid, D. G., W. E. Melquist, J. D. Woolington, and J. M. Noll. 1986. Reproductive effects of intraperitoneal transmitter implants in river otters. Journal of Wildlife Management 50:92–94.

Relyea, R. A., I. M. Ortega, and S. Demarais. 1994. Activity monitoring in mule deer: assessing telemetry accuracy. Wildlife Society Bulletin 22:656–661.

Rempel, R. S., and A. R. Rodgers. 1997. Effects of differential correction on accuracy of a GPS animal location system. Journal of Wildlife Management 61:525–530.

Rempel, R. S., A. R. Rodgers, and K. F. Abraham. 1995. Performance of a GPS animal location system under boreal forest canopy. Journal of Wildlife Management 59:543–551.

Renyi, A. 1970. Probability theory. North-Holland, Amsterdam, Holland.

Rettie, W. J., and P. D. McLoughlin. 1999. Overcoming radiotelemetry bias in habitat-selection studies. Canadian Journal of Zoology 77:1175–1184.

Rexstad, E., and K. Burnham. 1991. Users' guide for interactive program CAPTURE. Colorado Cooperative Fish and Wildlife Research Unit, Colorado State University, Fort Collins, Colorado, USA.

Reynolds, P. S. 1992. White blood cell profiles as a means of evaluating transmitter-implant surgery in small mammals. Journal of Mammalogy 73:178–185.

Reynolds, T. D., and J. W. Laundré. 1990. Time intervals for estimating pronghorn and coyote home ranges and daily movements. Journal of Wildlife Management 54:316–322.

Rice, W. R., and J. D. Harder. 1977. Application of multiple aerial sampling to a mark-recapture census of white-tailed deer. Journal of Wildlife Management 41:197–206.

Ricklefs, R. E., and G. L. Miller. 2000. Ecology. Fourth edition. W. H. Freeman, New York, New York, USA.

Ries, E. H., L. R. Hiby, and P. J. H. Reijnders. 1998. Maximum likelihood population size estimation of harbour seals in the Dutch Wadden Sea based on a mark-recapture experiment. Journal of Applied Ecology 35:332–339.

Riggs, M. R., and K. H. Pollock. 1992. A risk ratio approach to multivariable analysis of survival in longitudinal studies of wildlife populations. Pages 74–89 in D. R. McCullough and R. H. Barrett, editors. Wildlife 2001: populations. Elsevier Science Publishers, London, England.

Ripley, B. 1977. Modelling spatial patterns. Journal of the Royal Statistical Society, Series B 39:172–212.

Ritchie, M. E. 1998. Scale-dependent foraging and patch choice in fractal environments. Evolutionary Ecology 12:309–330.

Rivest, L. P., S. Couturier, and H. Crépeau. 1998. Statistical methods for estimating caribou abundance using postcalving aggregations detected by radio telemetry. Biometrics 54:865–876.

Robertson, P. A., N. J. Aebischer, R. E. Kenward, I. K. Hanski, and N. P. Williams. 1998. Simulation and jack-knifing assessment of home-range indices based on underlying trajectories. Journal of Applied Ecology 35:928–940.

Rodgers, A. R., and A. P. Carr. 1998. HRE: the home range extension for ArcView. Centre for Northern Forest Ecosystem Research, Ontario Ministry of Natural Resources, Thunder Bay, Ontario, Canada.

Rodgers, A. R., R. S. Rempel, and K. F. Abraham. 1995. Field trials of a new GPS-based telemetry system. Pages 173–178 in C. Cristalli, C. J. Amlaner, Jr. and M. R. Neuman, editors. Biotelemetry XIII. Proceedings of the Thirteenth International Symposium on Biotelemetry. Williamsburg, Virginia, USA.

Rodgers, A. R., R. S. Rempel, and K. F. Abraham. 1996. A GPS-based telemetry system. Wildlife Society Bulletin 24:559–566.

Rodgers, A. R., R. S. Rempel, R. Moen, J. Paczkowski, C. Schwartz, E. J. Lawson, and M. J. Gluck. 1997. GPS collars for moose telemetry studies: a workshop. Alces 33:203–209.

Rodgers, A. R., S. M. Tomkiewicz, E. J. Lawson, T. R. Stephenson, K. J. Hundertmark, P. J. Wilson, B. N. White, and R. S. Rempel. 1998. New technology for moose management: a workshop. Alces 34:239–244.

Roeder, K., B. Dennis, and E. O. Garton. 1987. Estimating density from variable circular plot censuses. Journal of Wildlife Management 51:224–230.

Rogers, K. R., and C. L. Gerlach. 1996. Environmental biosensors, a status report. Environmental Science and Technology 30:486–491.

Romesburg, H. C. 1985. Exploring, confirming, and randomization tests. Computers & Geosciences 11:19–27.

Rotella, J. J., and J. T. Ratti. 1992. Mallard brood movements and wetland selection in southwestern Manitoba. Journal of Wildlife Management 56:508–515.

Rotella, J. J., D. W. Howerter, T. P. Sankowski, and J. H. Devries. 1993. Nesting effort by wild mallards with 3 types of radio transmitters. Journal of Wildlife Management 57:690–695.

Rotenberry, J. T. 1986. Habitat relationships of shrubsteppe birds: even good models cannot predict the future. Pages 217–221 in J. Verner, M. L. Morrison, and C. J. Ralph, editors. Wildlife 2000: modeling habitat relationships of terrestrial vertebrates. University of Wisconsin Press, Madison, Wisconsin, USA.

Rotenberry, J. T., and J. A. Wiens. 1980. Habitat structure, patchiness, and avian communities in North American steppe vegetation: A multivariate analysis. Ecology 61:1228–1250.

Rotenberry, J. T., and J. A. Wiens. 1998. Foraging patch selection by shrubstreppe sparrows. Ecology 79:1160–1173.

Rotenberry, J. T., S. T. Knick, and J. E. Dunn. In press. A minimalist approach to mapping species habitat: Pearson's planes of closest fit. In J. M Scott, P. J. Heglund, M. Morrison,

M. Raphael, J. Haufler, and B. Wall, editors. Predicting species occurrences: issues of scale and accuracy. Island Press, Covella, California, USA.

Rowland, M. M., L. D. Bryant, B. K. Johnson, J. H. Noyes, M. J. Wisdom, and J. W. Thomas. 1997. The Starkey Project: history, facilities, and data collection methods for ungulate research. USDA Forest Service General Technical Report PNW-GTR-396.

Rowland, M. M., P. K. Coe, R. J. Stussy, A. A. Ager, N. J. Cimon, B. K. Johnson, and M. J. Wisdom. 1998. The Starkey habitat data base for ungulate research: construction, documentation, and use. USDA Forest Service General Technical Report PNW-GTR-430.

Rowland, M. M., M. J. Wisdom, B. K. Johnson, and J. G. Kie. 2000. Elk distribution and modeling in relation to roads. Journal of Wildlife Management 64:672–684.

Roy, L. D., and M. J. Dorrance. 1985. Coyote movements, habitat use, and vulnerability in central Alberta. Journal of Wildlife Management 49:307–312.

Ruggiero, L. F., D. E. Pearson, and S. E. Henry. 1998. Characteristics of American marten den sites in Wyoming. Journal of Wildlife Management 62:663–673.

Rumble, M. A., and F. Lindzey. 1997. Effects of forest vegetation and topography on global positioning system collars for elk. Resource Technology Institute Symposium 4:492–501.

Sain, S. R., K. A. Baggerly, and D. W. Scott. 1994. Cross-validation of multivariate densities. Journal of the American Statistical Association 89:807–817.

Saltz, D. 1994. Reporting error measures in radio location by triangulation: a review. Journal of Wildlife Management 58:181–184.

Saltz, D., and P. U. Alkon. 1985. A simple computer-aided method for estimating radio-location error. Journal of Wildlife Management 49:664–668.

Saltz, D., and G. C. White. 1990. Comparison of different measures of the error in simulated radio-telemetry locations. Journal of Wildlife Management 54:169–174.

Samietz, J., and U. Berger. 1997. Evaluation of movement parameters in insects – bias and robustness with regard to resight numbers. Oecologia 110:40–49.

Samuel, M. D. 1988. Wildlife software: new feature for the Wildlife Society Bulletin. Wildlife Society Bulletin 16:104.

Samuel, M. D., and M. R. Fuller. 1996. Wildlife radiotelemetry. Pages 370–418 in T. A. Bookhout, editor. Research and management techniques for wildlife and habitats. Fifth edition. The Wildlife Society, Bethesda, Maryland, USA.

Samuel, M. D., and E. O. Garton. 1985. Home range: a weighted normal estimate and tests of underlying assumptions. Journal of Wildlife Management 49:513–519.

Samuel, M. D., and E. O. Garton. 1987. Incorporating activity time in harmonic home range analysis. Journal of Wildlife Management 51:254–257.

Samuel, M. D., and R. E. Green. 1988. A revised test procedure for identifying core areas within the home range. Journal of Animal Ecology 57:1067–1068.

Samuel, M. D., and K. P. Kenow. 1992. Evaluating habitat selection with radio-telemetry triangulation error. Journal of Wildlife Management 56:725–734.

Samuel, M. D., D. J. Pierce, and E. O. Garton. 1985. Identifying areas of concentrated use within the home range. Journal of Animal Ecology 54:711–719.

Samuel, M. D., E. O. Garton, M. W. Schlegel, and R. G. Carson. 1987. Visibility bias during aerial surveys of elk in northcentral Idaho. Journal of Wildlife Management 51:622–630.

Samuel, M. D., R. K. Steinhorst, E. O. Garton, and J. W. Unsworth. 1992. Estimation of wildlife population ratios incorporating survey design and visibility bias. Journal of Wildlife Management 56:718–725.

Sargeant, A. B. 1980. Approaches, field considerations and problems associated with radio tracking carnivores. Pages 57–63 in C. J. Amlaner, Jr. and D. W. Macdonald, editors. A handbook on biotelemetry and radio tracking. Pergamon Press, Oxford, England.

SAS Institute. 1999. SAS/STAT software, version 8.0. SAS Institute, Cary, North Carolina, USA.

Sauer, J. R., and J. E. Hines. 1989. Testing for differences among survival or recovery rates using program CONTRAST. Wildlife Society Bulletin 17:549–550.

Sayre, M. W., T. S. Baskett, and P. B. Blenden. 1981. Effects of radio-tagging on breeding behavior of mourning doves. Journal of Wildlife Management 45: 428–434.

Scheaffer, R. L., W. Mendenhall, and L. Ott. 1986. Elementary survey sampling. Third edition. Duxbury Press, Boston, Massachusetts, USA.

Scheuring, I., and R. H. Riedi. 1994. Application of multifractals to the analysis of vegetation pattern. Journal of Vegetation Science 5:489–496.

Schmutz, J. A., and J. A. Morse. 2000. Effects of neck collars and radiotransmitters on survival and reproduction of emperor geese. Journal of Wildlife Management 64:231–237.

Schmutz, J. A., and G. C. White. 1990. Error in telemetry studies: effects of animal movement on triangulation. Journal of Wildlife Management 54:506–510.

Schnabel, Z. E. 1938. Estimation of the size of animal populations by marking experiments. Bulletin of the U. S. Bureau of Fisheries 69:191–203.

Schoen, J. W., and M. D. Kirchhoff. 1985. Seasonal distribution and home-range patterns of Sitka black-tailed deer on Admiralty Island, southeast Alaska. Journal of Wildlife Management 49:96–103.

Schoener, T. W. 1981. An empirically based estimate of home range. Theoretical Population Biology 20:281–325.

Schooley, R. L. 1994. Annual variation in habitat selection: patterns concealed by pooled data. Journal of Wildlife Management 58:367–374.

Schooley, R. L., B. Van Horne, and K. P. Burnham. 1993. Passive integrated transponders for marking free-ranging Townsend's ground squirrels. Journal of Mammalogy 74:480–484.

Schroeder, M. 1991. Fractals, chaos, power laws. Minutes from an infinite paradise. W. H. Freeman, San Francisco, California, USA.

Schulz, J. H., A. J. Bermudez, J. L. Tomlinson, J. D. Firman, and Z. He. 1998. Effects of implanted radiotransmitters on captive mourning doves. Journal of Wildlife Management 62:1451–1460.

Schwartz, C. C., and S. M. Arthur. 1999. Radio-tracking large wilderness mammals: integration of GPS and Argos technology. Ursus 11:261–274.

Scott, A. 1920. Food of Port Erin mackerel in 1919. Proceedings and Transactions of the Liverpool Biological Society 34:107–111.

Scott, D. W. 1992. Multivariate density estimation: theory, practice, and visualization. John Wiley & Sons, New York, New York, USA.

Seaman, D. E., and R. A. Powell. 1996. An evaluation of the accuracy of kernel density estimators for home range analysis. Ecology 77:2075–2085.

Seaman, D. E., B. Griffith, and R. A. Powell. 1998. KERNELHR: a program for estimating animal home ranges. Wildlife Society Bulletin 26:95–100.

Seaman, D. E., J. J. Millspaugh, B. J. Kernohan, G. C. Brundige, K. J. Raedeke, and R. A. Gitzen. 1999. Effects of sample size on kernel home range estimates. Journal of Wildlife Management 63:739–747.

Seber, G. A. F. 1982. The estimation of animal abundance. Second edition. Macmillan, New York, New York, USA.

Seber, G. A. F. 1986. A review of estimating animal abundance. Biometrics 42:267–292.

Sedinger, J. S., R. G. White, and W. G. Hauer. 1990. Effects of carrying radio transmitters on energy expenditure of Pacific black brant. Journal of Wildlife Management 54:42–45.

Seegar, W. S., P. N. Cutchis, M. R. Fuller, J. J. Suter, V. Bhatnagar, and J. S. Wall. 1996. Fifteen years of satellite tracking development and application to wildlife research and conservation. John Hopkins Advanced Physics Laboratory Technical Digest 17:305–315.

Seidel, K. S. 1992. Statistical properties and applications of a new measure of joint space use for wildlife. M.S. Thesis. University of Washington, Seattle, Washington, USA.

Shao, J., and D. Tu. 1995. The jackknife and bootstrap. Springer-Verlag, New York, New York, USA.

Shugart, H. H. 1998. Terrestrial ecosystems in changing environments. Cambridge University Press, Cambridge, United Kingdom.

Silverman, B. W. 1986. Density estimation for statistics and data analysis. Chapman & Hall, London, England.

Siniff, D. B., and J. R. Tester. 1965. Computer analysis of animal movement data obtained by telemetry. BioScience 15:104–108.

Skellam, J. G. 1951. Random dispersal in theoretical populations. Biometrika 38:196–218.

Slade, N. A., J. J. Cebula, and R. J. Robel. 1965. Accuracy and reliability of biotelemetric instruments used in animal movement studies in prairie grasslands of Kansas. Transactions of the Kansas Academy of Science 68:173–179.

Slauson, W. L., B. S. Cade, and J. D. Richards. 1994. User's manual for BLOSSOM statistical software. Midcontinent Ecological Science Center, U.S. Geological Survey, Fort Collins, Colorado, USA.

Small, R. J., and D. H. Rusch. 1985. Backpacks vs. ponchos: survival and movements of radio-marked ruffed grouse. Wildlife Society Bulletin 13:163–165.

Smith, C. C., and S. D. Fretwell. 1974. The optimal balance between size and number of offspring. The American Naturalist 108:499–506.

Smith, D. R. 1996. Improved access to software published in the Wildlife Society Bulletin. Wildlife Society Bulletin 24:339.

Smith, D. R., K. J. Reinecke, M. J. Conroy, M. W. Brown, and J. R. Nassar. 1995. Factors affecting visibility rate of waterfowl surveys in the Mississippi Alluvial valley. Journal of Wildlife Management 59:515–527.

Smith, G. J., J. R. Cary, and O. J. Rongstad. 1981. Sampling strategies for radio-tracking coyotes. Wildlife Society Bulletin 9:88–93.

Smith, I. P., and G. W. Smith. 1997. Tidal and diel timing of river entry by adult Atlantic salmon returning to the Aberdeenshire Dee, Scotland. Journal of Fish Biology 50:463–474.

Smith, L. M., J. W. Hupp, and J. T. Ratti. 1982. Habitat use and home range of gray partridge in Eastern South Dakota. Journal of Wildlife Management 48:580–587.

Snyder, N. F. R., S. R. Beissinger, and M. R. Fuller. 1989. Solar radio-transmitters on snail kites in Florida. Journal of Field Ornithology 60:171–177.

Sodhi, N. S., I. G. Warkentin, P. C. James, and L. W. Oliphant. 1991. Effects of radio-tagging on breeding merlins. Journal of Wildlife Management 55:613–616.

Sokal, R. R., and F. J. Rohlf. 1995. Biometry: the principles and practice of statistics in biological research. Third edition. W. H. Freeman, New York, New York, USA.

Solomon, D. J., and C. E. Potter. 1988. First results with a new estuarine fish tracking system. Journal of Fish Biology 33:127–132.

Solow, A. R. 1989. A randomization test for independence of animal locations. Ecology 70:1546–1549.

Sorenson, M. D. 1989. Effects of neck collar radios on female redheads. Journal of Field Ornithology 60:523–528.

Southern, W. E. 1964. Additional observations on winter bald eagle populations: including remarks on biotelemetry techniques and immature plumages. Wilson Bulletin 76:222–237.

Spencer, H. J., and F. P. Savaglio. 1995. An automatic small-animal radio-tracking system employing spread-spectrum concepts. Pages 185–191 in C. Cristalli, C. J. Amlaner, Jr. and M. R. Neuman, editors. Biotelemetry XIII. Proceedings of the Thirteenth International Symposium on Biotelemetry. Williamsburg, Virginia, USA.

Springer, J. T. 1979. Some sources of bias and sampling error in radio triangulation. Journal of Wildlife Management 43:926–935.

Stanley, H. E., and P. Meakin. 1988. Multifractal phenomena in physics and chemistry. Nature 335:405–409.

Stapp, P., and V. Van Horne. 1997. Response of deer mice (Peromyscus maniculatus) to shrubs in shortgrass prairie: linking small-scale movements and the spatial distribution of individuals. Functional Ecology 11:644–651.

Steidl, R. J., J. P. Hayes, and E. Schauber. 1997. Statistical power analysis in wildlife research. Journal of Wildlife Management 61:270–279.

Steinhorst, R. K., and M. D. Samuel. 1989. Sightability adjustment methods for aerial surveys of wildlife populations. Biometrics 45:415–425.

Stewart, B. S., and R. L. Delong. 1991. Diving patterns of northern elephant seal bulls. Marine Mammal Science 7:369–384.

Stickel, L. F. 1954. A comparison of certain methods of measuring ranges of small mammals. Journal of Mammalogy 35:1–15.

Storch, I. 1997. The importance of scale in habitat conservation for an endangered species: the Capercaillie in central Europe. Pages 310–330 in J. A. Bissonette, editor. Wildlife and landscape ecology: effects of pattern and scale. Springer, New York, New York, USA.

Stüwe, M., and C. E. Blohowiak. 1985. McPAAL, microcomputer programs for the analysis of animal locations. Conservation and Research Center, National Zoological Park, Smithsonian Institute, Front Royal, Virginia, USA.

Sutherland, W. J. 1996. From individual behaviour to population ecology. Oxford University Press, Oxford, England.

Swenson, G. W., Jr., and W. W. Cochran. 1973. Radio noise from towns: measured from an airplane. Science 181:543–545.

Swenson, J. E., K. Wallin, G. Ericsson, G. Cederlund, and F. Sandegren. 1999. Effects of ear-tagging with radiotransmitters on survival of moose calves. Journal of Wildlife Management 63:354–358.

Swihart, R. K., and N. A. Slade. 1985a. Testing for independence of observations in animal movements. Ecology 66:1176–1184.

Swihart, R. K., and N. A. Slade. 1985b. Influence of sampling interval on estimates of home range size. Journal of Wildlife Management 49:1019–1025.

Swihart, R. K., and N. A. Slade. 1986. The importance of statistical power when testing for independence in animal movements. Ecology 67:255–258.

Swihart, R. K., and N. A. Slade. 1997. On testing for independence of animal movements. Journal of Agricultural, Biological, and Environmental Statistics 2:48–63.

Swihart, R. K., N. A. Slade, and B. J. Bergstrom. 1988. Relating body size to the rate of home range use in mammals. Ecology 69:393–399.

Swingland, I. R., and J. G. Frazier. 1980. The conflict between feeding and overheating in the Aldabran giant tortoise. Pages 611–615 in C. J. Amlaner, Jr. and D. W. Macdonald, editors. A handbook on biotelemetry and radio tracking. Pergamon Press, Oxford, England.

Sykes, P. W., Jr., J. W. Carpenter, S. Holzman, and P. H. Geissler. 1990. Evaluation of three miniature radio transmitter attachment methods for small passerines. Wildlife Society Bulletin 18:41–48.

Tacha, T. C., W. D. Warde, and K. P. Burnham. 1982. Use and interpretation of statistics in wildlife journals. Wildlife Society Bulletin 10:353–362.

Taillade, M. 1992. Animal tracking by satellite. Pages 149–160 in I. G. Priede and S. M. Swift, editors. Wildlife telemetry: remote monitoring and tracking of animals. Ellis Horwood, West Sussex, United Kingdom.

Takeuchi, E. S. 1995. Developments in battery technology. Pages 49–55 in C. Cristalli, C. J. Amlaner, Jr, and M. R. Neuman, editors. Biotelemetry XIII. Proceedings of the Thirteenth International Symposium on Biotelemetry. Williamsburg, Virginia, USA.

Tanner, R. W., K. M. Kilbride, J. A. Crawford, B. A. Williams, and A. J. Kimerling. 1992. Automated mapping of radiotelemetric locations for wildlife studies. Wildlife Society Bulletin 20:232–233.

Taylor, I. R. 1991. Effects of nest inspections and radiotagging on barn owl breeding success. Journal of Wildlife Management 55:312–315.

Taylor, K. D., and H. G. Lloyd. 1978. The design, construction and use of a radio-tracking system for some British mammals. Mammal Review 8:117–141.

Temple, S. A., and J. R. Cary. 1988. Modeling dynamics of habitat-interior bird populations in fragmented landscapes. Conservation Biology 2:340–347.

ter Braak, C. J. F. 1995. Calibration. Pages 78–90 in R. H. G. Jongman, C. J. F. ter Braak, and O. F. R. van Tongeren, editors. Data analysis in community and landscape ecology. Second edition. Cambridge University Press, Cambridge, United Kingdom.

ter Braak, C. J. F., and C. W. N. Looman. 1995. Regression. Pages 29–77 in R. H. G. Jongman, C. J. F. ter Braak, and O. F. R. van Tongeren, editors. Data analysis in community and landscape ecology. Second edition. Cambridge University Press, Cambridge, United Kingdom.

Thirgood, S. J., and S. M. Redpath. 1997. Red grouse and their predators. Nature 390:547.

Thomas, D. L., and E. J. Taylor. 1990. Study designs and tests for comparing resource use and availability. Journal of Wildlife Management 54:322–330.

Thomasma, L. E., T. D. Drummer, and R. O. Peterson. 1991. Testing the Habitat Suitability Index for the fisher. Wildlife Society Bulletin 19:291–297.

Thome, D. M., C. J. Zabel, and L. V. Diller. 1999. Forest stand characteristics and reproduction of Northern Spotted Owls in managed north-coastal California forests. Journal of Wildlife Management 63:44–59.

Thompson, W. L., G. C. White, and C. Gowan. 1998. Monitoring vertebrate populations. Academic Press, New York, New York, USA.

Trent, T. T., and O. J. Rongstad. 1974. Home range and survival of cottontail rabbits in southwestern Wisconsin. Journal of Wildlife Management 38:459–472.

Trexler, J. C., and J. Travis. 1993. Nontraditional regression analysis. Ecology 74:1629–1637.

Tsai, K. 1996. Survival analysis for telemetry data in animal ecology. Ph.D. Dissertation. North Carolina State University, Raleigh, North Carolina, USA.

Tsai, K., K. H. Pollock, and C. Brownie. 1999a. Effects of violation of assumptions for survival analysis methods in radiotelemetry studies. Journal of Wildlife Management 63:1369–1375.

Tsai, K., K. H. Pollock, C. Brownie, and D. Nychka. 1999b. Smoothing hazard functions for telemetry data in animal studies. Bird Study 46(Supplement):47–54.

Tufto, J., R. Andersen, and J. Linnell. 1996. Habitat use and ecological correlates of home range size in a small cervid: the roe deer. Journal of Animal Ecology 65:715–724.

Turchin, P. 1991. Translating foraging movements in heterogeneous environments into the spatial distribution of foragers. Ecology 72:1253–1266.

Turchin, P. 1996. Fractal analyses of animal movement: a critique. Ecology 77:2086–2090.

Turchin, P. 1998. Quantitative analysis of animal movement: measuring and modeling population redistribution in animals and plants. Sinauer Associates, Sunderland, Massachusetts, USA.

Turcotte, D. L. 1997. Fractals and chaos in geology and geophysics. Second edition. Cambridge University Press, Cambridge, United Kingdom.

Turner, M. G., and R. H. Gardner, editors. 1991. Quantitative methods in landscape ecology. Springer-Verlag, New York, New York, USA.

Turner, M. G., G. J. Arthaud, R. T. Engstrom, S. J. Hejl, J. Liu, S. Loeb, and K. McKelvey. 1995. Usefulness of spatially explicit population models in land management. Ecological Applications 5:12–16.

Ueta, M., F. Sato, E. G. Lobkov, and N. Mita. 1998. Migration route of white-tailed sea eagles *Haliaeetus albicilla* in northeastern Asia. Ibis 140:684–686.

Unsworth, J. W., L. Kuck, and E. O. Garton. 1990. Elk sightability model validation at the National Bison Range, Montana. Wildlife Society Bulletin 18:113–115.

Unsworth, J. W., D. F. Pac, G. C. White, and R. M. Bartmann. 1999. Mule deer survival in Colorado, Idaho, and Montana. Journal of Wildlife Management 63:315–326.

Upton, G., and B. Fingleton. 1985. Spatial data analysis by example. Volume 1. Point patterns and quantitative data. John Wiley & Sons, New York, New York, USA.

Urban, D. L., and H. H. Shugart, Jr. 1984. Avian demography in mosaic landscapes: modeling paradigm and preliminary results. Pages 273–280 *in* J. Verner, M. L. Morrison, and C. J. Ralph, editors. Wildlife 2000: modeling habitat relationships of terrestrial vertebrates. University of Wisconsin Press, Madison, Wisconsin, USA.

U.S. Department of the Interior. 1996. Effects of military training and fire in the Snake River Birds of Prey National Conservation Area. BLM/IDARNG Research Project Final Report. U.S. Geological Survery, Biological Resources Division, Snake River Field Station, Boise, Idaho, USA.

Van Emon, J. M., and V. Lopez-Avila. 1992. Immunochemical methods for environmental analysis. Analytical Chemistry 64:279–288.

Vangilder, L. D., and S. L. Sheriff. 1990. Survival estimation when fates of some animals are unknown. Transaction of the Missouri Academy of Science 24:57–68.

Van Horne, B. 1983. Density as a misleading indicator of habitat quality. Journal of Wildlife Management 47:893–901.

Van Vuren, D. 1989. Effects of intraperitoneal transmitter implants on yellow-bellied marmots. Journal of Wildlife Management 53:320–323.

Van Winkle, W. 1975. Comparison of several probabilistic home-range models. Journal of Wildlife Management 39:118–123.

Varland, K. L., A. L. Lovaas, and R. B. Dahlgren. 1978. Herd organization and movements of elk in Wind Cave National Park, South Dakota. National Research Report, No. 13. Department of the Interior, National Park Service, Washington, D.C., USA.

Vekasy, M. S., J. M. Marzluff, M. N. Kochert, R. N. Lehmann, and K. Steenhof. 1996. Influence of radio transmitters on prairie falcons. Journal of Field Ornithology 67:680–690.

Venables, W. N., and B. D. Ripley. 1994. Modern applied statistics using S-Plus. Springer-Verlag, New York, New York, USA.

Venzon, D. J., and S. H. Moolgavkar. 1988. A method for computing profile-likelihood based confidence intervals. Applied Statistics 37:87–94.

Verner, J., M. L. Morrison, and C. J. Ralph, editors. 1986. Wildlife 2000: modeling habitat relationships of terrestrial vertebrates. University of Wisconsin Press, Madison, Wisconsin, USA.

Vickery, P. D., M. L. Hunter, Jr., and J. V. Wells. 1992. Is density an indicator of breeding success? Auk 109:706–710.

Viswanathan, G. M., V. Afanasyev, S. V. Buldyrev, E. J. Murphy, P. A. Prince, and H. E. Stanley. 1996. Lévy flight search patterns of wandering albatrosses. Nature 381:413–415.

Vo-Dinh, T., K. Houck, and D. L. Stokes. 1994. Surface-enhanced raman gene probes. Analytical Chemistry 66:3379–3383.

Voegeli, F. M. 1989. Ultrasonic tracking, position monitoring, and data telemetry systems. Pages 279–284 in C. J. Amlaner, Jr., editor. Biotelemetry X. Proceedings of the Tenth International Symposium on Biotelemetry. University of Arkansas Press, Fayetteville, Arkansas, USA.

Voight, D. R., and R. R. Tinline. 1980. Strategies for analyzing radio tracking data. Pages 387–404 in C. J. Amlaner, Jr. and D. W. Macdonald, editors. A handbook on biotelemetry and radio tracking. Pergamon Press, Oxford, England.

Wallingford, B. D., and R. A. Lancia. 1991. Telemetry accuracy and a model for predicting telemetry error. Pages 178–188 in Proceedings of the Annual Conference of Southeastern Association of Fish and Wildlife Agencies, White Sulphur Springs, West Virginia, USA.

Walls, S. S., S. Mañosa, R. J. Fuller, K. H. Hodder, and R. E. Kenward. 1999. Are dispersers pioneers or outcasts? Evidence from radio-tagged raptors. Journal of Avian Biology 30:407–415.

Wand, M. P., and M. C. Jones. 1995. Kernel smoothing. Chapman & Hall, London, England.

Wanless, S. 1992. Effects of tail-mounted devices on the attendance behavior of kittiwakes during chick rearing. Journal of Field Ornithology 63:169–176.

Ward, D. H., and P. L. Flint. 1995. Effects of harness-attached transmitters on prem-igration and reproduction of brant. Journal of Wildlife Management 59:39–46.

Warner, R. E., and S. L. Etter. 1983. Reproduction and survival of radio-marked hen ring-necked pheasants in Illinois. Journal of Wildlife Management 47:369–375.

Waser, P. M. 1985. Does competition drive dispersal? Ecology 66:1170–1175.

Waser, P. M. 1987. A model predicting dispersal distance distributions. Pages 251–256 in B. D. Chepko-Sade and Z. T. Halpin, editors. Mammalian dispersal patterns: the effects of social structure on population genetics. University of Chicago Press, Chicago, Illinois, USA.

Wasser, S. K., C. S. Houston, G. M. Koehler, G. G. Cadd, and S. R. Fain. 1997. Techniques for application of faecal DNA methods to field studies of Ursids. Molecular Ecology 6:1091–1097.

Wasser, S. K., K. E. Hunt, J. L. Brown, K. Cooper, C. Crockett, U. Bechert, J. J. Millspaugh, S. Larson, and S. L. Monfort. 2000. A generalized fecal glucocorticoid assay for use in a diverse array of non-domestic mammalian and avian species. General and Comparative Endocrinology.

Weatherley, A. H., P. A. Kaseloo, M. D. Gare, J. M. Gunn, and B. Lipicnik. 1996. Field activity of lake trout during the reproductive period monitored by electromyogram radiotelemetry. Journal of Fish Biology 48:675–685.

Weinstein, M., M. Conner, G. A. Hurst, and B. D. Leopold. 1997. HOMERUN: software for correctly calculating size of restricted minimum convex polygon home ranges. Presentation at the Forum on Wildlife Telemetry, Snowmass Village, Colorado, USA.

Welch, D. W., and J. P. Eveson. 1999. An assessment of light-based geoposition estimates from archival tags. Canadian Journal of Fisheries and Aquatic Sciences 56:1317–1327.

Welham, C. V. J., and R. C. Ydenberg. 1993. Efficiency-maximizing flight speeds in parent black terns. Ecology 74:1893–1901.

Wells, D. E., editor. 1986. Guide to GPS positioning. Canadian GPS Associates, Fredericton, New Brunswick, Canada.

White, G. C. 1983. Numerical estimation of survival rates from band-recovery and biotelemetry data. Journal of Wildlife Management 47:716–728.

White, G. C. 1993. Evaluation of radio tagging marking and sighting estimators of population size using Monte Carlo simulations. Pages 91–103 in J. D. Lebreton and P. M. North, editors. Marked individuals in the study of bird population. Birkhauser-Verlag, Basel, Switzerland.

White, G. C. 1996. NOREMARK: population estimation from mark-resighting surveys. Wildlife Society Bulletin 24:50–52.

White, G. C., and K. P. Burnham. 1999. Program MARK: survival estimation from populations of marked animals. Bird Study 46(Supplement):120–138.

White, G. C., and R. A. Garrott. 1984. Portable computer system for field processing biotelemetry triangulation data. Colorado Division of Wildlife Game Information Leaflet 110:1–4.

White, G. C., and R. A. Garrott. 1986. Effects of biotelemetry triangulation error on detecting habitat selection. Journal of Wildlife Management 50:509–513.

White, G. C., and R. A. Garrott. 1990. Analysis of wildlife radio-tracking data. Academic Press, San Diego, California, USA.

White, G. C., K. P. Burnham, D. L. Otis, and D. R. Anderson. 1978. User's manual for program CAPTURE. Utah State University Press, Logan, Utah, USA.

White, G. C., D. R. Anderson, K. P. Burnham, and D. L. Otis. 1982. Capture-recapture and removal methods for sampling closed populations. LA-8787-NERP. Los Alamos National Laboratory, Los Alamos, New Mexico, USA.

Whittingham, M. J. 1996. The use of radio telemetry to measure the feeding behavior of breeding European golden-plovers. Journal of Field Ornithology 67:463–470.

Wiens, J. A. 1972. Anuran habitat selection: early experience and substrate selection in Rana cascadae tadpoles. Animal Behaviour 20:218–220.

Wiens, J. A. 1989a. Spatial scaling in ecology. Functional Ecology 3:385–397.

Wiens, J. A. 1989b. The ecology of bird communities. Volume 2. Cambridge University Press, New York, New York, USA.

Wiens, J. A. 1997. Metapopulation dynamics and landscape ecology. Pages 69–92 in I. A. Hanski and M. E. Gilpin, editors. Metapopulation biology: ecology, genetics and evolution. Academic Press, San Diego, California, USA.

Wilson, K. R., and D. R. Anderson. 1985. Evaluation of a nested grid approach for estimating density. Journal of Wildlife Management 49:675–678.

Winterstein, S. R. 1992. Chi-square tests for intrabrood independence when using the Mayfield method. Journal of Wildlife Management 56:398–402.

Wisdom, M. J. 1998. Assessing life-stage importance and resource selection for conservation of selected vertebrates. Ph.D. Dissertation. University of Idaho, Moscow, Idaho, USA.

With, K. A. 1994. Using fractal analysis to assess how species perceive landscape structure. Landscape Ecology 9:25–36.

With, K. A., S. J. Cadaret, and C. Davis. 1999. Movement responses to patch structure in experimental fractal landscapes. Ecology 80:1340–1353.

Wolfinger, R., and M. O'Connell. 1993. Generalized linear mixed models: a pseudo-likelihood approach. Journal of Statistical Computation and Simulation 48:233–243.

Wong, C. N. 1996. Population size estimation using the modified Horvitz-Thompson estimator with estimated sighting probability. Ph.D. Dissertation. Colorado State University, Fort Collins, Colorado, USA.

Woodbridge, B., K. K. Finley, and S. T. Seager. 1995. An investigation of the Swainson's hawk in Argentina. Journal of Raptor Research 29:202–204.

Wooley, J. B., Jr., and R. B. Owen, Jr. 1978. Energy costs of activity and daily energy expenditure in the black duck. Journal of Wildlife Management 42:739–745.

Worton, B. J. 1987. A review of models of home range for animal movement. Ecological Modelling 38:277–298.

Worton, B. J. 1989. Kernel methods for estimating the utilization distribution in home-range studies. Ecology 70:164–168.

Worton, B. J. 1995. Using Monte Carlo simulation to evaluate kernel-based home range estimators. Journal of Wildlife Management 59:794–800.

Wray, S., W. J. Cresswell, and D. Rogers. 1992. Dirichlet tesselations: a new, non-parametric approach to home range analysis. Pages 247–255 in I. G. Priede and S. M. Swift, editors. Wildlife telemetry: remote monitoring and tracking of animals. Ellis Horwood, Chichester, England.

Yerbury, M. J. 1980. Long range tracking of Crocodylus porosus in Arnhem Land, Northern Australia. Pages 765–776 in C. J. Amlaner, Jr. and D. W. Macdonald, editors. A handbook on biotelemetry and radio tracking. Pergamon Press, Oxford, England.

Zahn, H. M. 1985. Use of thermal cover by elk in a western Washington summer range. Ph.D. Dissertation. University of Washington, Seattle, Washington, USA.

Zar, J. H. 1974. Biostatistical analysis. First edition. Prentice-Hall, Englewood Cliffs, New Jersey, USA.

Zar, J. H. 1984. Biostatistical analysis. Second edition. Prentice-Hall, Englewood Cliffs, New Jersey, USA.

Zar, J. H. 1996. Biostatistical analysis. Third edition. Prentice-Hall, Englewood Cliffs, New Jersey, USA.

Zimmerman, G. M., H. Goetz, and P. W. Mielke. 1985. Use of an improved statistical method for group comparisons to study effects of prairie fire. Ecology 66:606–611.

Zimmerman, J. W. 1990. A critical review of the error polygon method. Pages 251–256 in L. M. Darling and W. R. Archibald, editors. Bears – their biology and management: Proceedings of the Eighth International Conference on Bear Research and Management. Victoria, British Columbia, Canada.

Zimmerman, J. W., and R. A. Powell. 1995. Radiotelemetry error: location error method compared with error polygons and confidence ellipses. Canadian Journal of Zoology 73:1123–1133.

Zippin, C. 1956. An evaluation of the removal method of estimating animal populations. Biometrics 12:163–169.

Zippin, C. 1958. The removal method of population estimation. Journal of Wildlife Management 22:82–90.

SUBJECT INDEX

CPSIA information can be obtained at www.ICGtesting.com
Printed in the USA
LVOW07*0854280813

349991LV00008B/65/P

9 780124 977815